P9-EAN-958

A World of Rivers

A World of Rivers

Environmental Change on Ten of the World's Great Rivers

ELLEN WOHL

The University of Chicago Press   Chicago and London

ELLEN WOHL is professor of geosciences at Colorado State University.

The University of Chicago Press, Chicago 60637
The University of Chicago Press, Ltd., London

Printed in the United States of America

20 19 18 17 16 15 14 13 12 11   1 2 3 4 5

ISBN-13: 978-0-226-90478-8 (cloth)
ISBN-10: 0-226-90478-4 (cloth)

Library of Congress Cataloging-in-Publication Data
Wohl, Ellen E., 1962–
A world of rivers : environmental change on ten of the world's great rivers / Ellen Wohl.
p. cm.
Includes bibliographical references and index.
ISBN-13: 978-0-226-90478-8 (cloth : alk. paper)
ISBN-10: 0-226-90478-4 (cloth : alk. paper) 1. Rivers. 2. Nature—Effect of human beings on. 3. Global environmental change. 4. Stream health. 5. Stream ecology. I. Title.
GB1205.W64 2010
551.48'3—dc22   2010019713

♾ The paper used in this publication meets the minimum requirements of the American National Standard for Information Sciences—Permanence of Paper for Printed Library Materials, ANSI Z39.48–1992.

Contents

Preface

I have enjoyed the assistance of many people in researching and writing this book. My first river journey in association with this work was a memorable trip down the Amazon that Bob Meade made possible. Cheng Liu helped arrange the trip upstream through the Chang Jiang's Three Gorges. My journeys through the Murray-Darling and Ganges basins were made many years before I conceived the idea of this book, but numerous Australian and Indian colleagues helped make those travels possible. Colleagues who work on rivers were also generous in providing their time and talents to read drafts of the river chapters and help to reduce errors or omissions in the text. Bob Meade provided extensive commentary on the Amazon chapter, Klement Tockner reviewed the chapter on the Danube, Jürgen Runge reviewed the Congo chapter, and Gary Brierley reviewed the chapter on the Murray-Darling basin. Bob Meade, David Galat, and Martin Reuss thoroughly reviewed the Mississippi chapter. Andrew Goudie, Meg Lowman, three anonymous reviewers for Yale University Press, Nicole Gasparini, and an anonymous reviewer for the University of Chicago Press each provided many useful suggestions. I was unfortunately not able to visit every one of the river basins profiled in this book, but photographs provided by Jürgen Runge for the Congo basin and by Victor Baker and Pavel Borodavko for the Ob-Irtysh basin greatly enhanced the chapters on those rivers I could not visit. The enthusiasm shown for this book by Jean Thomson Black of Yale University Press and Christie Henry of the University of Chicago Press provided important encouragement. Finally, my mother, Annette Wohl, shared my river journeys on the Mississippi and the Mackenzie and patiently and carefully read the early drafts of the river chapters and the connecting interludes. The concern that my mother and my late father felt over human alterations of the world's ecosystems continues to shape my worldview and my life's work.

ONE

A Round River

Contemplating the lace-like fabric of streams outspread over the mountains, we are reminded that everything is flowing. JOHN MUIR

Along the networks of the world's rivers lives a disproportionate richness of plants and animals, all adapted to the rivers' yearly pulsing and occasional outbursts. Change either of the vital components—the supply of water or sediment—by altering the climate, by cutting the forests that stabilize the soil, or by damming the river, and the river itself changes in a cascading effect that influences all the rich life linked to the river corridor. We humans have been learning this fundamental yet highly complicated lesson over and over for millennia.

People too often view constantly changing rivers as inconveniences. We try to stabilize them by confining them in single straight channels that do not spill across the floodplain or migrate from side to side across the valley bottom. This confinement diminishes the complexity and diversity of habitat that nourish abundant and varied species of plants and animals. Where rivers regularly overtop their banks, floodwaters carry sediment and nutrients that rejuvenate the wetlands and forests of the floodplain. Fish disperse from the main channel across the newly flooded lands, where the warmer, shallower, slower-flowing waters that nourish the growth of microbes, algae, and insects also provide food and nursery habitat for vulnerable young fish. Floodwaters receding to the main channel carry organic matter that helps supply the river's food web.

Ecologists identify this "flood pulse" as one of the primary contributors to river health.

In addition to spreading water, sediments, and nutrients across the valley bottom, floods can also reconfigure the river channel. Their greater energy enhances bank erosion, shifting channels across the floodplain and leaving depressions that become ponds or lakes, as well as secondary channels that are fully connected to the main channel only during floods. The presence of swift, deep flow in the main channel, slower flow in the partially connected secondary channels, and still water in the floodplain depressions creates a variety of habitats. Animals that live primarily in upland environments also spend time in these riverine corridors where resources are abundant. When dams reduce the annual flood pulse and channelization eliminates the diverse habitats of the river corridor, the complexity that supports abundant life is lost. Jürg Bloesch, president of the International Association for Danube Research, refers to uniformity as the illness of rivers, not only because of lost complexity and abundance, but also because of lost function. The filtering effect of riverside vegetation and floodplains is reduced or lost when the microbes inhabiting water and soil lose habitat. In consequence, the river is more likely to pass downstream any contaminants that it receives, from excess sediment and nutrients to pesticides and PCBs, rather than storing them or breaking them down into biologically less harmful compounds. Contaminants passed downstream reduce water quality along the length of the river and into the coastal environment. The physical buffering provided by rivers is also lost to increasing uniformity. Rather than spreading across broad floodplains and moving slowly downstream, floodwaters remain concentrated in the main channel and create destructively large and fast-moving floods that are more likely to damage structures and communities near the river.

This book explores how the changes humans impose can impoverish the rivers, and by extension impoverish all the many creatures that rely on them. To illustrate the nature of rivers, I have chosen ten of the world's largest. Among these the Amazon, Congo, and Mackenzie remain relatively unaffected by humans. The Ganges and Chang Jiang exemplify rivers undergoing rapid change as a result of increasing human alterations. The Ob-Irtysh, Nile, Danube, Mississippi, and Murray-Darling represent the variety of heavily altered rivers present in the world today.

These ten also exemplify the climatic, topographic, and biological variety present among the world's largest river basins. The tropical

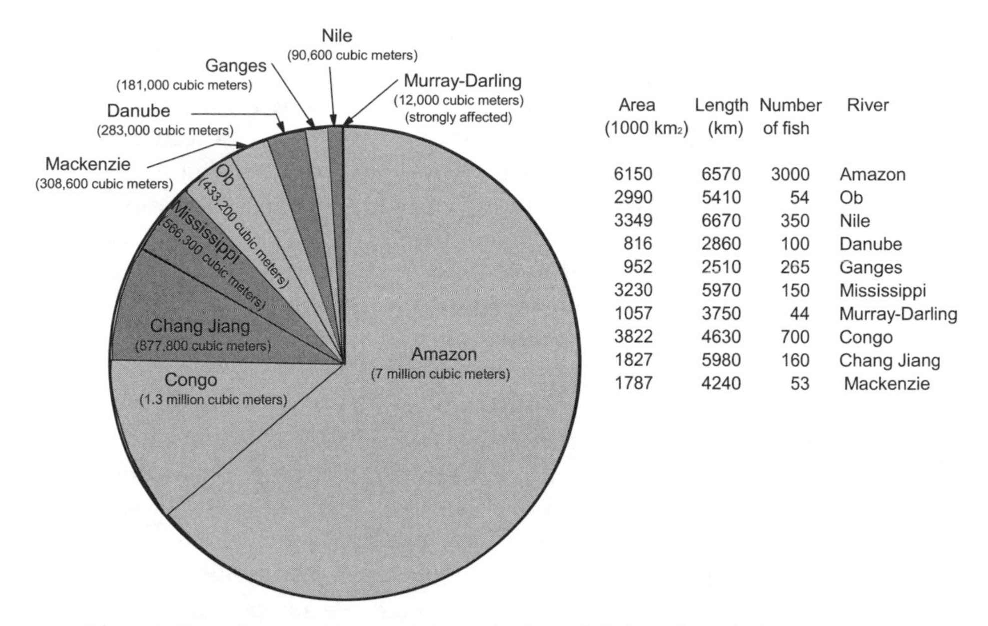

Area (1000 km^2)	Length (km)	Number of fish	River
6150	6570	3000	Amazon
2990	5410	54	Ob
3349	6670	350	Nile
816	2860	100	Danube
952	2510	265	Ganges
3230	5970	150	Mississippi
1057	3750	44	Murray-Darling
3822	4630	700	Congo
1827	5980	160	Chang Jiang
1787	4240	53	Mackenzie

1.1 Schematic illustration comparing average annual volume of discharge for each river discussed in this book. Darker gray shading indicates basins strongly affected by flow regulation as of 2005; lighter gray indicates basins moderately affected. The table at right lists rivers in the order they are discussed in this book, with drainage area in thousands of square kilometers, length of the mainstem river in kilometers, and approximate number of fish species in the drainage basin.

regions drained by the Amazon and the Congo pump enormous volumes of water and sediment into the adjacent oceans, whereas the Nile and the Murray-Darling emit only a comparative trickle from their arid drainages. The Amazon drops precipitously from the heights of the Andes, then meanders broadly for more than three thousand kilometers across an immense, nearly flat basin before reaching the Atlantic. The Ganges drops from the roof of the world in the Himalaya, then gradually turns and runs beyond the base of the mountains before making its way to the Bay of Bengal. The Congo traces a broad arc around its central depression before breaking through the Crystal Mountains along its lower course and tumbling down a series of cascades to the Atlantic. The Danube begins in gentle, hilly mountains, then alternates downstream between multiple channels as it flows across broad basins and through a single narrow gorge punched through the mountain ranges that cross its path to the Black Sea. The Amazon supports a tremendous diversity of fish species, whereas the Mississippi is particularly rich in species of mussels. Freshwater dolphins swim in the Amazon and the Ganges.

What the rivers share, besides their great size, is their crucial role in shaping the surrounding landscapes and biological communities. Rivers sculpt the lowlands they flow across by governing where sediment is removed through erosion and added through deposition. They also indirectly control the shape of adjacent uplands. Geologists working along the Indus River in Pakistan, for example, demonstrated that when the Indus cuts down rapidly, the neighboring valley sides become oversteep and unstable, triggering landslides that fill the valley bottom with sediment and temporarily slow the rate of downcutting.

Relative to the percentage of the landscape they occupy, river corridors host a disproportionately large number of plant and animal species. Rivers almost entirely dominate the transport of sediment to the oceans, with headwater regions producing most of the sediment. Biochemical processes along rivers and adjacent wetlands govern the amount of nitrogen, carbon, and other nutrients reaching coastal areas and thus strongly influence coastal and oceanic productivity. Rivers and wetlands also strongly influence groundwater storage by providing recharge zones where surface waters infiltrate to greater depths or by draining groundwater as channels cut downward and intersect the water table.

Despite the wealth of vital services and resources supplied by the world's rivers, human activities have impoverished most of the largest river basins. An assessment of fragmentation and flow regulation by dams published in 2005 indicates that of the ten large rivers profiled in this book, five (Mississippi, Nile, Danube, Chang Jiang, and Murray-Darling) are strongly affected catchments in which a substantial portion of the flow is regulated or the main channel and tributaries are highly fragmented by dams that effectively break the channel up into shorter segments. The remaining five rivers profiled here are moderately impacted catchments. Only 120 of the 292 catchments assessed for the study were judged to be unaffected as a result of fragmentation or flow regulation by dams, and most of these are smaller rivers.

Environmentalists sometimes quote John Muir, who wrote, "When we try to pick out anything by itself, we find it hitched to everything else in the Universe." Interconnections are particularly apt for describing rivers. Precipitation, windblown sediments, and atmospheric contaminants enter them from the air. Flying insects emerge from the river, and river water evaporates. Water carrying dissolved elements and compounds percolates down through the streambed to the groundwater, and groundwater seeps into river channels. Microscopic organisms and aquatic insects move back and forth between the river and the shallow subsurface, as do water and dissolved chemicals. Water, sediment, nutrients, and organisms flood across valley bottoms, then recede into river channels. Sediment and organic matter move from adjacent hill slopes and uplands into river corridors. And water, sediment, contaminants, and organisms moving downstream, as well as other organisms moving upstream, stitch together the uplands and oceans along the seams of rivers.

Aldo Leopold wrote of the functioning of an ecosystem as a "round river" to emphasize the cycling of nutrients and energy. This phrase could also aptly describe the planet, including the atmospheres, oceans, landmasses, and groundwater. I emphasize the importance of these global connections among seemingly individual river basins in the short "interludes" between river profiles. This book begins with precipitation over the headwaters of the Amazon. Following the river downstream across South America, chapter 2 ends where the Amazon enters the Atlantic. The intervening interlude traces the pathway of a hypothetical water droplet leaving the mouth of the Amazon until that droplet falls as precipitation on the headwaters of the Ob-Irtysh river system in Siberia. Each succeeding pair of chapters follows this format, tracing the path of the same water droplet down another river basin and then following it to the next one. In the interludes that trace pathways between river basins I have tried to estimate realistic rates of travel for a droplet moving with specific ocean currents and atmospheric circulation patterns. The travel times between river basins are underestimates, however, because many of the droplets moving down a river take much more circuitous pathways, with long periods of storage in groundwater, glaciers, or deep ocean currents. Polar waters take approximately a thousand years to circulate, for example, then hundreds of years more to surface at the equator, where they rise slowly from the ocean depths at only two to five meters a year. But water droplets do circulate around the entire globe, and contaminants, windblown sediment, and a wide variety of organisms, from microbes to whales, travel with them. It is this sometimes slow but always inexorable mixing that makes it ludicrous to think of any river basin in isolation. What people in the United States do in the Mississippi River basin does make a difference in the basins of the Congo, the Ob-Irtysh, and the Murray-Darling, and vice versa. We are all a part of one round river. The following chapters discuss some of the details of that river's ebb and flow using ten great rivers as pathways of learning and exploration.

My purpose in writing this book is to explore the natural history of some of the world's largest rivers and the history of human alterations to these river ecosystems by taking readers on a journey down each river and along the pathways that connect seemingly far distant rivers. My focus is on the science underlying our understanding of each river. Although I am fascinated by the approach of environmental historians, I do not explore the social history of people living in each river basin. The book is written to be accessible to nonspecialist readers interested in natural and environmental history, but the bibliographic sources for each chapter also provide reference material for those who wish to delve further into the topics discussed. Every river discussed here is the sole subject of several other books, as well as numerous scholarly articles. If the people of the world continue to impoverish the functioning of the greatest rivers, we cannot plead ignorance of the effects of our actions.

TWO

The Amazon: Rivers of Blushing Dolphins

The Andes Mountains form a wall more than six thousand meters high in Ecuador, Peru, and Colombia. The mountain wall separates two of the most different regions on Earth by only eight hundred kilometers. To the west of the divide, the absence of water defines a landscape with jagged edges of rock and topography. Across the continental divide, the mountains drop abruptly into a vast green plain. Elevation drops from more than six thousand meters to approximately three hundred meters over a distance of three hundred kilometers. The plain is contoured by immense rivers that wind back and forth across it, and the abundance of water defines the landscape.

Water sifts down onto the high peaks as snowflakes that feed glaciers and icefields, and as snowmelt in the mountain streams. The Amazon River heads in a small spring that seeps from spongy grassland high on Mount Huagra. This spring grows into a small stream called the Huarco. As other streams fed by melting snow and ice join the river, it becomes the Rio Toto, then the Rio Santiago, followed by the Rio Apurimac. (Spanish-speakers use an accent on Río.) By this point the river has dropped down into cloud forest, moist and cool, but below the frost level. The name Apurimac comes from an Indian language in which it means "Great Speaker," for the steep Apurimac is filled with rapids and waterfalls. The downstream point where the rapids end marks the boundary between the

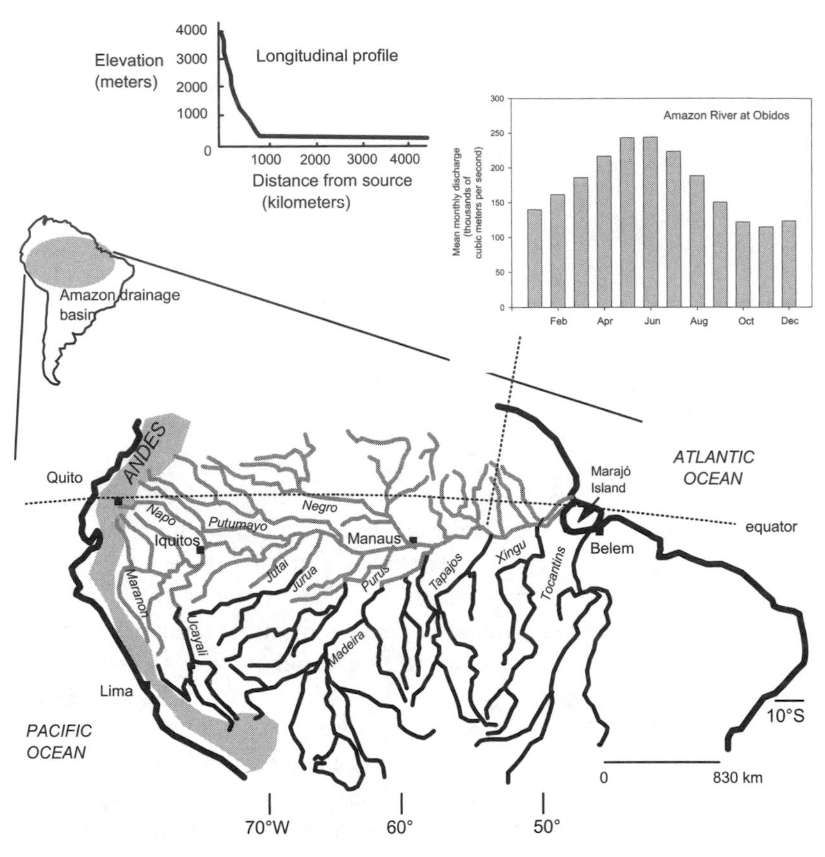

2.1 Map of the Amazon drainage basin in South America, showing principal tributaries and cities. The entire drainage basin is represented by the shaded area in the inset figure; the shading in the main figure represents the Andes Mountains. The flooding season in different parts of the Amazon basin is indicated by gray lines (June–August) and black lines (January–May) along the river courses (after Goulding et al. 2003, *The Smithsonian Atlas of the Amazon,* "The Flooding Season," 36). The hydrograph at upper right shows monthly average flow at Obidos, Brazil.

crystalline rocks of the Andes and the younger sediments of the Amazon basin.

The noisy Apurimac becomes the Rio Ene, then the Rio Tambo, which flows into a main tributary of the Amazon, the Rio Ucayali. The Peruvians begin calling the river Amazonas at the confluence of the Ucayali and the Marañon. Amazonas remains the river's local designation until it enters Brazil, where it is known as the Rio Solimões until it reaches the confluence with the Rio Negro, when it is known once more as the Rio Amazonas. (Portuguese-speaking Brazilians neither accent nor capitalize rio.) Victorian naturalist Henry Walter Bates coined the name the Rivers Amazon in recognition of the different local names given to the great river. The plural seems particularly apt in this land of rivers.

By the time the flowing water reaches the Ucayali, rains are abundant. Torrential "male" rains cascade from thunderheads over the feathery stands of palm trees. Gentler "female" rains moisten the leaves of the enchanted broccoli forest where hundreds of tree species create an undulating canopy of varied shapes and sizes. Water seeps into the ground and collects in the stream channels, swelling them to rivers that overflow their banks and flood for tens of kilometers across the surrounding lowlands, creating vast lakes and flooded forests. The water's force cannot be contained, and the rivers continually shift within their channels, eroding banks here, depositing sandbars there, jumping abruptly to a new channel. Each river sweeps back and forth in extravagant loops, doubling on itself, trailing secondary channels, chute cutoffs, and oxbow lakes in its wake.

The city of Iquitos, Peru, rises abruptly within the dense forest and network of rivers, without a gradual buildup of rural areas and suburbs. Iquitos remains unconnected by the network of roads so ubiquitous elsewhere at the start of the twenty-first century. Two hundred thousand people live in Iquitos, and everything they do not make themselves comes in by airplane or by boat. During the flood season, water rises up the stilts of houses on the floodplain. Each year the river can rise and fall here by nine meters vertically. Many structures are built on floating rafts.

People have been living in Iquitos since at least the fifteenth century, but the settlement became a city with the rubber boom that brought both intense suffering and lavish prosperity to the Amazon basin early in the twentieth century. Rubber has not boomed in Iquitos in many decades, but other resources can be extracted. The city has large sawmills where the giant kapok trees (*Ceiba pentandra*) that rear like massive umbrellas above the surrounding canopy are peeled to make plywood. A more limited harvest now occurs following a decade of intense logging that created a scarcity of kapok trees, but recent oil discoveries suggest to some residents the promise of another resource extraction boom.

Downstream from Iquitos the river alternately narrows and then widens into a broad inland sea. Birds screech loudly along the riverbanks. The turbid water resembles café au lait. The river surface holds a complicated topography of wrinkles and calms, and upwellings where big boils rise off submerged dunes. The water is not uniformly opaque. Particles of silt and clay churn within it like split pea soup being stirred, and the sediment forms subtle and complex honeycomb patterns in shades of brown and golden brown. The depths remain inscrutable.

The river flows dirty with the fecundity of life. Grasses, water hyacinths (*Eichhornia crassipes*), reeds, twigs, logs, and clumps of water lettuce (*Pistia stratiotes*) float by like bouquets dropped on the water. Herons and egrets perch on floating islands of grass. Butterflies form scraps of color on the breeze. Parrots pass overhead screeching loudly, their blunt-beaked silhouettes and short wingbeats as distinctive as their harsh cries. Along the banks, the globular nests of oropendola (*Psarocolius* spp.) colonies hang down from bare trees. *Ribereños* paddle small homemade dugouts that ride low in the water.

The ribereños are river people, as their name implies. Of mixed Indian and European ancestry, many of them are descended from people who migrated into the Amazon during the rubber boom and remained here after the boom ended. The ribereños live in isolated huts and small villages, supporting themselves through a mixture of fishing, hunting, gathering, small-scale agriculture, and trade.

The ribereños live within a landscape of impressive power, where oceans of air collide and precipitate huge quantities of water that in turn fuel the river's vast seasonal floods. The Amazon basin lies beneath the Intertropical Convergence Zone (ITCZ), the great meeting point of air flowing equatorward from both Northern and Southern hemispheres. As the warm, moist air masses collide the air rises upward, sending rain down from fierce, mobile convective storms.

The annual migrations of the ITCZ induce wet and dry seasons that are out of phase in the northern and southern sides of the Amazon basin. Flood season is January to May in the southern portion of the basin and June to August in the northern basin. Peak flows from northern tributaries lag those of southern tributaries by three months because of these seasonal differences in rainfall, as well as the large volumes of water that are stored and slowly drained from the basin's vast floodplains.

All the water falling over the basin originates in evaporation from the Atlantic Ocean. The water is then continually recycled, falling as precipitation and evaporating from water bodies or transpiring from plants. More than half of the rain falling on the interior parts of the basin has fallen at least once elsewhere in the Amazon. The cycling of water among myriad pathways defines the world of the Amazon: atmospheric flows of water vapor coming inland from the Atlantic; river flows back to the ocean; overbank flows from channels across the

2.2 *Above,* Canoa along the margin of the Amazon in Peru. The immense height and mass of vegetation, as well as the huge size of the river, dwarf humans, yet ribereños confidently paddle these dugouts with only a few centimeters of freeboard above river waters rushing along at more than a meter per second. *Below,* Ribereño house along the Amazon in Peru. The hut is roofed with palm thatch, and the house is built on stilts. The interior is simple, with cheap aluminum cooking pots and utensils, a table, and hammocks. A papaya tree may grow next to the house, and red peppers may grow in an old canoa filled with soil.

adjacent lowlands; and flows of soil moisture into plant roots, up stems, and out of leaves into the surrounding air.

The quantities contained in these flows dwarf those of other river basins. At Iquitos the Amazon flows with two times the average discharge of the Mississippi River. Even the Amazon's tributary Rio Negro and Rio Madeira each flow at about two times the Mississippi's volume. Unlike other large rivers that are fed primarily by snowmelt in their headwater mountains, by the time the Amazon reaches Iquitos, more than 95 percent of the flow is fed by rain falling below the mountains. This tremendous volume of rainfall results in the river's carrying approximately 15 percent of all the river water that is discharged into the world's oceans, or a flow equal to twelve Mississippi Rivers. The huge volume is the first of the Amazon's outstanding characteristics.

The relentless forces of this water have carved and deposited a flat landscape. Nothing but the tall trees lining the banks is visible from the river. In satellite imagery, the texture of the floodplain appears as intricate as the grain of wood where the river has repeatedly shifted back and forth to create secondary channels, oxbow lakes, backswamps, and long islands. Some of the islands are rimmed by natural levees, with depressions that form lakes in the center. Slight differences in surface elevation, and differing histories of deposition and erosion, equate to differences in subsurface grain size and moisture retention that in turn support different water levels throughout the period of low water and thus different plant and animal communities. The floodplain preserves a historical record of river dynamics that creates diverse habitats, but to the untrained eye the landscape beyond the river looks like an undifferentiated mass of greenery.

During flood stage, the Amazon may spread to a width of fifty kilometers as it spills across the adjacent floodplains. Where the floodwaters are turbid with sediment, the Brazilians call the flooded forest zone *várzea*. Where the floodwaters are clear and black, the flooded forest is *igapó*. Together these forests make up 2 percent of the total Amazon rain forest, an area of approximately a hundred thousand square kilometers.

During his 1848–52 explorations in the Amazon, English naturalist Alfred Russel Wallace adopted the local Indian practice of distinguishing whitewater, blackwater, and bluewater (which later became known as clearwater). This usage can be confusing to those in North America, where whitewater commonly refers to turbulent flow over rapids. In the Amazon basin whitewater is the brown, sediment-laden water of rivers such as the tributary Rio Madeira and the Amazon itself. The

2.3 *Above*, floodwaters from the Amazon backflooding the lower reaches of the tributary Rio Napo to form a várzea. Grama dulce extends above the water surface at right. *Below*, flooded forest (igapó) along the Rio Jutai. This site is more than a kilometer from the main channel of the Amazon, but the water is several meters deep.

whitewater is relatively rich in nutrients and solutes, as well as sediment, and of a nearly neutral pH.

Whitewater rivers head in the Andes and mainly occupy the western and southwestern portions of the Amazon basin. If the Andes are a wall, then Robert Frost's line "Something there is that doesn't love a wall" describes erosion. Glaciers, plants, wind-driven sand grains, flowing water, and the daily cycle of solar heating and cooling all break down the rocks of the Andes into sediment that eventually enters the rivers of the Amazon. Most of the sediment in the whitewater rivers comes from the Andes. The mainstem Amazon drains the Peruvian Andes, and the tributary Rio Madeira carries huge loads of sediment from the Bolivian Andes.

During the wet season, the turbid water of the main river mixes with clearer water coming down some of the tributaries and off the floodplain. More than 95 percent of the particulate material carried by the Amazon consists of tiny grains of silt and clay traveling suspended in the river water. The greatest amount of silt and clay movement occurs about midway through the rise of the annual flood.

As the turbid waters of the main channel flood across the bottomlands during the wet season, the water velocity decreases and sediment settles from suspension onto the floodplain. The floodplain thus decants sediment from the river water, returning clearer water to the main channel. As the annual flood recedes, the river water cuts at the exposed channel banks, returning some of the sediment to the flow.

When Leal Mertes studied the dynamics of sediment along the Amazon, she found that more sediment is deposited on the floodplain than moves downstream with the river water. More than 80 percent of the suspended sediment that flows onto floodplains from the main channel or tributaries settles from the water and remains on the floodplain at least temporarily. Sediment eventually returns to the main channel through bank erosion. Of the approximately 1.5 billion tons of sediment eroded annually along a two thousand kilometer reach of the Amazon, more comes from bank erosion than from tributaries or from transport down the channel. Most sediment thus repeatedly cycles between the floodplain and the main channel. Although how long the sediment stays on the floodplain is unknown, it probably is on the order of many hundreds to a few thousand years, with the longest residence times at sites farthest from the main river channels. Nonstop movement of water or sediment from the Andes to the Atlantic in the main channel would take only about two months, but such direct movement probably involves only a small proportion of the sediment particles being transported.

The Amazon is one of the few major rivers of the world where water and sediment are allowed to move through their own circuitous paths. On this great river engineers have not yet channelized and leveed the channels or regulated the flow. Not a single bridge constricts the river. Ship captains navigating the river annotate their charts as they proceed because the channel is continually changing. Water and sediment still surge across the floodplains of the Amazon each wet season, driving the life cycles of plants and animals, and the waters drain slowly back into the channels each dry season. The continued existence of natural rhythms of water and sediment movement, as well as the adaptations of plants and animals to these rhythms, is the second outstanding characteristic of the Amazon.

Above the level of the annual flood rise the uplands of terra firme, or solid earth. Many of these uplands are the remnants of a great wedge of sediment deposited along the eastern margin of the developing Andes. With time this wedge has been dissected by rivers and worn down, and the soil of the remaining areas is now colored burnt red from iron oxides. The constant heat and high precipitation of the tropics produce rapid mineral weathering. Mobile ions such as calcium and magnesium are quickly leached from the soil downward into the groundwater or taken up by plants and animals. Microbes and other organisms rapidly decompose any twigs or leaves shed by plants, freeing the nutrients for use by all the layered plant and animal life of the rain forest. Only stable oxides of iron and aluminum remain to form a soil that quickly loses fertility once the forest and its nourishing rain of plant litter are removed.

Winter provides a rest period in the temperate regions. During this rest more organic matter is added to the soil than is used by living organisms. Through centuries and millennia, dark, organic-rich layers accumulate in the upper few meters of temperate soils. There is no rest period in the tropics and no subsurface soil organic layer. Plants have shallow roots because there is no need to reach deep for nutrients. Yet from these shallow roots many trees grow exceptionally tall, developing buttress roots or stilt roots to increase stability, although there is still much treefall in the tropics.

Life in the Várzea

The plants and animals of the Amazon basin have adapted to the seasonal fluctuations present where the whitewater rivers overflow their banks and spread across the adjacent forest. Várzea like that along the tributary Rio Napo downstream from Iquitos hold a stunning diversity of life. Plants are densely layered, growing on and over and through one another in intense competition for sunlight. Although people of the temperate zones commonly regard deserts as a challenging environment for plants, the periodic superabundance of water, the limited nutrients, and the competition among individual plants in the tropical rain forests can present equally intense challenges. Woody vines hang like flattened, twisted fire hoses from tall trees in the várzea. Epiphytes sprout from the vines in little feathery tufts. Epiphytes, named for the Greek *epi*, "on top of," grow on host plants. Epiphytes such as bromeliads and orchids take nothing from the host plant, but their weight can break the host's limbs. Hemiepiphytes such as lianas grow on other plants but remain rooted in the ground for water and nutrients. The vines and leaves may grow flat along the trunk of a host tree to avoid being eaten by caterpillars or other herbivores, then fan out wherever the plant encounters a patch of light filtering through the canopy. Strangler figs start high in the canopy from a bird-deposited seed and grow downward. Some trees have evolved peeling bark as a defense against strangler figs. The sloughing bark sheds the figs like Houdini wriggling out of chains.

Aechmea and *Streptocalyx* bromeliads bloom in bursts of red high among the branches of kapok and other tall trees along the Napo. Trees leaning out across the water host their own aerial gardens thick with mosses, fungi, and epiphytes. Tufts of feathery palm fronds erupt in the canopy. Bulbous maroon fruits hang down like half-filled water balloons from the bare branches of false kapok trees (*Pseudobombax munguba*). Spiny black palms (*Bactris maraja*) grow along the riverbanks.

More than thirty thousand species of vascular plants live in this world of intertwined greenery along the Amazon. By comparison, approximately eighteen thousand species are present in the United States. The Amazon also hosts at least thirty thousand species of fungi and an unknown quantity of mosses, lichens, and algae. Trees are the most diverse type of plant, with the greatest number of species per hectare. Studies record approximately 300 species within a one-hectare plot near Iquitos, and 186 tree species at a plot near Manaus, midway down the Amazon. Instead of the limited number of species, each represented by large numbers of individuals, that can occur in higher latitudes, the tropics typically have large numbers of species with fewer individuals of each. The off-the-chart numbers of species of plants, fish, and other organisms are the third outstanding characteristic of the Amazon basin.

A three-toed sloth hangs like a hammock high in the crook of a bare tree along the Napo. Much of the sloth's diet here consists of leaves of the *Cecropia* tree, a fast-growing gap specialist common along the edge of rivers. Tiny poisonous *Azteca* ants protect the *Cecropia* by nesting in the extensive hollows of the trunk and branches among colonies of small insects that secrete a sugary substance the ants eat. The ants also obtain nutrients from special growths at the base of the leaf stalks. In return, the ants vigorously attack any animal other than a sloth that approaches the tree, thus limiting herbivory, and kill climbing plants that might otherwise strangle the *Cecropia*.

This symbiosis between ant and plant is only one of many ingenious strategies Amazonian plants have evolved to limit herbivory. Some trees produce dense wood by incorporating silica compounds into their tissue as a protection against wood borers and termites. Other plants sprout leaves that look unappetizing because, although perfectly healthy, they are splotched with yellow as though diseased, or lacy with holes as though some other herbivore had already eaten them.

Elsewhere in the várzea, ants boil up out of partly submerged logs at the slightest touch. Ant nests and termite nests form dark bulges among the tree trunks and limbs. Like other invertebrates, the termites move seasonally from nests in the soil during the dry season to nests among the tree branches during flooding. Scientists have described about 150 termite species from Amazonian forests, although many other species remain undescribed. Floodplain forests have about five times fewer species than terra firme forest, but the total number of termites is approximately equal in the two environments.

The turbid brown depths of the várzea, which appear inscrutable from the surface, also hold wonders. Tiny neon tetra fish (*Hyphessobrycon innesi*) swim among fruit-eating giants as big as a human. Plant parts plop into the water and roll downstream with the current. At least some of the plant parts will be eaten by fish, for the Amazon basin is unusual in that fish are the major dispersers of seed in the flooded forest. Many of these fish have large molars and jaws specialized for cracking hard seeds. In these turbid waters the fish have evolved specializations such as electroreception and the barbels of catfish to find their way about in the darkness of a world only centimeters below the intense sunlight.

Plants crowd in everywhere along the margins of the water, and thick stalks of grama dulce protrude above the water surface. This so-called floating vegetation is actually composed of many species of rooted grass, including *Paspalum repens* and *Echinochloa polystachya*, which can be twelve meters long but protrude less than a meter above the water.

Scientists studying *Echinochloa* have found that the grass uses a type of photosynthetic pathway common among plants in arid or semiarid regions but unusual in wetter areas. The high productivity of this type of photosynthesis allows the grass to keep pace as water levels in the Amazon rise approximately eight meters during each year's flood season. New shoots form at the nodes of old stems during low water levels, and these shoots root in exposed sediment. Each shoot forms a new plant when the old stems rot and die, and the stems of the new plants grow rapidly upward as rising water levels cover the sediment in late November or December. The rates of vegetative growth associated with this annual feat are among the highest values ever observed for a natural plant community.

Nutrients traveling attached to fine sediment in the whitewater rivers also promote this rapid growth. River waters carry dissolved nitrogen leached from the uplands down to the floodplains, where microbes and plants consume the nitrogen. The organic matter from these microbial and plant producers then fuels animal consumers in the water, river sediments, and floodplain soils.

Although the floodplain composes only about 3 percent of the Amazon basin, the extremely high rates of grass productivity indicate that floodplain vegetation forms a critical part of the nutrient cycle. Scientists studying *Echinochloa* conservatively estimate that the grass occupies five thousand square kilometers of the basin and consumes about seventy thousand kilograms of carbon dioxide each year. Much of this carbon is released back to the atmosphere as the stems and leaves die and decompose, and this contributes to the high fluxes of carbon gases measured in the floodplains. The stands of grass also absorb large quantities of nutrients such as phosphorus and nitrogen during the flood season, slowing the flow of nutrients through the river system. When the plants die and decompose during the dry season, the nutrients are released to the floodplain sediments. As long as the grass stands remain intact, nutrients within the floodplain system are conserved and enriched. When people burn the grasses during the dry season so they can plant the floodplain with short-lived food crops or use it as temporary pasture, the life cycle of *Echinochloa* is disrupted. After repeated burning the grass community may be lost, and with it the ability to store and recycle nutrients in the floodplain.

The floating meadows formed by grama dulce can be up to a kilometer long and several hundred meters wide. The meadows support shrubs, trees, and other plants as well as grasses, and a rich animal community lives among the plants. More than a hundred species of fish are specialized to live within the root mats of the meadows, which also provide good nursery habitat for fish and habitat for aquatic invertebrates. Among the animals living in the darkness below the plant roots are green knife fish (*Eigenmannia virescens*), which transmit an electrical discharge and have a modified vertebral column that transmits sound in freshwater. Ten to twenty species of electricity-generating fish can be present in a single floating meadow, creating what must be a chaos of electricity for those organisms sensitive enough to detect it. Water monkeys (*Osteoglossum bicirrhosum*) or silver arowana, another common name, also live beneath the floating meadows. While young, these insect-eating fish are brooded in the male's mouth, venturing out to eat and then returning to papa's mouth when danger threatens. This specialized care of the vulnerable young is a response to the large numbers of predators waiting hungrily in the flooded forest.

Insects, crustaceans, mollusks, birds, and at least a hundred species of frogs also thrive in the dense vegetation of the floating meadows. As the water level drops following the annual flood season, the floating meadows move downstream like arks carrying both plants and animals, until they eventually break apart.

The insect sounds forming a background to the varied birdcalls grow progressively louder as the light fades along the Rio Napo. Thousands of hidden frogs join the chorus. Of the four thousand species of amphibians known worldwide, sixteen hundred live in the Amazon. This region has the most diverse frog fauna in the world, and biologists regularly find new species here. Frogs of the Amazon occupy every conceivable niche. They live in the water, on the forest floor, and in the trees. They eat insects, fish, other frogs, snakes, and rodents. Some species lay their eggs on branches overhanging the water so that the tadpoles can drop in when ready. Others are carried to the water by a parent or grow up in puddles within hollow bamboo stems. Poison dart frogs eat ants and use the formic acid, along with other compounds, to produce the alkaloid poisons the frogs are famous for.

Francisco de Orellana was the first European to hear these sounds of insects and frogs along the Napo. Orellana was one of the conquistadores accompanying Gonzalo Pizarro in his conquest of the Incan empire west of the Amazon. Orellana left the main expedition in what is now eastern Ecuador and traveled down the Rio Napo to the Amazon with a smaller group in December 1541, intending to bringing provisions from downstream settlements back to the starving expedition. Neither

Orellana nor Pizarro accurately estimated the difficulties of traveling downstream on the Napo, let alone the impossibility, as it turned out, of returning back upstream.

Amid much difficulty, Orellana and his men traveled the length of the great river. As they approached the confluence of the Amazon and the tributary Rio Madeira, they heard stories of women warriors. The Spaniards immediately called these women "Amazons," taking the name from an old Greek story of warrior women in ancient Scythia. The tales of the Amazon warriors held a place of particular fascination among all the stories coming out of Orellana's adventures, and eventually Europeans came to know the main river and its entire drainage basin by this name.

Forest Tea

Farther downstream along the Amazon, tributaries such as the Rio Ampiyacu form an *igapó*, or blackwater wetlands, during the wet season floods. Blackwater in the Amazon is clear and slightly to strongly acidic, more the color of strong tea: as geologist Bob Meade describes it, "tea brewed from the forest litter." Ions of calcium, magnesium, and other minerals that are dissolved in and buffer the whitewater are absent from the blackwater, having long ago been leached from the forest soils through which this water moved as groundwater before entering a river. White beaches of quartz sand that invite swimming are prominent only along the blackwater rivers, which occur mainly in the northern part of the Amazon basin around the Rio Negro. The Amazon river basin is depauperate in mussels in part because the blackwater rivers lack enough calcium to support their shell building. Lack of calcium has also been hypothesized to explain the lack of large herbivores in the Amazon forest, where the tapir is one of the largest at 1.8 meters long.

Dozens of species of birds enliven the igapó. Yellow-rumped male caciques (*Cacicus cela*) call loudly as they display outside their funnel-shaped nests. The vivid yellows and blacks of the birds' plumage contrast sharply with the green vegetation. Greater kiskadees (*Pitangus sulphuratus*) and lesser kiskadees (*Pitangus lector*) utter the characteristic calls that give them their common name. Various members of the five species of kingfishers dart above the water, many uttering their own volley of sound as they move. Amazonian umbrella birds (*Cephalopterus ornatus*) sport a distinctive feathery headpiece like a pompadour. The staccato knocking of a lineated woodpecker (*Dryocopus lineatus*) sounds from the top of a dead tree. The highest bird diversity along the river corridor occurs at the meeting of the Andes and the Amazon. Some one thousand to fifteen hundred bird species are known in the Amazon; North America has seven hundred species.

The variety of birds is an index of the richness of flooded forest habitat. An area of less than a hectare is used during the year by an average of 125 bird species. Most species favor one or two of the four vertical levels of the forest. When the three lower levels of ground, lower, and middle tiers are inundated during the flood season, birds migrate to the upper level or to adjacent terra firme forest. Birds also migrate elsewhere during the dry season. Most of the bird species in the Amazon are migratory, and less than half are endemic species that are found only here.

Part of the richness of igapó habitat stems from the abundance of fruit. Most trees of the flooded forest have an intense fruiting period between the start and the peak of the annual floods. This abundance of fruit attracts monkeys and birds from adjacent upland areas and fish from the river channels. The dense veining of rivers across the basin means that only fifty to a hundred kilometers separate individual patches of flooded forest, aiding animal migration between patches.

Although the várzea and igapó host a variety of animals, the annual outpouring of water from the river challenges plants. Water in the flooded forest can reach nine meters deep, and individual sections of forest can be flooded for three months to eleven months at a time. Botanists still do not fully understand how trees can survive the absence of oxygen in the root zone created by this prolonged deep flooding, yet trees here can be up to forty meters tall and several hundred years old. Many trees grow mainly during periods of low water, then drop their leaves during periods of inundation. Other plants time their life cycles differently with respect to flooding. Floating algae known as phytoplankton living in floodplain lakes throughout the Amazon grow rapidly after river water enters the floodplain and supplies nutrients.

Most of the Amazon lakes are isolated during low water, merging into nearby flowing rivers or side channels during high water. Aquatic animals that spread into the flooded forest during high water migrate back into the lakes or river channels during low water, but many of the lakes do not provide good low-water habitat. The lake water is too muddy or poor in nutrients to support much phytoplankton, and low levels of dissolved oxygen can make breathing difficult. Many of the floodplain lakes are stratified, with an oxygen-depleted lower layer in which anaerobic bacteria accelerate the decomposition of organic matter. The surrounding waters of the flooded forests are also low in dissolved oxygen. As plant litter and other organic materials decay in these oxygen-poor waters, they release methane. The Amazon floodplains and lakes release 5 to 10 percent of the total global methane flux from wetlands during the annual inundation. In response to low levels of dissolved oxygen in the lake and floodplain waters, South American lungfish, swamp eels, electric eels, catfish, and pirarucu—in total at least twenty fish species—surface periodically to breathe atmospheric oxygen. Predators from piranhas and caimans to vultures and cormorants also concentrate in and around the lakes during the dry season.

Along the edges of the floodplain, the mixing zone between the clear waters of the Ampiyacu and the main channel is abrupt as a fence, though churned by large vortices. Flashes of pink appear and disappear, revealing the presence of botos (*Inia geoffrensis*), freshwater dolphins that are pink because subcutaneous blood vessels expand with exertion, making their color more vivid. This is one of the wonders of the Amazon: pink dolphins up to 2.5 meters long and 160 kilograms in weight, often swimming upside down, clicking as they go to echolocate their prey in the murky water. And some of the prey is hard to eat: the back teeth of botos are commonly badly worn and ground down, probably by the tough coverings on the armored catfish.

These freshwater dolphins may be one of the species remnant from the time when the Amazon basin was a shallow ocean. The geologic provinces of the Brazil Shield and Guiana Shield, in the southern and northern portions of the Amazon, respectively, were once highlands from which rivers flowed westward. As the Andes began to rise to the west about fifteen million years ago, the region between the Andes and the highlands became a shallow ocean. Continuing uplift of the Andes established the present eastward drainage approximately ten million years ago, but dolphins and rays living in the shallow ocean apparently evolved to remain in the freshwaters that replaced the sea.

Five of the world's great rivers have or recently had their own species of freshwater dolphin: the Amazon, the Chang Jiang, the Indus, the Ganges, and the Mekong. The Amazon is unique in having two species; botos divide up the river habitat with smaller tucuxi (*Sotalia fluviatilis*), gray dolphins up to 1.8 meters long.

Stories about the boto are numerous. Legend has the dolphin as a male seducer of humans who comes ashore at night, and some birth certificates along the Amazon still list the father as a boto. Real botos, however, are declining in number as a result of the combined effects of accidental dolphin kills during commercial driftnet fishing and use of dolphin parts for medicine and ceremony by people living along the lower Amazon. Dams on some of the tributaries restrict the distribution of dolphins, and pesticides washing into the rivers from agricultural fields accumulate in their tissues. Mercury moving downstream from placer gold mining in the northern Amazon also poisons the dolphins.

Caboclos

The Amazon enters Brazil downstream from the riverside city of Leticia, Colombia's only port on the Amazon. This city of thirty thousand people, with its dock full of boats, jet aircraft flying in, and clearings in the forest spreading up and downstream, interrupts the otherwise largely unbroken expanses of forest along the river downstream from Iquitos. Occasional clearings for small cattle ranches lie along the river below Leticia, always on the burnt red uplands of the terra firme. The name terra firme provides a nice distinction from the watery, changeable world just a few meters lower in elevation. The distribution of terra firme controls the distribution of people, for only seasonal, temporary camps exist beyond the terra firme. Although most of the cattle ranches along the main river are small, family-run operations, ranching is the biggest threat to floodplain fisheries and overall biodiversity in the Amazon and is the main economic force behind destruction of primary forest.

Small palisades of wooden fish traps protruding above the water along the river's edge create a technological contrast with the motorboats docked upstream. The traps are used for subsistence fishing by the ribereños, who are known as *caboclos* in Brazil. Technologies more advanced than those they use overshadow the life of the caboclos in many ways. The size of individual fish and the numbers of fish caught are declining with time, for example, perhaps as a result of commercial driftnet fishing. Manatees have also been heavily hunted here and are now rare. Turtles and land tortoises are declining because both adults and eggs are heavily hunted. The giant Amazon river turtles in particular once laid hundreds of millions of eggs each year, but now the turtles are nearly gone.

The word caboclo has many connotations, including non-Amazonian ones, but it is widely used to describe indigenous, rural inhabitants of Amazonia who are not tribal peoples. The caboclos practice small-scale, diversified agriculture, using manioc as a staple crop. Fish caught with bow and arrow or small nets and traps form their major protein source, but the caboclos' diet is deficient in vitamins because they lack fruits and vegetables. Manioc provides a starchy staple.

Like the plants and animals around them, the caboclos adjust their lives to the rivers that dominate the Amazon region. As the floodwaters recede during the dry season, the caboclos cultivate beans or corn on the newly exposed river margins. Zebu cattle and water buffalo introduced from India stand in water up to their chests as they graze on emergent vegetation. During rising and high water the caboclos remove their cattle to terra firme or keep them on rafts or elevated corrals and feed them grass. Most caboclos live along the whitewater rivers, because blackwater rivers lack the rich alluvial soil that can support crops and because fewer fish live in blackwater rivers.

Although contemporary Brazilians and people living elsewhere commonly think of the Amazon basin as a huge wilderness unaltered by humans, soils across the basin record a long history of land use. Archaeologists estimate that the river corridor was already well populated

2.4 Floating cowboy along the Rio Guajará during the annual flood. Corrals built on platforms above the level of the yearly flood appear in the background. Water buffalo introduced to areas such as this one churn up the floodplain mud and prevent some of the native plants from germinating.

when Orellana journeyed downriver in 1541; more than three million Indians lived in the Amazon basin in 1500. The indigenous peoples had altered the ecology through the use of fire and selective cultivation. Some scientists hypothesize that repeated burning released sufficient nutrients to the soil to create rich, dark soils known as *terra preta do Indio* (Indian dark earth), which are still present on the terra firme. Others have hypothesized that anthropogenic burning beginning eleven thousand years ago altered forest ecology and regional climate sufficiently to cause the otherwise unexplained band of lower average rainfall through the central Amazon basin.

Terra preta typically occurs in small patches of a few hectares on low hills overlooking rivers, the kind of terrain favored by Indians. Together these small patches cover about 10 percent of the Amazon, an area the size of France, reflecting the extent and stability of human communities in the Amazon before contact with Europeans. The black soils are typically half a meter deep and full of broken ceramics. The oldest terra preta soils, in the central and lower Amazon, date back more than two thousand years. Contemporary farmers prize terra preta for its great productivity, and scientists have recently become intrigued by its existence. Fertile patches of terra preta exist under the same conditions of warm, wet climate that create adjacent red and yellow soils of low fertility. Warmth and rainfall leach nutrients and organic material from most tropical soil once it is cleared of vegetation. The terra preta, however, has more phosphorus, nitrogen, calcium, and other nutrients that are available to plants, as well as more organic matter. It retains moisture better and is not rapidly exhausted by agricultural use when managed well. Somehow, the ancient inhabitants of the Amazon were able to continue farming for centuries, creating and maintaining soil fertility in a manner that modern farmers have not been able to mimic. Traditional knowledge was lost with the massive population decline caused by diseases brought by Europeans. Only about fifty thousand Indians remain in the Amazon today.

Contemporary slash-and-burn farmers shift from one plot to another, newly cleared plot every two to four years. Archaeologists hypothesize that clearing the rain forest with stone tools was so difficult that such rapid movement among farm plots would have been impractical in the past. Instead, the ancient farmers learned to practice slash-and-char by burning the forest only partially, creating charcoal that was then stirred into the soil. Nutrients from excrement and waste such as turtle, fish, and animal bones were added to the charcoal-rich soil, and these nutrients helped to support a richer diversity and abundance of soil microorganisms. In addition, the ancient farmers practiced a type of agroforestry, growing at least thirty species of useful trees as well as annual crops. Scientists now hope to mimic some of these ancient techniques to help Amazonian soils support crops.

Manaus

Ranches and towns become more common along the riverbanks approaching Manaus, and swaths of the dense native vegetation are cleared for grazing lands and banana groves. The cuts of roads form reddish scars on the terra firme, and soon high-rises, jet aircraft, huge oil tankers, and smoke rising from fires on the bank are visible along the river. Moving from the headwaters downstream is a progression forward in time and downward in ecosystem degradation as a result of human impacts.

Manaus is a city of a million and a half people incongruously located in the middle of the jungle. The city is a major industrial and manufacturing center established with the status of an international free-trade zone in part because the Brazilian government fears losing sovereignty over the Amazon basin if the region is internationalized as a biosphere reserve. The Portuguese established Manaus in 1660, calling it São Jose do Rio Negro. The name subsequently changed to Manáos to recognize an Indian tribe that used to live here. Many of the cities along the Amazon experienced a period of growth and prosperity associated with the rubber boom, but Manaus is the preeminent boom city.

People have used Amazonian rubber for as long as written records of the region exist; early European visitors describe indigenous peoples coating their feet in sticky sap, then drying the sap over a campfire to create temporary waterproof footwear. The English-language name rubber comes from early use in pencil erasers, but Europeans were quick to appreciate the versatility of the substance derived from the sap of *Hevea brasiliensis*. More than three hundred million of these trees grow in the Amazon, mostly singly rather than in groves, and mostly south of the main river. The tree's natural geographic distribution is restricted to the Amazon basin.

The rubber boom of 1880 to 1912 represented the meeting of the Amazon and the Industrial Revolution. All of the industrial world's rapidly growing demand for rubber came from trees in the Amazon that were tapped for latex by incising the bark. The Amazon boom ended when the British successfully cultivated rubber trees in Asia and developed large commercial plantations. The economy of Manaus slumped when the rubber boom ended, remaining stagnant until the free-trade zone was declared in 1967. Now the city manufactures electronics, Honda motorcycles, jewelry, and watches as well as processing locally derived raw materials such as rubber and timber and refining oil from Peru. Roads built during the 1970s link Manaus to Venezuela and to southern Brazil, but the city's most important connection to the outside world remains the Amazon and, twenty-three hundred kilometers downstream, the Atlantic Ocean.

The fish market in Manaus hints at the fecundity of the surrounding waters. Gleaming bodies striped and speckled with the colors of the sun-dappled flooded forest lie stilled on the counters. Fish in the markets are smaller than they were in the 1970s, owing to commercial fishing. Pirarucu (*Arapaima gigas*) formerly reached more than 140 kilograms and nearly three meters long, but now such big specimens are rare. These fruit-eating fish are among the largest freshwater fish in the world. Because they breathe air, pirarucu must surface every ten to twenty minutes, which makes them vulnerable to fishermen. Other species are caught en masse in driftnets.

Fish and the Flood Pulse

In an age when a few keystrokes can call up detailed lists from electronic databases, we do not know how many fish species live in the Amazon basin. Approximately three thousand are known thus far, including about eighteen hundred endemic species that are found only here. This total is about ten times the number of fish species in Europe and two to three times the number in the Congo River. Some fifty new species are identified in the Amazon each year. Ichthyologists are just now beginning to access the fish of the deepwater channels as new technologies become available for catching them. The Amazon basin supports high fish diversity in part because it is so large. Some of the diversity of fish species in the Amazon also derives from the fifty million years during which South America remained an island after separating from Africa approximately sixty-five million years ago. This long isolation allowed species to evolve to fill all the specialized niches of river and floodplain.

The Amazon basin does not just have catfish, for example: it has more than twelve hundred species of catfish. These include fish that can "walk" on their ventral fins between patches of water; piraíba

2.5 Fish of the Amazon. *Above*, a fruit eater. *Below*, fish on sale in the Manaus fish market.

2.5 (continued) Fish that has been nibbled by piranha. Some species of piranha are specialized to feed on the fins of other fish, thus ensuring a continuing food supply without killing their prey.

(*Brachyplatystoma filamentosum*) up to 1.8 meters long and weighing more than ninety kilograms; fish that change their diet between high- and low-water periods; and fish that disperse seeds by eating fruits dropping from trees in the flooded forest. Other species of catfish, including those known as the piramutaba (*Brachyplatystoma vaillantii*) and dourada (*Brachyplatystoma flavicans*) in Brazil, migrate five thousand kilometers upstream to spawn, more than the east-west span of the continental United States. Amazonian catfish have the broad, flattened head and barbels characteristic of catfish, but otherwise they defy expectations with their sleek, muscular bodies striped like a tiger or gleaming silver. The catfish need strong muscles to hunt in the swift waters of the main channels; the Amazon basin has more predatory catfish species larger than seventy-six centimeters long than any other freshwater system in the world. The catfish also need senses other than eyesight to find their prey in the murky depths. Their main competitors, botos and tucuxi, use sonar. The catfish use a highly developed organ that links the inner ear and swim bladder to let them hear underwater, as well as olfactory cells on their barbels and bodies. Although most of the fish species in the Amazon are less than ten centimeters long when mature, the catfish actively seek out the larger species for their prey. The fish found in catfish stomachs also indicate that they are hunting throughout the entire water column rather than just remaining near the riverbed.

The turbid waters of the Amazon present special challenges for fish navigation and communication. Approximately sixty types of "electric" fish are present, finding their way through the murky waters by sending out electrical pulses that they receive back as signals bouncing off surrounding objects. Catfish use barbels as sensitive as a cat's whiskers. Lateral lines on bony fish help them sense changes in pressure, allowing them to school. Stridulation with hard structures, a process akin to crickets' "calling," aids communication among some fish species. Using these adaptations for communication, the fish school at tributary mouths. Here predators such as botos gather too, and the pink flash of a boto breaching to breathe provides a brief hint of all the activity going on below the surface.

Some of the Amazon fish are anadromous, spending a portion of their lives in the Atlantic Ocean but ascending rivers for breeding. Steep rapids and waterfalls punctuate the Amazon's headwaters where the rivers rush down from the Andean foothills to the vast lowland basin, or where the lowland tributaries flow off the edge of the Guiana Shield in the northern Amazon basin or the Brazilian Shield in the southern basin. Migrating fish jump up these rapids and falls as energetically as salmon migrating up the rivers of western North America.

The river's flood pulse sets the tempo for the life of most species of fish. More than two hundred species have evolved to eat fruits dropping into the waters of the flooded forest. These fish eat well and build up stores of fat that may see them through the dry season. Another hundred species of fish specialize in grazing on the algae, protozoans, and floating organic matter that collect around the flooded tree trunks. Many fish species spawn on the floodplain edges, and the currents carry the young fish into floodplain nurseries. The young fish feed well in the warm, shallow, nutrient-rich waters of the floodplain, gaining size and strength before retreating with the falling water back to the main channel and the perils of botos and other predators. In parts of the Amazon with a single dominant high-water season, fish species spawn once a year. In parts of the basin with two rainy seasons, some fish have more than one breeding season each year.

Most important, the floodplains are a critical source of primary production, the base of an ecosystem's food web, where plants transform the sun's energy into a form that animals can use. In the Amazon, primary production occurs in the flooded forests, in the floating meadows, and through plankton growth in floodplain lakes. Because limited sunlight penetrates the muddy whitewater rivers, primary production is minimal, and the rich flush of organic detritus, plankton, floating meadows, and newly grown fish that returns to the main channels from the floodplain during falling water is critical to supporting the catfish and other predators of the larger rivers. Primary production on forested blackwater rivers starts with plant parts falling from into the water overhanging vegetation. In other Amazonian environments, rising water levels trigger growth of plants, including the *Paspalum* and *Echinochloa* grasses. Invertebrates, fish, and other animals feed on the plants and in turn provide food for predatory species.

Decades of work on the plant dynamics of the Amazonian floodplains, and the exchanges of fish and other organisms between the main channels and the floodplains, led German ecologist Wolfgang Junk to develop the concept of the flood pulse. This now forms a powerful conceptual model for understanding the dynamics of large floodplain rivers around the world. Tropical South America, and especially the Amazon basin, provides the perfect location for developing this conceptual model. Junk estimates that the flat landscape and pronounced periodicity of a large amount of annual rainfall result in 20 percent of tropical South America being covered in wetlands, with most of these wetlands lying on river floodplains.

Plus or Minus One Mississippi

The large scale of everything on the Amazon makes it a challenging place for scientists to conduct field measurements. Geologist Bob Meade described the specially designed sediment sampler scaled up for the Amazon and carefully lowered over the side of a research vessel using a strong steel cable. The surging currents of the Amazon's depths broke the cable, and the sampler the geologists had named "the jolly green giant" was lost forever on its inaugural deployment.

Obidos is the farthest downstream point where the discharge of the Amazon can be gauged with any accuracy. At Obidos the Amazon flows at approximately ten times the Mississippi's flow rate. Striving for plus or minus 10 percent accuracy in their measurements of river discharge at Obidos, U.S. Geological Survey scientists and their Brazilian colleagues would unwind with cold beers at the end of a successful day, congratulating one another on measuring the Amazon's flow to plus or minus one Mississippi.

Downstream from Obidos, the corridor of the Amazon remains a vast watery world, but the margins appear more firmly bounded by terra firme uplands. Water hyacinths bloom in feathery clusters of white and pale lavender along quietly flowing tributaries. Each blossom lasts only a day and wilts after sunset. This apparent fragility is deceptive, however, for water hyacinths are superplants in other ways. Once a mother plant is established, it quickly sends stolons branching out in all directions. On the tip of each stolon new plants rapidly form and send out stolons of their own. The stolons can break away or decay when the young plants become established, but they can also remain to form a huge mat of interconnected plants within a few months. Petioles on the plants swell into spherical bladders that keep the mat floating. Water hyacinths also propagate rapidly by seed. Each seed can remain dormant for up to twenty years and then give rise to sixty-five thousand new plants in a single season. In its native Amazon, the water hyacinth is a lovely but limited addition to the flooded forest. Elsewhere in the world, the superplant is an invasive exotic that chokes waterways.

The Amazon divides into many narrow distributary channels as it approaches the ocean. Low tide exposes clay cutbanks riddled with crab holes. This portion of the Amazon is a depositional area of many small islands separated by narrow creeks. As the daily rise and fall of the ocean tides affect water level and current speed here, sediment drops from suspension, creating the start of the Amazon's delta zone. The river does not build a classic delta into the ocean, in the manner of the Mississippi or the Ganges-Brahmaputra river system. Instead, about three-quarters of the sediment plume from the Amazon is carried onto the continental shelf, and another quarter is carried north by ocean currents to join the sediment plume coming off the Orinoco River. Even this contribution is enough to overwhelm the Orinoco sediment, and there is more Amazon sediment than Orinoco sediment on the continental shelf off the Orinoco River.

The Amazon remains fresh rather than saline all the way to the Atlantic. Five hundred years ago the Spanish explorer Vicente Pinzón encountered a spot where the sea tasted fresh as he traveled up the coast of Brazil. He named the spot the sweet sea, or *mar dulce*. Despite this plume of fresh Amazon water intruding into the ocean, the tidal influence on water level fluctuations equals that of the annual flood pulse from upstream at this point along the river. This daily tidal cycle extends a thousand kilometers up the river, although the water level rises and falls by only a few centimeters that far upstream.

Exploiting the Amazon

Among the islands in the Amazon's delta is Marajó, one of the world's largest river islands at approximately fifty-two thousand square kilometers. The approximately forty thousand people of the city of Breves, at the western end of the island, are mostly involved in the lumber industry and in commerce. The clear-cuts around Breves create an exception: the deforestation of the Amazon, which has received so much international attention, is mostly not apparent along the river corridor. The most extensive deforestation at present occurs on a south-southeast axis along the existing highways, and Brazil is colonizing the Amazon from the south. Future construction of roads will likely determine the pace and location of deforestation and settlement.

Road construction is partly driven by pressure for land as people migrate from other regions of Brazil. The population of Brazil currently doubles approximately every fifty years, although at 170 million people the country still has a much lower population density than the United States. Those 170 million people and their descendants, however, seek higher living standards and more space.

Approximately 85 percent of the 3.6 million square kilometers of the Amazon basin in Brazil are forested, and approximately 80 percent of that forest is intact. This forest stores the equivalent of about twenty years of industrial output of carbon dioxide, so its role in global climate, as well as its biodiversity, is of international concern. A 2005 study based on computerized climate modeling, for example, predicts that Amazonian deforestation will shift patterns of tropical air circulation, allowing summer moisture to travel farther north and strengthening the summer monsoon in the southwestern United States.

The potential local effects of deforestation on hydrology and soil fertility also are worrisome. Reasoning by analogy from the history of land use in river basins of temperate regions, scientists predict that the severity and frequency of floods will increase in the Amazon lowlands as a result of deforestation. The mechanisms of such changes are well documented in more densely settled regions, where removing trees leads to erosion of hill slopes. Eroding hill slopes cause sedimentation in river channels, which then lose capacity to transport large volumes of water and flood overbank more frequently. Deforestation also results in loss of hill slope infiltration capacity and consequent increases in rapid delivery of precipitation from hill slopes to stream channels, further exacerbating flooding. Reduced transpiration as plant roots extract less moisture is also likely to be an important contributor to higher, faster rises in floodwaters within the Amazon basin. Water and sediment rushing off hill slopes carry carbon and nitrogen along with them. The secret to sustaining life in the Amazon basin is that nutrients that elsewhere are stored in the soil are here stored in the organisms of the forests and rivers. One of the tragedies of contemporary deforestation in the Amazon is that it releases these stored nutrients to be carried down the intricate network of rivers and released into the sea.

Other human enterprises besides deforestation now affect the Amazon basin. Development of oil and gas reserves has enhanced road building and caused spills of crude oil into the Amazon's tributary, the Urucu River. Principal gold mining areas form a belt through the south-central Amazon to the southwestern headwaters and along the northern margins of the basin. Mercury was used to process placer gold in some of the headwater streams, and the highly toxic mercury gradually made its way down tributaries and into the main Amazon. Somewhere between nine hundred and twelve hundred tons of mercury were released into the Brazilian portion of the Amazon basin during the 1990s as part of gold processing. Water, sediment, fish, and the people who eat fish along the Tapajós and Madeira rivers now show increased levels of mercury.

Although the main river remains undammed, twenty dams on tributary rivers fragment the connectivity of the river ecosystem and regulate the downstream movement of water. More than 1 percent of flow in the river basin is regulated, which is enough to earn the Amazon a moderately impacted rating on a 2005 evaluation of flow regulation on rivers around the world. As of 2003, seven hydropower dams existed in the basin, primarily on southern tributaries. By 2008, the Brazilian government proposed to build ten more dams within a few years, including some extremely large ones. One of the world's largest hydroelectric plants, the Tucuruí, was inaugurated on the Tocantins River in 1984. Power generated by this dam has enabled large-scale industrialization in the eastern Amazon. Brazil also proposes to build what would be the world's second largest hydroelectric dam along the Xingu. The primary purpose of this dam is to provide electricity for Chinese companies involved in metal mining and manufacture in the eastern Amazon.

Brazil is at present the third or fourth largest producer of hydroelectric power in the world, although only 4 percent of its hydropower potential is developed (compared with 24 percent development of potential hydropower sources in the United States). Study after study has demonstrated how dams have impoverished the diversity and health of rivers in the United States. Studies of Brazil's Tucuruí Dam show a 50 percent reduction in fish diversity downstream, yet the Brazilian government continues to support the construction of new dams.

Plans to divert water from the Tocantins basin into Brazil's semiarid Nordeste region largely ignore the potential for regional water shortages in the Amazon. This potential exists, however. By December 2005, the worst drought to occur since records began to be kept nearly a century ago had evaporated floodplain lakes, resulted in forest fires and the deaths of fish and crops, and stranded caboclos who rely on boat travel along smaller rivers. Scientists attributed the drought to the rise in water temperatures in the tropical Atlantic Ocean that also triggered the devastating hurricane season of 2005. More ominously, they speculated that if global warming is involved, this may be the start of more severe and frequent droughts in the Amazon basin. Thus, although less altered than most of the world's large rivers, the complex world of the Rivers Amazon has been altered and continues to change in response to human actions.

The Ultimate River

The immensity of water flowing down the Amazon represents the world's ultimate river: the largest drainage area, the greatest volume of discharge, and according to some geographers, the longest channel. This river creates its own world. Every living thing within the floodplains of the Amazon basin must adapt to the rise and fall of the great river's seasonal floods. Trees grow buttresses and stilt roots to remain upright in the waterlogged soil and send seeds floating out on the currents. Fish migrate thousands of kilometers to reach the abundance of the flooded forest, feasting on the fruit and seeds dropped by the inundated plants, then fasting during the dry season. Other fish time their spawning so their young can use the flooded forest as nursery habitat. Freshwater sponges in the flooded forest flourish during the wet season, then become dormant through the dry season. Spiders, centipedes, ants, and other small creatures move up the trees of the flooded forest as available habitat contracts with the rising of the waters, then move back down as the waters ebb. Forty species of monkeys and more than nine hundred species of birds move between the terra firme uplands and the flooded forest to take advantage of food and habitat as they become available.

The river world overflows with abundance. Some thirty thousand species of flowering plants grace the flooded forest, the terra firme, and the floating meadows. This is three times the number found in Europe. The rivers contain the greatest variety of predatory fish of any freshwater system on Earth. The tributary Rio Negro alone contains an estimated seven hundred fish species, six to seven times as many as in all of North America. In all, the region contains at least two million species of living organisms, relatively few of which have been studied or are well understood by scientists. This environment that is so poor in nutrients is nonetheless rich in life.

Now, unfortunately, the Amazon is also unique for the limited degree to which humans have altered the dynamics of the river basin. The Amazon still floods thousands of square kilometers each year, creating a complex topography of levees and secondary channels and carrying vast quantities of sediment down to the ocean. Countless fish still migrate up and down the river corridor. The Amazon provides a glimpse into the way many of the world's great rivers functioned before they were dammed, diverted, and polluted. But now the Amazon basin approaches a critical point in history. Commercial utilization of the uplands, the flooded forest, the power of the seasonally flooded rivers, and the abundance and diversity of the fish fauna will inevitably alter the Amazon. It remains to be seen whether this alteration takes the form of impoverishment, leaving the Amazon basin drained of the complexity that makes it such a wonderland at the start of the twenty-first century, or whether the alteration represents a milder loss of abundance. The evolutionary fate of hundreds of thousands of species, and of an entire river ecosystem, is in human hands. If in the future people in the Amazon can no longer hear the soft whoosh of botos surfacing to breathe, or see squirrel monkeys flinging themselves across gaps in the forest canopy like trapeze artists, or follow green tunnels into the secret world of the flooded forest where submerged microphones record the crackling and buzzing of electric fish beneath mats of grama dulce, we may have to begin children's fairy tales with, "Once, long ago, there was a world full of wonders. . . ."

Interlude

After a journey of nearly four months down the Amazon River, the water droplet that fell on the upper basin as late winter rain reaches the Atlantic Ocean. The Amazon enters as a gushing, unified flow with enough mass and momentum to lower the salinity more than a hundred kilometers out into the Atlantic. The floating islands of grass carrying insects and larger creatures, the big logs, the masses of sediment suspended in the warm brown water—everything gets rushed into the open ocean to mix with the warm surface current in the Atlantic that has flowed westward across the ocean south of the equator and then turned north and accelerated up the east coast of South America.

Approximately one-third of the mass of sediment brought down by the river enriches the delta of the lower Amazon. Another one-sixth is carried northwestward by the North Brazil Current and deposited along the coastlines of French Guiana, Suriname, and Guyana. The remaining half accumulates on the continental shelf adjacent to the Amazon's mouth at a rate that scientists estimate at half a billion tons each year. This tremendous accumulation has created the Amazon Fan, a wedge of sediment with a surface area larger than the state of Florida (more than 700 kilometers long and 250 kilometers wide) and eleven thousand meters thick that has sufficient mass to cause local flexure of Earth's crust. Sediment accumulating at the upper part of the fan periodically reaches an unstable mass and abruptly fails, flowing down the steeper offshore topography and out on to the abyssal plain of the deep ocean. These abrupt failures known as turbidity currents are the submarine equivalent of a debris flow on a terrestrial hill slope, and as they flow downslope at speeds of two to four meters per second they carve a canyon into the submarine fan. This canyon is the lesser known remainder of the great river. The terrestrial portion of the Amazon River is approximately sixty-three hundred kilometers long. The submarine portion of the river's channel extends another nine hundred kilometers beyond the edge of the continental shelf and has a much steeper gradient than even the Andean headwaters of the terrestrial portion of the river.

When the Amazon is flowing high, as it is in early June, intense vertical mixing enriches the water in its estuary with radium-228. Scientists use this distinctive isotopic signature to trace the Amazon water as it moves north and west along the coast of South America as part of the North Brazil Current. The naturally occurring radioactive isotope of radium that provides a tracer for the Amazon water is overshadowed by the many artificial radionuclides measured in low concentrations throughout the global oceans as a result of discharges from nuclear facilities and weapons testing. Fallout from weapons testing is the single largest source of radionuclides in the ocean, but industrial plants and military activities create locally high concentrations.

As the Amazon's freshwater mixes with the seawater, the unified flow separates into lobes. The lobes in turn separate into freshwater lenses three hundred kilometers in diameter that are twenty to fifty meters deep. These lenses of freshwater move like rafts within the North Brazil Current, beyond the border of Brazil to the coast off French Guiana. Here some of the water separates from the coastal flow and doubles back to the east and then south in the North Equatorial Countercurrent. The rest of the North Brazil Current continues northwestward past Suriname to form the Guiana Current.

In company with the rest of the Guiana Current, the Amazon water droplet enters the Caribbean between the Windward Islands and the coast of South America. The nutrients leached from the lush rain forests of the Amazon and Orinoco basins nourish a rich Caribbean ecosystem of giant corals and sponges. The Guiana Current can extend five hundred kilometers offshore during spring, but the highest velocities occur along the edge of the continental shelf. By September the current slows considerably as the Intertropical Convergence Zone shifts north. This atmospheric shift weakens the winds along the French Guiana coast of South America, which change from northeasterly to southeasterly. Combined with the decrease in Amazon discharge after

June, these changes also weaken the Guiana Current. But in June the Guiana Current is strong, and the Amazon water droplet is rushed into the Caribbean. Here, fifteen hundred kilometers from its source, the Guiana Current still retains the unique radium signature generated in the Amazonian estuary, despite being diluted by the discharge of the Orinoco River.

Infinite possibilities exist for the water droplet that has moved northward from the Amazon. It could be recycled among the reef communities many times, taken up by a zooxanthella, one of the millions of tiny, single-celled plants within a coral; ingested by a parrotfish grazing on the coral; passed on to a barracuda that eats the parrotfish and then to a large shark that in turn eats the barracuda; expelled by the shark and taken up by a tiny plankton; and so on. But in early September the water droplet is evaporated and moves into the upper atmosphere as part of the Hadley cell. This portion of the equatorial circulation pattern was described in 1735 by English lawyer and amateur meteorologist George Hadley (1685–1768).

Warm, moist air converges at the equator, where the Coriolis force is negligible and the air moves primarily upward. This is the Intertropical Convergence Zone, which mariners know as the doldrums, because wind-driven ships often made little progress among the light, constantly shifting breezes near the equator. Much of the upward movement in the doldrums occurs through thunderstorms that transport heat and moisture through the troposphere to the boundary with the stratosphere at twelve to fifteen kilometers above Earth's surface. At this point the air stops rising and flows either northward or southward. Air moving toward the poles at these heights is acted on by the Coriolis force, creating a clockwise rotation in the Northern Hemisphere and a counterclockwise rotation in the Southern Hemisphere. As the air moves toward the poles it loses heat and moisture and eventually descends back toward the surface at about 30° north and south of the equator. The descending air is compressed and warmed, and twisted by the Coriolis force. Well over a kilometer above the surface, the dry, subsiding air enters the boundary layer and flows back toward the equator at the surface in the form of the trade winds. When the trade winds reach the equator they once again rise, completing the Hadley cell.

The water droplet from the Amazon, having moved northward in the Hadley cell, descends toward the surface in the subtropical high-pressure belt at 30° north, where it is incorporated into the Ferrel cell, named for William Ferrel (1817–91). Ferrel developed a model of global atmospheric circulation based on three cells: the Hadley cell at low

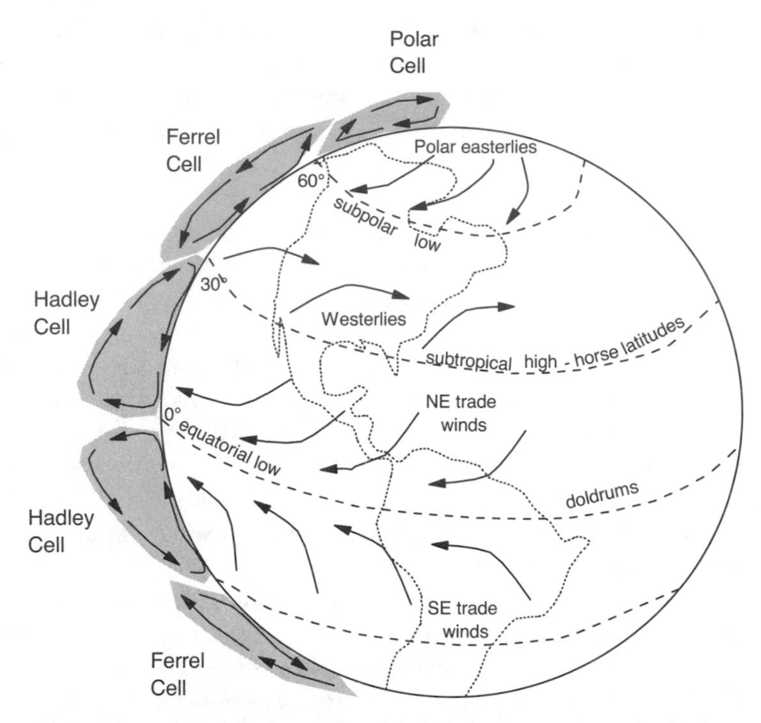

I.1 Idealized illustration of the three-cell model of wind patterns on a rotating Earth.

latitudes, the Ferrel cell at intermediate latitudes, and the polar cell at the highest latitudes. Air descending at the subtropical high-pressure belt can return toward the equator at the surface as part of the Hadley cell, or it can flow toward the pole at the surface as part of the Ferrel cell. Surface flow in the Ferrel cell extends to the subpolar lows, which are large bands of low pressure at 50° to 60° north and south, where the air rises again as a result of convergence with the polar cell. As it rises, the air diverges. Some returns to the subtropical high-pressure belt and descends once more, completing the Ferrel cell. Some air moves toward the poles before descending and flowing back at the surface as the polar cell.

Despite the intuitive appeal of the three-cell model of global air patterns, the Hadley cell is the only portion of the circulation that is well defined. Land-ocean interactions and the effects of surface topography distort the Ferrel and polar cells so dramatically that they are only vague abstractions superimposed on the complexities of real circulation patterns. Scientists use the Ferrel and polar cells almost as shorthand for the general patterns of movement in the higher latitudes.

Moving with the Ferrel cell, the Amazon water droplet travels across the Atlantic toward the low-pressure zone between 50° and 60° north, where its path is skewed east over Europe by the polar front jet stream. Climatologists describe the subtropical and polar front jet streams as rivers of air. Nothing could be more appropriate. Like a river of water, the jet stream branches and rejoins yet remains a distinct entity nine to twelve kilometers above Earth's surface. Moving along at speeds of 110 kilometers an hour during winter and 50 kilometers an hour in summer, turbulent currents roil its flow. And, like a river of water, a variety of substances other than air and water travel with the jet stream.

POPs are one of the travelers riding the jet stream, the Hadley cell, and other circulation systems around the globe. POPs sounds rather cute, and even the full phrase—persistent organic pollutants—doesn't sound too bad, thanks to that "organic" in the middle. In this case, organic can be very misleading, because it simply means that these compounds include carbon atoms. Indeed, POPs are among the most toxic materials synthesized by chemists. POPs include polycyclic aromatic hydrocarbons (PAHs), polychlorinated biphenyls (PCBs), dioxins, and other compounds that, although originally developed for purposes such as insulating or cooling transformers and capacitors, have proved to be both highly persistent in the environment and extremely toxic to nearly all forms of life. From their primary sources in urban areas, the POPs are spread globally through both local transport and deposition and long-range oceanic and atmospheric transport that mainly deposits them at high latitudes. The Amazon water droplet moving across Europe with the jet stream is joined by a variety of POPs released from European cities, growing more sullied with each kilometer of its journey.

As the water droplet moves toward Siberia, it is passed by a counterflow of migratory ducks, geese, and birds of prey returning to lower latitudes for the winter. The POP concentrations in these birds will depend on their gender (males usually have higher concentrations than females) and on what and where they eat.

The air mass carrying the water droplet moves across the vast flat, watery expanse of Siberia. Only the lines of rivers that have not yet frozen create perspective within the gray-and-white plains. The air mass and all its freight—water, dust, PCBs, mercury, and so forth—moves northeast across Siberia until it collides with the polar front. The Amazon water droplet falls as snow on the headwaters of the Ob-Irtysh River late in the autumn.

THREE

The Ob: Killing Grandmother

On a map, the channels of the Ob River basin look like the generic drainage network used in textbooks to illustrate river catchments. Dozens of smaller channels that spread across the broad upper basin join as they gradually concentrate into the slender lower basin. The rivers follow irregular, twining paths on even a very small map, and their sinuous traces join like the branches of a tree to form the perfect dendritic channel network. Except that the Ob network seems upside down. By convention, illustrations show rivers funneling down toward an outlet at the bottom of the drawing. The Ob flows north from the Altai and Ural mountains into the long Gulf of Ob that eventually opens into the Arctic Ocean.

Rivers in the Ob drainage form a dendritic pattern in part because most of the basin is flat and homogeneous. Once the rivers leave the Ural and Altai headwaters, they cross nearly a thousand kilometers of vast Siberian plains. No outcrop of resistant bedrock or rising landmass interrupts their course. This flat land is nonetheless a region of harsh extremes. Air temperature can drop to −55°C along the Ob in January but rise above 40°C in July. Much of the treeless plains of the middle and lower drainage is underlain by permanently frozen ground, so that winter snows melt into waters that form vast marsh complexes rather than soaking into the ground. Small portions of the drainage receive a hundred centimeters of precipitation each year, but most of the Ob's catchment averages only

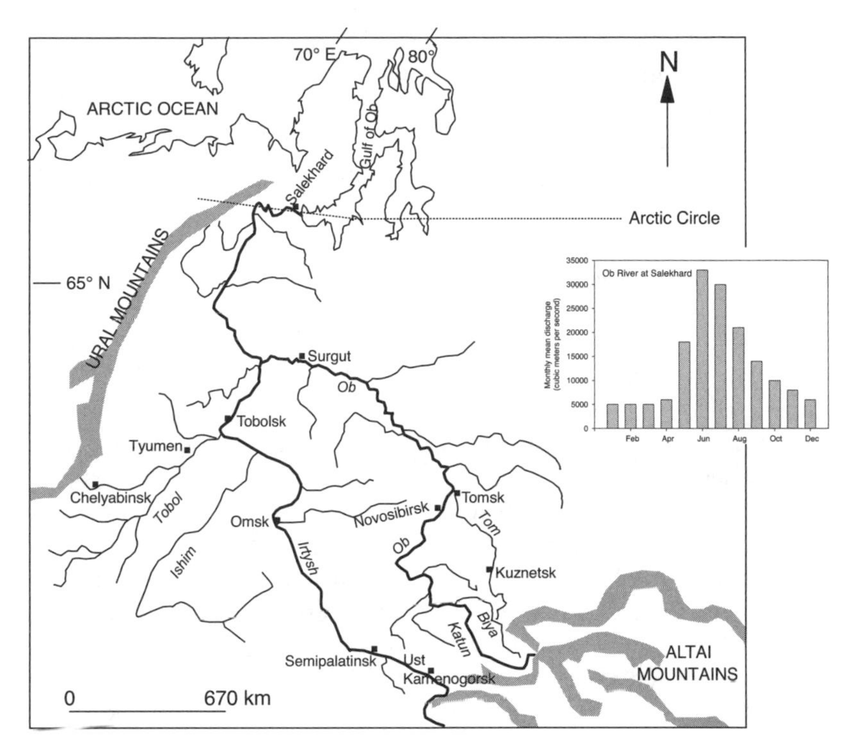

3.1 Schematic map of the Ob drainage basin, showing the major stream channels and cities mentioned in the text. Inset shows mean monthly discharge at Salekhard.

forty to sixty centimeters. Precipitation is also unpredictable from year to year. When averaged over a century, portions of the Ob's headwaters are in drought nearly half of the time.

Much of the precipitation that does fall on the Ob drainage comes as summer rain because a high-pressure zone dominates Siberia from November to March in response to strong cooling of the ground surface. This cooling occurs each year as snow cover increases reflectivity of Earth's surface and inhibits heat flow from the ground. Fronts of moister air coming off the North Atlantic periodically bring winter snowfalls to the Ob basin, and moist, warm air coming from the tropics will occasionally make it as far northeast as the Ob estuary. But mostly winter is dominated by cold, dry air flowing across Siberia from the Arctic Ocean and northern Europe. In summer, solar heating of the ground allows absorbed energy to be returned to the atmosphere, promoting turbulent heat and moisture exchanges that give rise to rainfall. Cool, humid air masses also move in from the Atlantic and Pacific oceans despite the blocking effect of the Ural and the Altai mountains along the margins of the Ob basin.

The Ural Mountains that separate Europe and Asia appear on topographic maps as a long, slender, slightly sinuous ridge. The Altai Mountains are more complex; skeins of ranges run east-west, then diverge to angle northwest or southeast. The mountains divide the closed, arid river drainages of Kazakhstan, Mongolia, and northwestern China from the huge rivers that flow north through Siberia to the Arctic Ocean. At a closer view, the Altai resolve into a rugged landscape. Rocky peaks that remain snow-streaked throughout the summer rise steeply above broad valleys stepped with river terraces. This is a geologically active landscape. Forces beneath Earth's crust continue to push the Altai upward along faults. Sediment-laden glaciers still grind along many of the upper valleys.

The headwaters of the Ob reach deep into the Altai. Rocky little creeks twine steeply down among the ridges, carrying sediment that varies in size from large boulders to silty "rock flour" ground from the bedrock by the abrasive glaciers. Waterfalls and rapids punctuate the course of broad rivers the color of milky tea. The rivers continually cut downward as they wind through canyons, leaving inside each meander bend a series of crescent terraces like those that farmers create on steep hillsides.

In some respects the Altai Mountains remain a nineteenth- or even eighteenth-century landscape. Asian and Slavic peoples meet here, and herdsmen live in yurts as they follow their herds. Men wield pikes to break up enormous logjams that form on the braided rivers as cut logs are floated downstream to collection booms. Hunters and trappers stalk snowshoe hare, lynx, deer, and black bears. Eagles fish the rivers. Mining and logging provide employment little changed from the past century, but then the next bend of the river reveals a hydroelectric dam or an industrial plant.

These contrasts reflect struggles over land use in the Altai region during the past few decades, as exemplified by the Katun River. One of the Ob's major headwaters, the Katun is one of Siberia's least impacted rivers, and its drainage harbors complex and little understood plant and animal communities. Relentless tectonic energy forcing mountains upward and great tongues of ice extending and retreating along the mountain valleys have periodically limited migration and genetic exchange, creating relict populations and endemic genera.

The Katun originates in high mountain lakes. Montane tundra (treeless barrens) at the highest elevations of the Katun drainage gives way

3.2 *Above*, terraces along the Chuya River, a tributary of the Katun River, in the Altai Mountains. Each terrace represents a former floodplain of the river, which has cut downward through time as the surrounding mountains have been elevated by tectonic forces. *Below*, the gorge of the Chuya River in the Altai Mountains. (Both photographs courtesy of Victor Baker.)

to shortgrass steppe, taiga (swampy coniferous forest), wetlands, and meadows at lower elevations. Small and unpredictable rains fall on the well-drained sandy soils of the steppe, which are often underlain by frozen ground. The short growing season and lack of rain preclude crops, so the human inhabitants herd horses, cattle, sheep, goats, and yaks. Fragile soils require that these animals be constantly moved over great distances, creating a seminomadic lifestyle for at least part of the year.

A Dam Defeated

Contemporary Altaians are a Turkic people. Waves of people have advanced and retreated over this region through the centuries as numerous empires have waxed and waned. Landscape has formed a constant throughout the political changes, and landscape features form central elements of the Altaian animistic belief system. Individual lakes, rivers, springs, and mountains are understood to have spiritual owners who must be acknowledged and honored. Shamans negotiate between the natural and the spiritual worlds on behalf of human communities, and subjugation or domination of the natural world is considered unacceptable. The Altaians consider the free-flowing Katun River a living being. The forested slopes that compose the watersheds of rivers and springs are also protected as living beings.

Into this belief system came the Soviets, with their Project of the Century, which was part of Stalin's post–World War II industrialization of Siberia. The project was designed to reverse the great northward-flowing Siberian rivers and make them flow south so that massive dams could harvest energy and irrigation canals could enable crops in arid Central Asia. Electrification of Siberia for industrialization and competition with the West continued to be a major goal for decades following Stalin's death. Dams, dubbed "temples of kilowatts" by the Russian poet Yevgeny Yevtushenko, became a Communist article of faith.

Reversing the drainage of half a continent went slowly, and opposition built by the 1980s. Scandinavia began to protest when predictions of the project's impacts included climate change in northern Europe. Public opinion throughout the industrialized world turned against dams and massive river diversions. Soviet leader Mikhail Gorbachev responded by establishing a special commission from the USSR Academy of Sciences to review the project. This commission, the first successful independent technical review of government projects in modern Russia, issued a negative report, and the project was dropped from the 1986 planning cycle. The Ministry of Reclamation and Water Management proceeded as planned, however, expending funds to initiate construction on individual projects. The resulting public outcry led to termination of the overall project in 1988 and has been credited with being central to perestroika and to fundamental changes in Russia's legislative process. But dam builders are determined people. Efforts to construct hydroelectric projects in the Altai Mountains continued despite the 1988 cease and desist order, triggering yet another wave of protest that became international in scope and incorporated indigenous peoples.

At the center of the Katun Dam protest were plans to construct dams along nearly seven hundred kilometers of the Katun River and its major tributaries. The main dam would be 180 meters high and capable of generating nearly two million kilowatts of electricity, most of which would go to downstream industrial centers. The dam would flood about 5 percent of the available arable land of the Altai Republic's prime agricultural bottomlands, in a country where less than 2 percent of the land is considered arable. Proposed dams would also have flooded over eight hundred hectares of riparian forest, which provide the main source of fuel and building materials for villages in this semiarid landscape, as well as important habitat for wildlife and for plants that people use for food and traditional medicine.

Russian scientists were concerned at the potential for mercury concentration in the reservoir sediments. In the headwaters of the Katun River lie metal-rich rocks. As the lesser component ranges that together make up the Altai Mountains were uplifted by movements along deep faults, molten rock from Earth's interior rose into the crust, carrying in trace amounts the elements that humans prize. Silver, gold, and copper have been mined in the Altai for centuries. Mercury associated with the ore belts has been dispersed throughout the environment as a result of mining, raising mercury levels in sediments of the Katun and Ob rivers. Because mercury travels adsorbed to fine sediment particles such as clay, any collection point for sediments, such as a dam reservoir, can build up dangerously high levels of this extremely toxic substance. Russian scientists were concerned at the threat to water supplies both at the Katun Dam reservoir and downstream.

People protesting the Katun Dam did not have to go far to find a precedent for the dire effects predicted to result from the dam. The Ob River was dammed at Novosibirsk in 1956 to create a hydroelectric power station. The resulting "Ob Sea" flooded some of the region's most productive agricultural land, retaining water that once flowed through the huge areas of wetland on the middle reaches of the Ob. Numerous fish species spawned and foraged in these wetlands, including eighteen commercially exploited species. Reduction in flow decimated the fish once the dam was completed; annual fishing yields abruptly dropped from six or seven thousand tons to only one thousand tons. The power stations cut off 40 percent of the spawning sites used by sturgeon and 40 percent of the spawning grounds for nelma (*Stenodus leucicthys nelma*), which is now endangered. Fish biomass is relatively high in the lakes that feed the headwaters of the Katun River, and herders supplement their summer diets by fishing for grayling (*Thymallus arcticus*) and osman (*Oreoleuciscus* spp.). These fish would also be severely affected by dam construction.

Construction on the Katun Dam, begun in 1982, was temporarily halted by early public protests. When work resumed after 1988, protests increased as unprecedented domestic and international news coverage heightened public awareness. Perhaps most impressive was that concerns over water quality went well beyond the usual discussion of known health risks associated with mercury and other contaminants. Many of the Altaian people who testified at public hearings gave a higher priority to the need to maintain environmental purity, especially the purity of water, than to either employment or material wealth. Describing the Katun River as a "blood artery that is essential to the survival of the Altaian people," more than four thousand inhabitants of the Altai Mountains signed an appeal calling for recognition of the unique riches of the region. Imagine a world where everyone felt this way about water and rivers.

The Katun Dam proposal was defeated. A national park was established around the river's headwaters in 1991, and in 1998 the region was added to the UNESCO system of World Heritage sites. The Katun still flows freely to its junction with the Biya River, where together they form the Ob. Unfortunately, this example of protecting rivers in the Ob drainage is the exception.

Exploiting the Ob

Indigenous hunter-gatherers have lived in the Ob catchment for thousands of years, but there is no evidence that they altered the rivers of the region. It was not until Russians entered the catchment that resource use became sufficiently intensive to affect river form and function. Although not topographically high, the long north-south line of the Ural Mountains effectively formed a barrier to shifting human populations for millennia. Lands west of the Urals were settled and Russianized between the twelfth and the sixteenth centuries, but lands east of the mountains came under Russian control only very gradually after the sixteenth century. By the start of the eighteenth century, Siberia was inhabited by approximately three hundred thousand Russians. At the start of the nineteenth century, more than a million Russians lived in Siberia, along with thirty distinct indigenous peoples. The indigenous peoples, estimated at one and a half million at the start of the twentieth century, had declined to half a million by 1990. Frederick Kempe quotes an indigenous Siberian woman as saying in 1991, "The Russians came like grasshoppers. And like grasshoppers, they didn't do as much good as harm."

Siberia formed a vast storehouse of natural resources for the Russians. Furs were the first magnets drawing them into Siberia. Black fox, white fox, ermine, beaver, squirrel, wolf, wolverine, and sable furs were exported to western Europe through the Siberian towns of Tara and Tomsk. Enormous quantities of grain, potatoes, hides, tallow, bristles, flax, and hemp eventually joined the steady flow of raw products from Siberia to Europe, but the fur trade dominated the Siberian economy for 150 years until displaced by mining. Mines along the eastern Urals began to assume importance early in the eighteenth century. Copper mines were established near Tomsk in 1726. Mines in the Altai led to smelters and supported industries that the Russians began trying to foster during the early twentieth century.

The prevalence of mining in Siberia reflects the geological history of the Altai and Ural mountains. The Urals form one of the largest belts of folded rock on Earth, stretching for almost six thousand kilometers and reaching widths of nearly two thousand kilometers. The contemporary Eurasian continent is a collage of smaller subcontinents that crashed into one another and stuck together during the hundreds of millions of years over which tectonic plates have moved across the planet's surface. The deformation produced by two colliding plates is messy and prolonged. Normally rigid rocks fold plastically under pressure or rupture in faults. Molten rock from Earth's interior migrates upward, metamorphosing the overlying rock and depositing mineral veins as superheated water saturated with metals is injected into fractures in the surrounding rocks. More than a hundred iron ore deposits are concentrated in the Urals, as well as hundreds of copper pyrite deposits and deposits of nickel, cobalt, and aluminum.

Gentle deformation of sedimentary rocks also created broad subsurface flexures in the Urals that effectively trapped oil and gas particles moving through permeable rocks, eventually forming economically valuable petroleum deposits. The first oil was extracted from the Urals in 1929 and from the Siberian Plate east of the Urals in 1959. Numerous deposits have been found since the first strikes, with oil production especially intense today in the Surgut region of the middle Ob basin and gas development focused on the Nadym region near the Ob's mouth. Coal is mined along the entire length of the Urals, as well as from basins spread across Siberia.

Indigenous peoples discovered and used many of Siberia's richest coal, iron, and gold mines before Russians reached the region. Among these peoples are the Shors, who live at the northwestern base of the Altai around the Tom and Katun rivers. The Cossacks, who conquered the Shors, knew them as the Kuznetsk Tatars. The Kuznetsk basin, the richest industrial basin of Siberia, was subsequently named for these people, but now the Shors make up less than a fraction of 1 percent of the population in the Kuznetsk basin.

Historically, Shors were either smiths or hunters. When the Mrassa River thawed each spring, the Shors thanked the river god for the miracle that fish were still alive after the long, hard winter. When American Frederick Kempe visited the Shors in 1991, he noted that the Mrassa River had fallen along with Shor society. The Shors attributed the declines in water level and river fish to Russian lumberjacks who floated logs down the river.

Timber was floated along many of the headwater tributaries of the Ob River until very recently. Many of the trees sank to the bottom, where the Shors believed they destroyed the breeding grounds for fish and poisoned the water with their sap. However, sunken logs would be likely to create a stable framework that would trap gravel and provide spawning habitat for fish, and leachates from logs are natural dissolved organic carbon that would not harm fish. Overfishing, toxic contaminants from industry, and deforestation are more likely causes of the recent changes in the Mrassa River. Deforestation and removal of other natural ground cover reduced the soil's ability to act as a sponge that gradually released water to the river throughout the summer. Lacking this filtration, snowmelt and summer rains produce short-lived floods, but the Mrassa is low and impassable by late summer. One of the Shors reminisced to Kempe about the steamship that plied the river during his childhood. "Now you can only take a canoe down the river. . . . It was once known as Fish River, but now there are practically no fish at all."

As the rivers of the upper Ob flow beyond the Altai Mountains, they increasingly encounter the obstacles that make the survival of any fish seem miraculous. One of the most significant of these obstacles is the huge dam at Novosibirsk. This dam flattens the river's annual hydrograph by increasing low flows and reducing flood peaks. It also traps much of the sediment that used to be carried downstream each year. As usually occurs downstream from dams, the reduction in sediment load caused erosion of the riverbed and riverbanks. Normally this process eventually produces a new stable channel form, but extraction of sand and gravel for use in construction exacerbated channel erosion along the Ob. Between 1966 and 1984, more than forty million cubic meters of sediment were removed from a thirty kilometer stretch of the river channel within the city limits of Novosibirsk. As a result of these changes, river levels fell by an average of a meter, and fish habitat vanished.

The Novosibirsk dam alters the hydrological regime of the upper and middle Ob River for more than a thousand kilometers and helps to earn the entire basin a rating of "moderately impacted" in a 2005 global overview of flow regulation on the world's large rivers. The main effect of the Novosibirsk dam is the loss of the spring flood peak. The Ob historically provided one of Russia's most important fisheries, with particularly rich fish stocks as a result of the well-developed floodplains lining the main channel. The Ob typically has a small flood in late April, immediately after the ice covering the river melts. Ice breakup is accompanied by heavy ice jams that accentuate the spring flooding. A more substantial flood occurs about a month later as snow melts throughout the basin. Many species of fish spawn in the flooded meadows during May, and the young fish grow rapidly in the relatively warm, quiet shallow waters of the floodplain during June and early July.

The reliance of fish on a regular annual flood resembles the situation described for the Amazon, but the high-latitude Ob has far fewer fish species than the tropical Amazon. Despite the favorable combination of a large floodplain system, an extensive delta, and a freshwater bay, the Ob has a fish fauna of only fifty-four species and subspecies. The Ob's fish, however, include several species that together account for 70 percent of the total catch of freshwater fish for Siberia and 40 percent of the total for Russia. Although near the bottom in a ranking of fish species richness relative to drainage area for great rivers of the world, the Ob has large numbers of individuals in its relatively few species. As on the Amazon, these fish depend on access to broad floodplains for their spawning and growth. When the floods occur, how long they last, and how far they spread are vital, and it is just these factors that are most altered by flow regulation associated with the Novosibirsk dam. A minimum of twenty days of flooding are needed for fish to spawn, hatch, and grow each year in the Ob. Floodplain spawning and feeding areas are reduced by half in years of average flow because of the dam and are now essentially nonexistent during years of drought.

Similar but smaller dams exist on the Irtysh River at the Bukhtarma, Ust'-Kamenogorsk, and Shul'binsk hydroelectric stations. A total of eight major dams impede the flow of rivers in the Ob drainage. None of these dams allows adequate passage for migrating fish. This creates at least two problems, by limiting genetic exchange among populations and by constraining each population's ability to find suitable habitat during periods of stress such as low water. By reducing flood peaks, the dams also disconnect the river from its floodplains. Formerly productive floodplain areas along the Irtysh downstream from Bukhtarma Dam are now arid lands no longer useful for fishing or agriculture.

The losses of river productivity at existing dams are not discouraging new proposals to alter flow in the Ob catchment. At the start of the twenty-first century China is pushing hard for construction of a three hundred kilometer irrigation channel to divert some of the flow in the Irtysh headwaters. The Chinese propose to initially draw 5 percent of the river's flow but then steadily increase the diversion to 40 percent in subsequent years. The first 500 kilometers of the Irtysh flow through China, the next 1,700 kilometers through the Kazakh Republic, and the final 2,000 kilometers through Russia to the Irtysh's junction with the Ob. Kazakh scientists estimate that diverting only 6 percent of the annual flow would damage the ecology of the Irtysh and potentially drain the Kazakh Zaisan Lake in the manner of the Aral Sea, which is disappearing through evaporation and lack of river inflow. Water coming down the Irtysh from China is already severely polluted by heavy metals, petroleum compounds, and nitrates, and decreasing flow will reduce the river's ability for self-cleansing. The outcome of negotiations between the Kazakh Republic and China is being kept secret, but China does not have a history of giving way to neighboring countries.

The contemporary status of the Siberian sturgeon (*Acipenser baerii*) reflects the challenges faced by fish in the Ob River today as a result of dams, pollution, and other changes. Like the piraíba of the Amazon, sturgeon are physically impressive fish that typically weigh sixty-five kilograms but can grow to be more than two hundred kilograms and two meters long. Growing slowly in the cold Arctic waters, sturgeon can live up to sixty years. They look vaguely prehistoric, with long, flattened snouts from which barbels hang like lanky mustaches and with ridged lines along their backs and sides. Swimming along the streambed, sturgeon suck up the bottom sediments like piscine vacuum cleaners, filtering out chironomid larvae, amphipods, isopods, and other creatures too small to see. Sturgeon ascend rivers to spawn during summer, and their sticky eggs adhere to the streambed until the larvae hatch. Sturgeon historically ranged throughout Siberia, from the Ob drainage in the west to smaller rivers east of the Lena. In the Ob, they swam from the sea up to the Katun, Biya, Irtysh, and other headwater tributaries. The Ust-Kamenogorsk (1952) and Novosibirsk (1957) dams cut the sturgeon off from about 40 percent of their upstream spawning grounds, and the Shulbinsk hydroelectric dam (1985) downstream from Ust-Kamenogorsk sliced off another length of previously accessible river. These losses are not unique to the Ob basin. The range of the sturgeon throughout Siberia is constantly being reduced, and the number of individuals in each population is decreasing.

The catch of Siberian sturgeon has always been small relative to other commercially fished species. Catches from the Ob basin were usually the highest recorded, reaching a peak of 1,400 tons during the late 1930s. Catches declined severely after that point, reaching essentially zero in the mid-1990s. The earlier twentieth-century catch rates were probably unsustainably high, particularly when coupled with loss of habitat and water pollution. Contaminants in Siberian rivers do not necessarily have the obvious and immediate effect of floating dead fish. Instead, many of the contaminants gradually erode the ability of fish populations to survive, leading to declines in numbers over decades that can be easy to ignore unless someone specifically studies the toxicology of river fish.

Examination of sturgeon in the Lena River between 1964 and 1977 revealed only single cases of degeneration of eggs in females during critical periods of egg growth. But in 1986 the number of females with such defects was an astonishing 59 percent of the total population. Autopsies of female sturgeon essentially revealed a widespread breakdown in reproductive ability. The scientists who detected these abnormalities attributed them to "a high level of water pollution, especially by pesticides." What happened between the 1970s and the 1980s?

The city of Tomsk lies near the junction of the Ob and the tributary Tom River. When Charles Wenyon visited Tomsk in 1893 he noted, "Almost all the various industries of Siberia are represented in . . . Tomsk; its rivers abound with fish, and its forests with game; in the hilly regions of the south there is mining for gold and silver and precious stones; in the wide valleys farmers have tens of thousands of acres of

3.3 Tom River near Tomsk. The river here is severely polluted by discharges from metallurgical and chemical industries and no longer "abounds with fish," as described by Charles Wenyon in 1893. (Photograph courtesy of Pavel S. Borodavko.)

fertile lands under cultivation; and other settlers rear great herds of horses and cattle on the plains."

The short answer to what happened between the 1970s and 1980s is that all of these "industries" grew up, as well as adding new enterprises such as chemical manufacturing and oil and gas drilling. As mining, agriculture, and the other industries became more extensive, they also became more intensive, often using the new synthetic chemicals, including pesticides, developed after World War II. Wenyon described the Tom as "a fine, navigable river." A 1998 report on the status of the Tom River noted the presence of "several environmentally-devastated urban and industrial areas. . . . River stretches receive heavy pollutant loads and sometimes experience shortages of water" (Vasiliev 1998).

The Tom basin includes the Kuzbass Industrial Complex, which is now the largest center of coal mining in Russia, as well as a center of metallurgical and chemical industries. The Tom River supplies the water for these industries and collects the wastewater from over a thousand industrial plants and human settlements, as well as waste from cattle breeding complexes and the runoff from agricultural lands. As the 1998 report summarized, "The river waters contain practically all types of pollutants (a variety of organic compounds, heavy metals and others)." Major contaminants include petroleum compounds, phenols, formaldehyde, naphthalene and its derivatives, and heavy metals such as cadmium, zinc, chromium, and copper. All of these are extremely toxic to all living organisms at the levels detected in the Tom, which "often drastically exceed the national standards for water quality."

Many of these compounds accumulate during winter periods of low flow. Although the spring and summer floods dilute the pollutants, runoff from adjacent agricultural lands carries pesticides into the river, and they increase in concentration during the growing season. Only about 10 percent of wastewater in Russia is subject to any type of purification. Severe pollution of many drinking water supplies results in a high rate of intestinal diseases and hepatitis, as well as increasing the risk of carcinogenic and mutagenic effects on humans. As might be expected, these loads of contaminants are also associated with dramatically depressed numbers of river organisms downstream from large industrial centers. Most downstream aquatic species have elevated levels of bioaccumulated heavy metals and other chemicals.

It is not only those organisms unfortunate enough to live downstream from Kuzbass that are affected by a variety of contaminants. A study of surface waters throughout the Tom and upper Ob drainages found widespread contamination by chemicals such as dialkyl phthalates. These chemicals appear to be literally falling like rain. Atmospheric precipitation spreads the chemicals throughout entire watersheds and, through infiltration, into groundwater. Water collected from a spring high in the Altai Mountains had extremely high concentrations of hydrocarbons, despite no obvious source of water pollution.

By the late 1990s, Russian scientists were publishing papers in international journals that included statements such as "The quality of the Ob water deteriorates constantly, and this results in an intense transformation of the ecological and geochemical conditions of the environment of the entire river basin," and "The majority of the tributaries [of the Ob] are classified as contaminated and highly contaminated water bodies." Scientists also pointed to specific sources of contaminants, such as the phthalic ester leaching from plastic wastes at the Tomsk Petroleum Chemical Plant. Although statements like these have been made in Western industrialized societies for more than thirty years, they are new with respect to Siberian rivers.

In a paper published in 2000 (Zhulidov et al. 2000), Russian scientists noted that although the Russian water quality monitoring network was one of the most extensive in the world during the Soviet era, the data collected were treated as state secrets. Most scientists, whether Russian or foreign, were unable to access, analyze, or publish water quality data before the 1990s. The restrictions have now been lifted, but many of these records remain inaccessible or of such poor quality as to be unreliable for longer-term assessments. Most papers detailing water pollution in the Ob basin deal only with conditions from the late 1990s onward. Lack of information on longer-term trends in water quality limits scientists' ability to identify thresholds beyond which a river ecosystem or individual species cannot be sustained or to understand how contaminants are dispersed through a river network and how they persist or degrade with time.

The Nuclear Ob

Stalin initiated a crash program to develop the USSR's nuclear capabilities and succeeded in detonating a bomb by the end of the 1940s. But one bomb, or a few bombs, was not enough for this man who helped rush the world toward the philosophy of mutually assured destruction. Stalin set in motion a vast array of mining, processing, manufacturing, and disposal sites across the Soviet Union, all developed in the greatest possible secrecy. As Philip Pryde wrote in 2002, "Even entire secret cities were built with impressive speed, but often with far from impressive safety procedures" (Pryde 2002, 448).

Uranium ore must be mined and then concentrated because, like most ores, the economically valuable material is disseminated widely through large amounts of host rock. The concentrated ore then travels to enrichment plants, where the percentage of fissionable isotopes is increased. The enriched uranium is processed into fuel rods for reactors or into bomb-grade material that can sustain a chain reaction. Used fuel rods must be processed again so they can be stored long term in designated repositories. Nearly everything the uranium touches also becomes radioactive and must be stored in some type of isolation from air, water, soil, and living organisms, commonly for thousands of years.

Much of the Soviet Union's nuclear development occurred in Siberia, which the government viewed as a remote area suitable both for sacrifice to radioactive contamination and for maintaining secrecy about what was being done. Atmospheric bomb tests, nuclear waste storage sites, commercial nuclear reactors, mining, fabrication, and reprocessing sites, and underground bomb tests have all been located in different parts of the Ob basin. Among the most significant sites were the city of Tomsk-7 and the Mayak facility in the southern Urals.

Tomsk-7 is a former secret city near the historical city of Tomsk. Nuclear fuel production reactors, fissile (bomb) materials production, and injection and other high-level radioactive waste storage facilities were located at Tomsk-7. Today this site has the dubious distinction of having one of the world's most extensive deposits of uncontained radioactive wastes, as well as releasing the largest amount of radioactive wastes into the environment. An estimated 127,000 tons of solid radioactive wastes and thirty-three million cubic meters of liquid radioactive wastes are stored underground at Tomsk-7. Liquid wastes have been discharged directly into the Tom River. Storage ponds contain radioactive wastes at levels comparable to those of the most contaminated place on Earth—the Mayak site. Liquid wastes have been injected into the subsurface at Tomsk-7 at depths of up to 365 meters, and although these wastes do not pose an immediate threat, they are uncontained and may eventually migrate into the aquifers that feed the Ob and Yenisey rivers.

Tomsk-7 is known to most non-Russians mainly because of a widely reported accident at the Siberian Chemical Combine on April 6, 1993. A chemical explosion in a large uranium processing tank destroyed the tank and the building that housed it, sending a radioactive plume tens of kilometers to the northeast, contaminating more than 110 square kilometers that included two villages.

The 1993 accident indicates the ongoing problems with large accidental radioactive contamination in the Ob basin, which began at the Mayak facility during September 1957. Russian biochemist Zhores Medvedev first brought Mayak to the world's attention in an article titled "Two Decades of Dissidence," published in late 1976. Medvedev explained that a nuclear accident in the Urals contaminated more than 970 square kilometers. Several hundred people died immediately, thousands more were evacuated and hospitalized, and an extensive area in an industrially developed region was declared dangerous and off-limits. Scientists did extensive research in the contaminated zone for years after the accident, documenting the die-off of organisms from soil microbes to trees, but the results of this research were kept secret until very recently.

The Mayak accident was actually three separate incidents of high-level radiation releases over the period 1949 to 1964 near two secret cities together known as Mayak. During the 1950s this complex covered more than two hundred square kilometers. Scientists now estimate that at least 130 million curies of radioactivity were discharged into the environment from operations in this region. To put this in perspective, this is about 2.6 times the amount of radiation released by the Chernobyl explosion and fire in 1986.

Problems initially resulted from depositing millions of cubic meters of nuclear wastes directly into the Techa River, whose source is near the Mayak plant. High-level discharges into the river occurred mainly from 1949 to 1952, but contamination continued at a reduced rate until at least 1964. The Techa continues downstream to join the Iset, the Tobol, the Irtysh, and finally the Ob, and these wastes moved downstream in the water and adsorbed to fine sediment. Sediment sampling throughout the Ob basin in 1994 and 1995 revealed contamination from the atmospheric fallout to which the entire globe has been subjected. The sediment samples also revealed persistent contamination at four sites: in the Ob above its junction with the Irtysh, coming from Tomsk-7; in the Tobol above its confluence with the Irtysh, coming from Mayak; in the Irtysh above its confluence with the Tobol, coming from the Semipalatinsk weapons test site; and in the Ob delta by at least two of these sources that now pollute the entire length of the Ob River.

Radioactive contaminants are insidious. They create no odors or tastes that might warn organisms of their toxicity, yet they can spread everywhere. As many as 124,000 people who reside near the Techa River were exposed to radiation as a result of waste dumping into the river. Not surprisingly, statistically elevated rates of infant mortality and malignant tumors characterize these residents, as do average life spans substantially lower than those for Russia as a whole. Estimates of the victims of radiation exposure at Mayak rose past one million people by the end of the twentieth century. Perhaps the greatest tragedy is the longevity of this contamination. Strontium contamination persists for hundreds of years. Highly toxic radionuclides of plutonium have half-lives of tens to hundreds of thousands of years. The contamination from Mayak is effectively permanent.

The Petro-Ob

As the Ob and its tributaries flow farther from the Altai and the Ural mountains, the vast, flat steppes dominate the landscape. Every traveler who writes of Siberia comments on the immensity of this landscape where the absence of tall trees creates long vistas. Patches of forest scat-

3.4 Ob River. The extensive low-lying lands on either side of the river drain poorly because of underlying frozen ground, and this promotes the formation of wetlands that formerly supported large populations of fish, birds, and other animals. Dams that store floodwaters and limit overbank flooding, combined with severe pollution and destruction of native vegetation by oil and gas development, have substantially reduced the numbers and vitality of wetland plants and animals. (Photograph courtesy of Pavel S. Borodavko.)

tered throughout the middle and lower river basin increasingly give way to wetlands. The lack of summer warmth limits the northward extent of trees because photosynthesis occurs only when plant tissues are warm enough. The short growing season of the highest latitudes allows plants to produce leaves for photosynthesis and roots for storing food but does not allow sufficient time to grow tall, woody stems. Species of birch (*Betula*), willow (*Salix*), and conifers present in the Arctic are typically dwarfed plants less than a meter tall. Shallow soils underlain by ice also limit rooting depth, and roots must be able to tolerate high moisture levels when the soil thaws in summer.

The steppes surrounding the Ob are a soggy place despite the relatively low precipitation. Wetlands cover over a million square kilometers of western Siberia, particularly in the tundra and taiga zones, where underlying frozen ground and clay-rich soils prevent infiltration and flat terrain limits lateral water movement. Once the river enters the region of permanently frozen ground, no vegetation penetrates the soil deeper than fifteen centimeters or grows upward more than thirty centimeters. The wetlands formed here once supported huge flocks of birds.

Arctic plants must withstand not only cold but also a short growing season, drought, strong winds, and infertile soil. Most hunker down with their leaves close to the dark ground that absorbs the sun's heat as soon as the snow melts. Leaves that develop late in summer and survive the winter without withering jump-start photosynthesis the following spring. Dead and withered leaves that remain attached to the plant help cushion living tissue from drying winds and abrasion by wind-driven snow. Numerous species of woody shrubs, nonwoody herbs, mosses, and lichens nonetheless manage to grow under these challenging conditions, and the Arctic tundra along the middle and lower Ob presents a miniature canopy as varied in color, texture, and form as that of the rain forest.

Historically, the middle regions of the Ob drainage hosted rich agricultural lands from which grains, butter, and honey were exported to Europe. This land of milk and honey also gives richly of oil and gas. Northern Eurasia as a whole contains 13 percent of the world's known oil reserves and 35 percent of its gas reserves. About 3.3 million square kilometers of land in western Siberia are potential oil and gas territory. In the Ob drainage these deposits are concentrated along the middle and lower course of the river. These concentrations have proved problematic for the health of the river ecosystem as the Soviets pushed for maximum industrial output in the shortest possible time, regardless of environmental costs. Only 209,000 tons of oil were produced from the Tyumen region along the Ob in 1964. Then the western Siberian oil fields boomed during the 1970s and 1980s, and by 1984 the Tyumen output had risen to one million tons a day. When the Soviet Union collapsed, the regional economy collapsed with it, although oil and gas production continue. Little money has been available to repair and update the oil and gas infrastructure, which is now prone to accidents and waste.

Satellite images record the extent of destruction. Vegetation cover and soil horizons are completely destroyed within 5 percent of the oil and gas regions and damaged to varying extents in another 40 percent. Much of this destruction arises when drilling derricks and heavy equipment are relocated during summer by connecting a few tractors to a single vehicle that plows up the tundra, forming deep furrows that become sites of soil erosion and thawing. These lower portions of the Ob drainage basin are underlain by permafrost, a layer of permanently frozen ground anywhere from a few centimeters to a few meters below the surface. Clearing vegetation and disturbing the soil changes the ground's thermal regime, melting the permafrost and causing changes in the infiltration and runoff that affect not only the rivers but also the many wetlands in this area. When existing vehicle tracks become filled with water and mud, new routes are plowed alongside them. Each drilling pad or production site generates its own network of access roads, pipes, and power lines. One observer estimates that more modern technology could replace up to twenty of these drilling pads with a single drilling point.

In addition to surface disruption, drilling generates significant contamination. Chemical and oil pollution result from the spreading of solutions used to ease drilling. Oil escapes from boreholes or in spills. Each oil well produces approximately thirty-nine thousand cubic meters of wastes, which are usually buried in nearby pits. In western Europe disposal pits are commonly lined with impermeable materials to prevent seepage into groundwater and soil. Disposal pits in Russia are commonly unlined. Leaking storage tanks that stand on the floodplain release contaminants during high-water periods. The rivers, lakes, and marshes of the Ob floodplain form an interconnected waterway. Surface runoff from marshy areas, which accounts for up to 80 percent of the total drainage area in some portions of the Ob, moves laterally and downhill toward the rivers because the underlying frozen ground and clay-rich soils limit infiltration. Tributaries also carry in loads of contaminants. Two or more small rivers flow through most of the oil and gas fields now being worked in western Siberia, and scientists have documented increases in a wide range of contaminants within these waters.

The background concentration of hydrocarbon contaminants in the middle reaches of the Ob averages four to six times the maximum permissible under Russian law (Shvartsev et al. 1996). When a pipeline breaks or an accident occurs, concentrations of hydrocarbons in the water increase to hundreds or thousands of times the maximum permissible levels, and streambed sediments become heavily and persistently polluted. An average of 120,000 tons of petroleum products are transported into the Arctic Ocean by the Ob River each year, and about 35 percent of the petroleum that enters the river in spills and effluents settles to the streambed and gives rise to chronic secondary pollution. Pollution of the middle Irtysh by petroleum products now averages 2.4 times greater than in the mid-1960s. In the middle Ob, such contamination has increased by a factor of forty-five.

Harsh weather and poor maintenance contribute to frequent oil spills along Siberian pipelines, which are the largest source of contaminants associated with oil and gas drilling. At the beginning of oil field development, the decision was made to send oil, gas, and water together in transport pipes from the oil fields to the shipment points and refineries, in contrast to the practice used elsewhere in the world of separating these elements at the source. The consequence is a highly sulfurous compound that prematurely erodes pipes. As of 2002, more than six hundred large oil pipeline breaks occurred annually in the Ob basin, spilling an estimated three to ten million tons of oil. Individual spills can be extremely damaging. An estimated 420,000 tons of oil spilled into the Ob River in the spring of 1993. Smaller accidents are even more frequent. In total, anywhere from 7 percent to 20 percent of the oil extracted in Russia each year (between fifteen and twenty million tons) is lost through accidents on the pipelines. Up to half of these accidents are caused by steel corrosion, which increases as a pipeline ages. At the start of the twenty-first century, about half of all oil pipelines in Russia were over twenty years old and in need of replacement.

The effects of oil pollution on fish and waterfowl are well documented. Oil pollution reduces the dissolved oxygen content of water, which is particularly damaging at these latitudes because Arctic fish species require high levels of oxygen concentration. Contamination of bottom sediments damages insects and fish larvae, destroys overwintering habitat and spawning grounds, and decreases the possibility of normal reproduction. Petroleum products are rapidly accumulated by organisms such as filter-feeding insects, then passed on to their predators in concentrated doses. Accumulation of petroleum products in the brain of fish depresses respiration, produces a narcotic effect, disrupts spatial orientation and chemoreception, and weakens vision. A large proportion of fish from polluted zones have diseased or deformed internal organs. Migratory fish such as the sturgeon pass through the most highly polluted region of the middle Ob during sexual maturation, when they are particularly vulnerable to contaminant-induced disruption of their own bodies and their eggs and sperm. Single accidents can result in massive fish kills. A pipeline break that spilled petroleum directly into the Ob killed an estimated 114,000 whitefish and other fish. Elsewhere, the fish accumulate so many toxins in their bodies that they are not fit to eat. The Siberian Institute of Fishery estimates that 1,200 streams and 250 rivers have lost their marketable fish resources because of pollution. The commercial fish harvest on the lower Ob in the vicinity of the Tyumen oil and gas region declined by 50 percent between 1970 and 2000.

Diving birds drown from oil on their feathers or die from ingesting oil while preening or eating. Contamination of nesting birds' eggs kills the embryos through asphyxia. Lemmings are trapped in small oil-filled depressions. Oil and petroleum products carried to the Arctic Ocean by the river spread even farther through coastal and marine food webs. Loss of fish and other wildlife in turn affect indigenous peoples who rely on these animals for some portion of their food supply, as well as destroying the reindeer pastures that are vital to indigenous peoples' subsistence.

Local pollution created by oil and gas development may also accelerate global climate change. As worldwide concern grows over global warming, scientists look to the high-latitude regions of the world as the most sensitive sites that will change first. One of the concerns is how changes in snow and ice cover and melting of permafrost may affect the Arctic Ocean and the global water balance. Natural gas in the rock layers above the oil deposits is commonly burned in torches onsite. This burning produces artificial heat comparable to the annual radiation budget of the area and causes periodic mass deaths of migrant birds. Oil films reduce evaporation by approximately 50 percent and ice growth rate by a factor of 1.5. They also decrease the reflectivity of snow and ice by 10 to 35 percent, causing earlier and more intensive melting. Deposition of dust and hydrocarbons exacerbates early melting of snow and degradation of permafrost. Snow melts five to ten times faster in polluted areas than in clean areas. At present, snow covers the Siberian wetlands for about one-third of the year. Following snowmelt, the wetland is flooded by meltwater, and the rising water table changes the water and heat balances. When the water table is above the surface, energy is stored in the wetlands. When the water table falls below the surface, the surface temperature increases and there is a large transfer of energy to the air. If warming and pollution cause early melting of each year's snowfall and permanent melting of frozen ground, the extensive Siberian wetlands could become dry during late summer and autumn, transferring more heat to the air and exacerbating regional and global warming. Increases in air temperature might also alter the carbon balance of Siberian wetlands, changing these vast areas from a carbon sink into a carbon source and further increasing global warming. Extensive areas of oil development and contamination thus have the potential to alter the hydrologic cycle of Siberia and with it that of the Arctic Ocean and the entire planet, but scientists cannot yet predict exactly how.

The dynamics of high-latitude lakes provide an example of why it is so difficult to precisely predict the effects of global change on these ecosystems. Most high-latitude lakes have nutrient-poor waters containing very low concentrations of phytoplankton that form the base of the aquatic food web. In lower-latitude lakes with more abundant algae, incoming solar radiation is absorbed primarily by the algae. In high-latitude lakes, incoming solar radiation is attenuated by the water itself and by dissolved organic acids carried into the lake from the surrounding soils and vegetation. These dissolved acids give the lake water distinctive colors of green, yellow, brown, or black, depending on their concentrations. They also substantially reduce the damaging effects of ultraviolet radiation on aquatic biota. The acids generally occur only at low concentrations, however, and therefore provide only poor protection against ultraviolet radiation. Even small changes in the dissolved organic material entering high-latitude lakes could give rise to large shifts in the transparency of the water column to ultraviolet radiation. This suggests that high-latitude lakes are especially sensitive to climate-related changes in hydrology or catchment vegetation that alter their dissolved acids.

Where these lakes are in developed areas, their ecology is already altered by contamination or by overfishing. Frederick Kempe quotes a woman of the Khanty, an indigenous people of the lower Ob basin, who describes how the Russians have depleted all the important lakes in her region of fish. Where once there had been enough fish in each lake to feed several families for many years, "Now there are only the names left. We have a pond called Fish Lake that has no fish. There is Pike Lake, but it has no pike in it. There is Perch Lake, but you can't find perch there."

Although the high-latitude Ob drainage has only a fraction of the species diversity of the tropical Amazon, both rivers historically hosted an abundance of individual fish, waterfowl, and other organisms nourished by annual floods that spread nutrients across the diverse habitats of the riverine lowlands. As a result of dams, overfishing, and pollution, the diminished ecosystems of the Ob now stand in stark contrast to the still vibrant Amazon.

The River and the Ocean

Having traveled more than fifty-five hundred kilometers from its headwaters, the Ob at length enters a delta nearly a hundred kilometers long that includes fifty islands. The delta, like other parts of the Ob, is the setting for a dangerous mixture of freight traffic to the inland oil and gas fields plus a commercial fishery of great importance. Arctic regions experience an intense pulse of life during the short summer growing season. Plants and animals must breed and grow to maturity during a much shorter time span than organisms at lower latitudes. Similar conditions govern ship traffic. The Ob is navigable for only 190 days a year in the south and 150 days in the north. The freeze begins in mid-October, and the river is not ice-free until the end of May or the start of June. This concentrated season of great shipping and industrial activity coincides with the most biologically productive times of the year. Waters off the Ob estuary are more productive than other coastal areas of the Kara Sea because of the large fluxes of living and dead organic matter that the Ob brings into the sea. These fluxes fuel spring and summer algal blooms of phytoplankton that in turn feed fish, seabirds, and marine mammals.

By the time it reaches the Arctic Ocean today, the Ob is a very sick river. From the nearly three million square kilometers of its drainage basin come more than four hundred cubic kilometers of water each year. And with the water comes a toxic brew of petroleum products, radioactive wastes, pesticides, and heavy metals. People in Russia and around the world acknowledge severe contamination problems within the Ob drainage. The necessity of cleaning up these sites and reducing contaminant input to the river is complicated by the economic problems of Russia at the start of the twenty-first century and by the lack of any seriously implemented environmental standards or mediation. Cleanup is also complicated because the Ob drainage includes fourteen administrative districts in Russia, as well as parts of Kazakhstan (14.5 percent of the basin) and China (1.5 percent of the basin). The basin is home to about thirty million people.

At present the Russian environmental movement primarily focuses on urban areas and on defending people against immediate, severe hazards rather than on broader issues of widespread, persistent environmental degradation. Individuals who speak out too effectively still risk personal attacks that people in Western societies might associate with organized crime. Until dramatic changes occur in social and political conditions in Russia, the attention and pressure other countries bring to bear on Russian environmental degradation will continue to be the major impetus for any improvements.

International attention increasingly focused on the Ob and other Siberian rivers during the 1990s as scientists began to understand the importance of the Arctic Ocean to global climate and the importance of the great Siberian rivers to the Arctic Ocean. The Arctic Ocean contains only 1 percent of the world's ocean water, but it receives 11 percent of global river discharge from the Ob, the Yenisey, the Lena, Canada's Mackenzie, and other rivers. The freshwater coming from these rivers helps to maintain a surface layer of low salinity in the Arctic, allowing sea ice to form readily. The Kara Sea into which the Ob River flows is almost entirely covered by ice from October to May. Because the ice cover reflects solar radiation and separates relatively warm ocean water from the colder overlying atmosphere during winter, the ice helps reinforce the role of the Arctic region as a heat sink for the Northern Hemisphere. During 2008, Arctic sea ice reached the lowest extent recorded since 1978. Continuing changes in the extent of sea ice—estimated at a further 20 percent reduction by 2050—and terrestrial snow cover are expected to amplify the response of the Arctic to global warming, making this region particularly sensitive to climatic changes projected to occur within the next century.

Freshwater that flows into the Arctic Ocean is estimated to stay one to three years, but eventually it must go somewhere. Freshwater leaves mainly as sea ice and low-salinity water that moves through the Fram Strait and into the North Atlantic Ocean. All the ocean basins of the world are connected by an intricate network of both surface and deep ocean currents. The deep waters tend to be colder and more saline, hence denser, than waters at the surface. The production of deep water in the seas around Labrador and Greenland-Iceland-Norway, for example, is crucial to maintaining existing patterns of oceanic circulation, which warm the climate in northwestern Europe through transfer of heat by the Gulf Stream. The production of deep water appears to be sensitive to even a modest change in the salinity of the upper ocean driven by changes in outflow in the Fram Strait. As a result, scientists have recently given much attention to potential changes in river discharge into the Arctic Ocean.

Analysis of flow from the six largest Eurasian rivers to the Arctic Ocean indicates an increase of 7 percent from 1936 to 1999. This translates to an average annual increase of two cubic kilometers and a total flow of about 128 cubic kilometers more water per year now than when routine measurements began in the late 1930s. The question whether a 7 percent increase in freshwater discharge to the Arctic Ocean is sufficient to alter broader oceanic and atmospheric circulation patterns remains open. Scientists do not yet understand global circulation patterns sufficiently to precisely model the effects of such a change.

What is carried into the ocean along with the river water is also of great concern to scientists and to those who make their living from the fisheries of the Arctic. Water quality records suggest anomalously high concentrations of nitrogen in the form of ammonium entering the Arctic from the Siberian rivers, but as noted previously, the reliability of these data is suspect. Relatively little sediment reaches the Arctic from the Siberian rivers, reflecting partly long-term sediment storage on the extensive floodplains along the rivers and partly reservoir storage. The Novosibirsk reservoir, for example, traps up to 90 percent of the inflowing sediment. The sediment that does reach the ocean, however, is likely to include adsorbed contaminants.

Radioactive isotopes of cesium and plutonium appeared in bed sediments in the floodplain lakes and estuary of the lower Ob River starting in about 1952. Concentration of these isotopes peaked, on average, in 1966. This chronology matches the characteristics of global fallout deposition from atmospheric nuclear weapons tests, given the lag between testing and deposition in a sedimentary environment. The ratios of these isotopes in the Ob sediments suggest that about three-quarters of the radioactive nuclides come from the combined sources of global fallout and downstream transport of contaminated sediments from sites such as Tomsk and Mayak within the Ob basin.

In addition to these sources of radioactive contamination, from 1964 to 1986 the Soviet Union directly deposited highly radioactive wastes into many locations in the Barents and Kara seas. These wastes included damaged cores and other radioactive materials from nuclear icebreakers and submarines. Each year operating ships produce approximately 20,000 cubic meters of new liquid radioactive waste and 6,000 tons of solid radioactive waste, and as of 1998 more than a hundred nuclear submarines were in need of decommissioning. When Russia in 1991 officially acknowledged its history of radioactive dumping, the eight nations bordering the Arctic seas formed an assessment and monitoring program to evaluate radioactivity and chemical and metal pollution in Arctic ecosystems. One result of this effort is the *Arctic Environmental Atlas*, a beautiful collection of maps showing everything from animal and bird migration routes to sources and concentrations of various contaminants in the Arctic region.

Radioactively contaminated particles of silt and clay are incorporated into Arctic Ocean sea ice when these particles remain suspended in the water column, which is common in the shallow shelf zone where wave dynamics keep the sea bottom agitated. The contaminants then travel long distances with the ice throughout the Arctic basin and into the North Atlantic. An estimated 209,000 square kilometers of sea ice is produced each year in the Kara Sea and, along with the 256,000 square kilometers formed in the Laptev Sea, accounts for at least 20 percent of the total area of ice flux of a million square kilometers through the Fram Strait. Samples of dense pack ice can contain more than 260 metric tons of sediment per square kilometer of ice, indicating the great importance of ice transport in redistributing sediment in the Arctic. Scientists tracing the movement of ice masses have tracked high levels of radioactive contamination in the Beaufort Sea off the Alaskan coast back to sources in the Kara Sea. Self-purification of bottom sediment in the Arctic basin through degradation of organochlorine and petroleum contaminants has been estimated to take anywhere from five years to fifty years or more. Self-purification from radioactive materials will not occur for thousands of years.

Elevated levels of heavy metals such as lead and cadmium, arsenic and mercury from industrial activities, polyaromatic hydrocarbons from oil products, and organochlorine compounds such as PCBs and DDT are also present in Arctic sediments and sea ice originating off Siberia. Up to half of the Arctic ice cover can be discolored by impurities incorporated in the ice. In addition to pollutants adsorbed to fine sediments, pollutants concentrated in the oceanic surface microlayer and pollutants deposited from the atmosphere also accumulate on the snow and ice surface. Ice melts most extensively along the margins, a region of intense biological activity where the pollutants can easily enter the food web. Migratory organisms such as beluga whales, harp seals, birds, and fish can then distribute the contaminants even more widely in their tissues. Polar bears in Svalbard, along the Fram Strait, now have levels of PCBs high enough to cause reproductive damage.

Health of a River Grandmother

Siberia derives from the Mongolian *siber* ("beautiful, wonderful, pure") and the Tatar *sibir* ("the sleeping land"). The native peoples who lived along the Ob in this beautiful sleeping land for centuries called the river "grandmother." The Ob River dominates the vast, flat landscape it flows over. Spilling from the confines of its channel across the adjacent wetlands each year, the river has nourished a rich fishery that sustained the people living along its banks for millennia. The Ob has been the transport artery for those traveling in Siberia and exporting its riches of furs, timber, minerals, crops, butter and honey, oil and gas. The Ob has indeed been a grandmother to this landscape, sustaining all the creatures that live on it. But now the grandmother is dying, long before her time.

Does this seem too metaphorical or anthropomorphic? Perhaps in personifying the Ob the indigenous peoples incorporated an important insight. A river is more than simply a physical conduit for water and sediment. The processes that create and maintain a river's forms also create and maintain the habitat on which millions of individual organisms rely, from microscopic algae to 1.8 meter sturgeon. The river is an ecosystem, and although it is not human, it possesses the human attributes of vitality and continuity. The Ob will not fully cease to live. Even the radiation of Mayak and the oil and pesticides of Kuzbass are unlikely to render the river completely biologically sterile. But the Ob is becoming impoverished, its vitality and continuity lost as species after species is decimated or destroyed by contamination and loss of river form and function. Unfortunately, the pollution, flow regulation, and loss of fish in the Ob reflect those of many other great rivers, including the Yenisey and Lena in Siberia.

Form and function can be restored if historical conditions can be restored, if the flow regulation of the Novosibirsk and other dams can be ameliorated or the thawing of frozen ground and the early loss of snow cover stopped. Contamination can in some cases be mitigated by repairing the oil pipelines and implementing modern drilling technology to eliminate spills. Species can recover if overharvest ceases and the contaminants that render fish reproduction largely impossible are removed from the system. These actions are difficult and expensive, but not impossible. The global human community can save the Ob and each of the river grandmothers on which our existence depends.

Interlude

At the mouth of the Ob, the far-traveling water droplet that once fell as rain on the upper Amazon reaches the Arctic Ocean in early July. The droplet joins the low-salinity surface water of the Arctic Ocean, and there it remains, very slowly mixing with the Arctic deep water in the Beaufort Gyre. Along with the sediments and nutrients and uncountable billions and billions of other water droplets flowing from the Ob come the radioactive strontium from the Mayak accident, hydrocarbons from the oil and gas fields, DDT, PCBs, lead, cadmium, and other contaminants, as well as water, sediment, and contaminants from the Yenisey and several smaller rivers.

During summer, when the Ob water droplet joins the sea, the prevailing winds create a cyclonic, or clockwise, circulation of surface waters in the Kara Sea. Some of the Kara water moves north into the greater Arctic Ocean, some leaves through the southwestern end at the narrow gap between Novaya Zemlya and the mainland, and some simply remains in the Kara Sea, circling and circling.

As the flow from the Siberian rivers decreases with the approach of winter, some of the seawater freezes. Frozen into the ice are contaminants that have been resuspended in the water column by convective currents, as well as contaminants introduced from the atmosphere. The poisoned ice drifts out of the Kara Sea and in melting releases the contaminants into the biologically productive surface waters, where they are readily taken up in the Arctic food web.

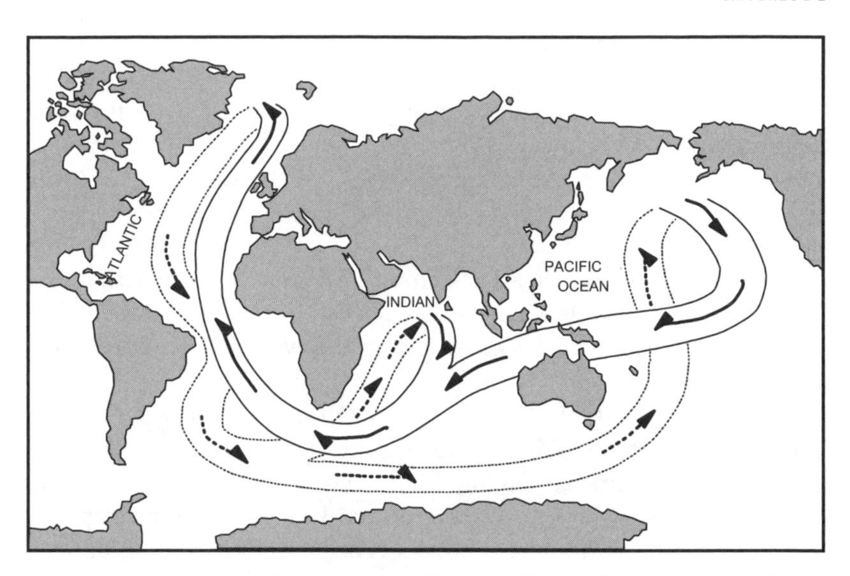

I.2 Schematic illustration of the Great Ocean Conveyor. Warm surface currents are indicated by solid lines, cold deep currents by dashed lines.

The ice forming in the Kara Sea is fresh, so the unfrozen seawater grows more saline and denser. The dense water sinks and flows out of the Kara Sea as part of a cold, deep, salty current flowing toward the equator in the depths of the Atlantic Ocean.

Now the Ob water droplet has joined the slow but immense Great Ocean Conveyor. Moving ten centimeters per second at its fastest, the circulation system includes a flow equivalent to one hundred Amazon Rivers, with more than thirty million cubic meters of water entering the conveyor each second. Warm, shallow water chilled in the far northern reaches of the Atlantic grows saltier and sinks. This dense water flows southward at depth around the southern tip of Africa and breaks into two branches, one moving northward in the Indian Ocean and the other continuing south around Australia, then north into the Pacific. Rising back to the surface as warm, less salty water, the two branches flow south and west, rounding Africa again and flowing north in the Atlantic to complete the cycle.

The Great Ocean Conveyor is critical in regulating global climate because the heat released into the atmosphere in the North Atlantic keeps northern Europe warmer than equivalent latitudes far from the ocean in Canada and Siberia. The conveyor is also critical to contemporary marine life; three-quarters of all marine organisms are maintained by the nutrient-rich waters from the deep ocean that rise to the surface as part of the conveyor.

The Ob water droplet moving slowly southward in the deep waters of the Atlantic is beginning a long journey. A droplet requires about a thousand years to complete one full cycle in the Great Ocean Conveyor. The Ob water droplet would probably take a few hundred years to complete its leg south and east around Africa, up to the surface in the Indian Ocean, and then back around Africa and northward again into the Atlantic. This is where "poetic license" comes in and we ignore the true duration of travel in order to keep the water droplet in contemporary times.

As it surfaces in the Indian Ocean, the Ob water droplet encounters a suite of synthetic organic chemicals. The atmosphere and waters of tropical Asia contain particularly high concentrations of DDT, which is still used in these regions. Chlordane and PCB emissions also appear to be increasing in tropical countries. The movement of the droplet northward in the Atlantic is only one of several mechanisms by which these contaminants are globally transported toward the polar regions.

The Ob water droplet is withdrawn from the Great Ocean Conveyor when it finally reaches the subtropical southern Atlantic Ocean. The high-pressure zone that forms over this region each summer gradually draws water from the Atlantic via moist, shallow equatorial westerly winds that move across Africa. The droplet is evaporated from the Atlantic and swept along on the westerlies.

Leaving the blue oceans behind, the water droplet moves over the lush green forests of the northern Congo basin. The forests breathe out, transpiring moisture recycled from the water that has fallen on them in brief, intense tropical rains. The greenery below grows sparser as the droplet continues east and slightly north. A long, slender lake appears where the crust of Earth is rupturing as the African Plate pulls apart. The land has appeared sparsely settled to this point, but now the hills are cleared for agriculture and are bleeding topsoil from a thousand wounds of landslides and gullies. As the moist air mass in which the water droplet travels is forced to rise over the Ethiopian highlands, the droplet condenses from vapor to liquid and falls as part of the abundant rains that drench the highlands from March to September. The droplet infiltrates into the soil and slowly moves downslope as groundwater that eventually feeds a headwater tributary of the Nile River.

FOUR

The Nile: Lifeline in the Desert

The Nile River forms northern Africa's tree of life. Nearly two hundred centimeters of rain fall each year over the highlands of Ethiopia, where tall mountains force water from the passing clouds. Tributaries of the Blue Nile swiftly collect these waters and rush them down toward Khartoum. To the southwest, the western mountains of the Lake Plateau collect five hundred centimeters of rain each year, filling the great depressions of Lake Victoria, Lake Kyoga, and Lake Albert (also known as Lake Mobutu Sese Seko). These lakes feed the Bahr al-Jabal, which flows northward into the vast swamplands of the Sudd, emerging on the other side as the White Nile. The White Nile flows a thousand kilometers northward to Khartoum through a desert that contributes little to its flow. Beyond Khartoum the two great branches of the Blue Nile and the White Nile unite to form the Nile, the trunk of the tree. The Atbara River forms a low-growing branch, but otherwise the trunk extends uninterrupted to its "roots" in the Nile delta 170 kilometers upstream from the Mediterranean Sea.

Water streaming northward through this long, narrow river basin nourishes everything that lives in the vast deserts of the Sudan and Egypt. A map of the channel network in the Nile catchment looks strangely unfinished, as though the cartographer had forgotten to delineate the channels of the lower, northern half of the catchment. This northern portion constitutes more than 40 percent

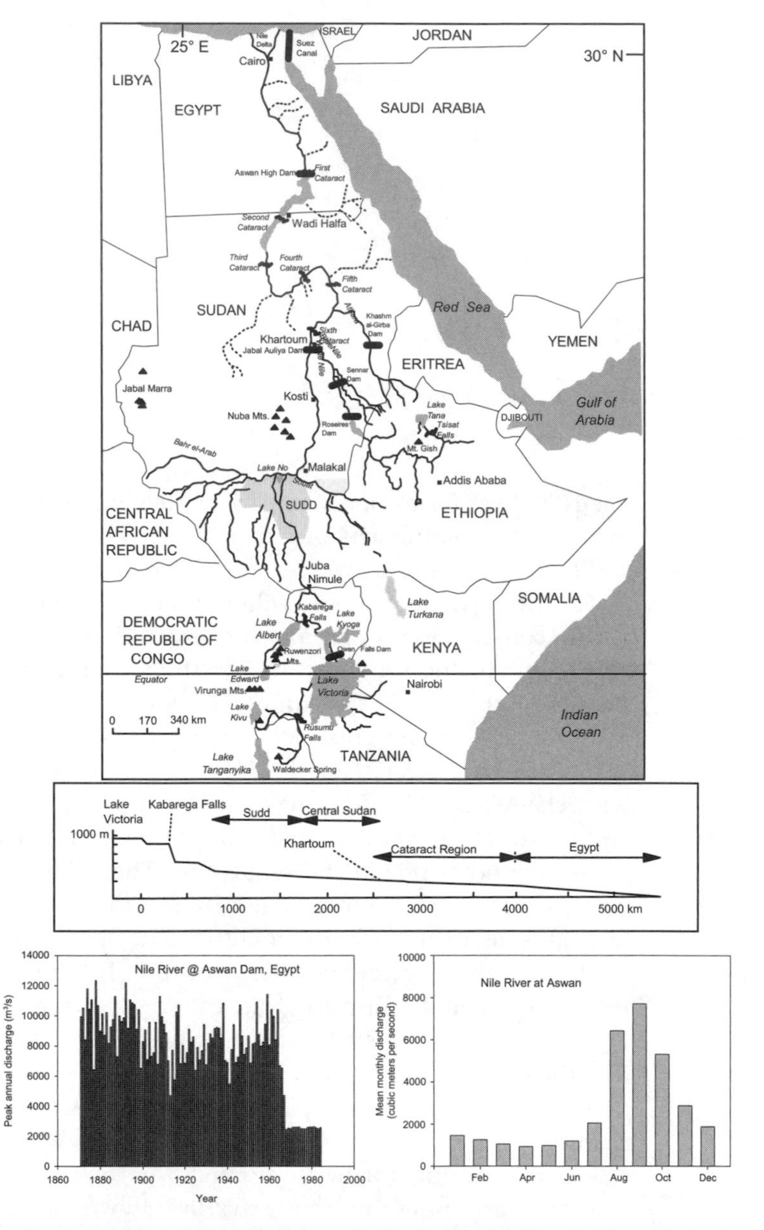

4.1 *Top:* Location map of the Nile drainage basin, showing localities mentioned in the text. Slender lines indicate national boundaries; dashed lines indicate ephemeral channels. *Middle:* Simplified longitudinal profile from Lake Victoria to the Mediterranean Sea. *Bottom:* Mean monthly discharge and peak annual discharge through time below the Aswan Dam in Egypt. The abrupt drop in discharge during the 1970s reflects completion of the dam.

of the Nile's entire drainage area and usually contributes no flow to the river. A delicate tracery of channels crosses these northern drylands, but on a map the channels show as dashed lines, indicating streams that come to life only briefly after rare rainfalls and seldom contribute water to the main Nile. An even more obscure network of channels lies below the sands of the desert, relicts of a wetter period thousands of years ago when rivers as mighty as the Nile flowed over streambeds of gravel. No one knew these ancient rivers existed until satellite imaging radar carried by the space shuttle *Columbia* in 1981 revealed the network of channels below the thick sands of modern desert dunes.

Because the northern deserts receive too little rainfall to produce flowing tributary streams, the Nile's discharge is small relative to its vast length and drainage area. The river crosses thirty-two degrees of latitude and eight degrees of longitude as it descends from the Lake Plateau to the Mediterranean. Waters from nearly three million square kilometers of Africa feed the Nile. Yet the river's average yearly flow is less than 2 percent of the mighty volume that gushes into the Atlantic each year from the Amazon. Where the Amazon produces an average of 730,000 cubic meters of water for each square kilometer of its drainage basin, the Nile manages less than 30,000 cubic meters per square kilometer. But this seemingly poor flow is everything in a land as dry as northern Africa.

People have clung to the lifeline of the Nile for at least seven thousand years. During periods of warmer, drier climate, the Nile valley formed a refuge for humans from the surrounding deserts. When climate ameliorated, the valley formed a reservoir for repopulation of the surrounding area. Archaeological sites containing shells, edible wild plants, and the bones of birds, large mammals, and fish give some indication of the diet of these first people. The earliest humans to live along the Nile relied on the yearly flood, just as do modern occupants of the river valley. The first agriculturalists planted crops where the waters flooded the land from July to September, but artificial irrigation began as early as 3000 BC.

Some practical genius invented basin irrigation, in which water diverted from the river into canals flowed into a series of basins eight hundred to sixteen hundred hectares in area that gradually stair-stepped downstream. High flow from the Nile flooded each basin to a depth of one to two meters, soaking the soil and leaving a residue of rich mud and organic material before draining into the next lower basin and, eventually, back into the Nile. Where croplands adjacent to the river were too high above the channel to employ basin irrigation, a single hard worker using a *shaduf*—a bucket counterpoised on a beam that pivoted through a pair of uprights—could raise 2,200 liters of water a day.

Artificial irrigation increased the area that could be devoted to crops. A rich civilization developed as food production and population grew along the river valley. Archaeological sites along the Nile index the wealth and complexity of dynastic Egypt, with its intricate religious and social systems, hieroglyphic writing, beautiful art, and towering monuments of stone. For three thousand years, from approximately 3050 to 332 BC, the periods of prosperity and trouble in this society corresponded to the times of good flow and of drought on the Nile. The Egyptians worshiped the river as the god Hapi, and they remained thoroughly aware that the Nile provided their only source of water for drinking, irrigating crops, and manufacturing goods. For these desert dwellers, the powerful river was clearly the source of life.

The civilizations that followed the Egyptian dynasties remained cognizant of the Nile's critical role. Greeks, Romans, Byzantines, Arabs, and Europeans came to the region as successive "conquerors," but each in turn was forced to work within the limitations imposed by the river. If the Nile flowed poorly, famine came. And no towns grew up beyond the reach of the river's waters. The Greek historian Herodotus wrote that "Egypt is the gift of the Nile."

Herodotus received a map of the Nile basin in 460 BC. This map looks remarkably accurate by modern standards. The river's source is depicted as springs between two tall mountains. Yet the source of the Nile remained obscure until the exact location of the springs on Herodotus's map was finally discovered by German explorer Burkhart Waldecker as he explored the country of Burundi in 1937, twenty-four hundred years after Herodotus received the map.

Lacking detailed knowledge of the Nile's source, the people whose lives depended on its flow downstream prayed to their gods to keep the river flowing and paid careful attention to its behavior. The modern science of hydrology began along the Nile in AD 622, when Arabs built the first nilometer on the island of Roda at Cairo. A series of these stone pilings engraved with depth scales were erected along the river over many years to gauge the annual high flow. From these records eventually emerged a picture of how the river fluctuated over periods of decades to centuries.

Charting general patterns of river flow through time did not give the ability to predict the Nile's flow in any particular year. All societies are vulnerable to the climatic fluctuations that limit the growth of food, but the so-called Nilotic peoples of the northern African deserts lived closer to the margin of survival than many societies and hence were particularly vulnerable to the vagaries of this single river that rose from obscure sources somewhere to the south. The only means of reducing this vulnerability was to store some of the water flowing swiftly downstream during periods of high flow, but efforts to create storage did not begin until early in the twentieth century.

Documented exploration of the upper Nile waited nearly as long. Arab and Indian traders reached Lake Tanganyika in the late 1830s, but the first real progress in charting the course of the upper Nile occurred when Egyptian and Turkish captains passed through the Sudd in the 1840s. The vast swamps of the Sudd, named from the Arabic word *sadd,* for obstacle, had previously formed the major barrier to upstream travel along the river.

Bypassing the Sudd, most British explorers eager to locate the Nile's source entered the region from the east, through the African seaport of Zanzibar. Richard Burton and John Hanning Speke, Samuel and Florence Baker, David Livingstone and Henry Morton Stanley all explored the region during the 1850s to 1870s. The preponderance of British discoveries in this region helped set up a proprietary attitude toward the Nile among the British. With the opening of the Suez Canal in 1869, and the opportunities for foreign political intervention created by the financial collapse of Egypt in 1876, Egypt appeared both strategically important and ripe for British colonization. British international policy focused on the Lake Plateau region as the key to controlling the Nile, which in turn was the key to controlling Egypt. The British did not realize that nearly 90 percent of the Nile's flow originated in Ethiopia. Satisfied that they understood the source of the Nile, the British imperialists set out to control the White Nile by regulating and storing its floodwaters from the lakes to far downstream at Aswan.

The Blue Nile

By the time the annual flood reaches Aswan, the waters of the river have followed a long and varied transect across Africa. The Nile flows from many sources. The Blue Nile flowing from Ethiopia contributes nearly twice the annual flow of the White Nile, which descends from the Lake Plateau of equatorial Africa. When the Blue Nile is combined

4.2 The Semien Mountains in the Ethiopian headwaters of the Nile drainage. The steep terrain and seasonally heavy rains of this region promote naturally high sediment yields, but agricultural land use has increased sediment loads in recent decades. (Photograph courtesy of Martin Williams.)

with Ethiopia's Atbara River, the Ethiopian sources contribute about three-quarters of the Nile's total flow. This contribution rises to nearly 95 percent of the main river's discharge during the summer floods.

The great floods that have nourished the lower Nile civilizations for millennia are the product of monsoon rains over the Ethiopian highlands. A high-pressure zone forms over the subtropical southern Atlantic Ocean during the summer months. Simultaneously, a low-pressure zone forms at about 20 degrees north latitude over the Sahara. The resultant pressure gradient draws water from the Atlantic by moist, shallow equatorial westerly winds that release abundant rains as they rise over the Ethiopian highlands. The rains replenish the spring at Sakala at the foot of the mountain of Gish, and here the Blue Nile begins.

Europeans reached this source of the Nile much earlier than Waldecker's 1937 charting of the White Nile source. Portuguese Jesuit Pedro Paez described the spring at Sakala in April 1613. From the spring, a stream called the Little Abbai flows steeply down rapids and waterfalls to Lake Tana.

The sharply dissected mountains surrounding the lake are drenched by violent thunderstorms from March to September. Then the monsoonal circulation reverses and dry winds cool the mountains from October to February. The region was once covered in primary tropical rain forest, but deforestation has replaced much of the rain forest with secondary growth and shrublands. This rugged landscape supplies 98 percent of the sediment entering the Nile. Removal of the forests that help to hold the soil in place is increasing the volume of sediment eroded from the hill slopes. The World Resources Institute estimated in 1998 that 91 percent of the original forest cover had been removed throughout the Nile basin as a whole and that deforestation continued at a rate of 6 percent of the remaining forest each year (Revenga et al. 1998).

Shallow Lake Tana lies where a lava flow dammed the valley of the Blue Nile, which the Ethiopians call the Abbai. The lake contributes less flow to the Abbai than do the numerous tributaries cascading down the adjacent mountains. Twenty-six kilometers downstream from the lake, another lava dam creates Tisisat Falls, "smoke of fire," where mist rises as the Abbai drops over a cliff into a basalt gorge that runs at right angles to the upper river and in places narrows to four meters. When the river finally encounters softer sedimentary rock, it forms a wider canyon and gradually changes its direction of flow from south to west and eventually to northwest as it makes a huge bend around the highlands of Gojjam.

The countryside opens out as the river leaves the Ethiopian Plateau. The river's blue waters take on a heavier load of silt as they pass through the coffee-growing country at the border of Ethiopia and Sudan. English explorer Samuel Baker wrote in the 1860s that during the dry season "the water is beautifully clear, and, reflecting the cloudless sky, its colour has given it the well-known name of Bahr el Azrak, or Blue River" (Baker 1869). The river would be unlikely to receive that name today, since it carries progressively more of the Ethiopian highlands downstream as millions of tons of sand and silt.

The last major rapids occur just before the river enters Sudan, when the turbid waters churn into a thundering white mass as they cut through the resistant rocks of the Blue Nile's western gorge. During the wet season the floodwaters spread out across an irregular bedrock bench, and a thousand little vortices grind sand and cobbles against the tough rock. The river contracts to an inner channel less than fifteen meters wide during the dry season, leaving the adjacent bedrock polished to a gloss and sculpted with potholes.

A thousand kilometers downstream from Lake Tana, the waters of the Blue Nile encounter their first human-built obstacle: Roseires Dam. From this point to its outlet in the Mediterranean more than thirty-three hundred kilometers downstream, the river will be steadily dammed and diverted and used for all the thousand things that humans require of water. Completed in 1966, Roseires was designed to supply irrigation to agriculture in the Gezira region around the confluence of the Blue Nile and the White Nile. In satellite images of the region, the dense network of irrigation canals now resembles the intricate tunnels insects bore in wood. Roseires Dam was also built to supply hydroelectric power to downstream cities including Khartoum. But deforestation in Ethiopia has allowed the torrential monsoon rains to steadily strip the hill slopes of soil, so that the reservoir has lost capacity to store water and to generate power as the sand and silt pile up behind the dam. These huge quantities of sediment have smothered and destroyed large beds of the Nile oyster (*Etheria elliptica*). The reservoir was also the scene of heavy fish mortality during 1967 when water stratification depleted the newly filled lake of oxygen.

Downstream from Roseires the Blue Nile enters the drylands that characterize the rest of its course. Tributary streams such as the Dinder and the Rahad may contribute 10 percent of the Blue Nile's annual flow, but this contribution is heavily weighted toward the flood season. Once the rains cease each year, the tributaries shrink back to pools separated by lengths of sandy streambed. Beyond the stream channels stretch savannas covered by grasses, bushes, and small trees. The soil here is fertile, and water diverted from the Blue Nile supports fields of cotton and sugarcane. Some of this diverted water comes from the reservoir behind Sennar Dam. The Sennar was the first dam on the Blue Nile, built by the British in 1925 primarily to ensure a plentiful supply of cotton for Lancashire textile factories. From Sennar the Blue Nile flows uninterrupted to its meeting with the White Nile downstream at Khartoum.

Darwin's Dream Pond

The White Nile comes into its own during the low-flow period of March to June, when it contributes three-quarters of the total discharge to the main Nile. This contribution is greatly overshadowed by the volume of water pouring forth from the Blue Nile during the flood season. Despite the relative unimportance of the White Nile's flow, it was this tributary that captured the imagination of Arabic and European explorers seeking the source of the Nile.

The White Nile catchment penetrates much farther south into Africa than does the Blue Nile drainage; the White Nile originates from the Waldecker Spring, nearly 5° south of the equator. Water from the spring flows northward along the Luvironzia River, which joins first the Ruvuvu River draining the highlands of northwest Burundi and then the Kagera River draining the hill country of Rwanda. In Tanzania the Kagera pours over Rusumu Falls, enters an unusually straight reach as it flows north through swampy lakes along a fault line between Tanzania and Rwanda, and then turns abruptly eastward into Lake Victoria.

Victoria, the greatest of African lakes, was named by John Speke for an English queen who never saw the lake. People have inhabited the shores of Lake Victoria for thousands of years. The changing composition of pollen grains among the lake sediments, in combination with historical and archaeological records, indicate human influences on the environment through time. Sudanic-speaking peoples settled the western shores of the lake at least three thousand years ago. These people fished, hunted, herded cattle, and grew crops, but settlements were small and dispersed. Pollen records from that time indicate gallery forest along the lake, woodlands on the adjacent plateau, and savanna woodland inland. Bantu agriculturalists arrived by 500 BC. The Bantus started clearing the gallery forests along the lake and experimenting with more intensive agriculture. Between 200 BC and AD 0 they began to smelt iron along the lakeshore. Settlements grew larger, the gallery forests were increasingly cut, and timber harvest extended to smaller trees and swamp forests. Population and resource use expanded steadily for the next five hundred years as the gallery forest and adjacent woodlands disappeared, leading to soil erosion and loss of fertility.

Population declines attributed to disease, malnutrition, and environmental degradation began about AD 600. People moved away from the lakeshores, leaving behind a drier, deforested landscape. The forests gradually recovered during the next four centuries, only to be cleared once more as expanding agricultural populations resettled the region beginning about AD 1000. Pastoral and agricultural groups competed for the land during succeeding centuries, and by the early twentieth century the landscape was once again denuded of trees. Today farmers attempting to wrest a poor living from the stripped landscape battle malnutrition, poverty, illness, and lack of schools for their children. The landscape of peaceful beauty Speke described is now the setting for what the United Nations Under-Secretary-General for Humanitarian Affairs in 2004 called the world's "largest neglected humanitarian emergency." As in nearby Rwanda and Sudan, resource scarcity has helped to foster civil unrest in Uganda, which forms the northwestern shore of Lake Victoria.

The most dramatic ecological changes of the past two to three decades have played out within the lake itself, which is infamous among ecologists as the scene of the largest mass extinction in modern times resulting from an introduced species. Among bodies of freshwater, Lake Victoria is exceeded in size only by Lake Superior and the Caspian Sea. Although less than one hundred meters deep, the lake covers about sixty-seven thousand square kilometers. Perched on the Lake Plateau between the two valleys that form the great branches of the African Rift, Lake Victoria is bisected by the equator. In the early twentieth century the lake had some 350 native fish species, including more than 200 species of cichlids, almost all found only in this lake. Ecologist Tijs Goldschmidt called the lake "Darwin's dream pond." Studying the small, perchlike cichlids during the early 1980s, Goldschmidt found an aquatic ecosystem that very little was known about. The lake harbored cichlids that fed off organic waste on the lake bottom; cichlids that ate algae scraped from rocks; cichlids that tore off pieces of aquatic plants; and cichlids that crushed and ate snails. Other cichlids ate zooplankton, insects, prawns, other fish, or fish embryos. Some cichlids were specialized to rasp scales, which are rich in protein, off other fish. Less than 750,000 years old, Victoria dried up partially into isolated smaller lakes—or perhaps even dried up completely—14,000 years ago. In the evolutionarily short span of time between then and now, the numerous fish species rapidly evolved to exploit the many feeding niches within the huge lake.

Writing in the early 1860s, John Speke noted that "fish and crocodiles are said to be very abundant in the lake." The most popular fish species in Lake Victoria were already threatened by overfishing in the 1920s, as evinced by steadily shrinking gill net mesh designed to catch the smaller, less mature fish. As mesh sizes decreased and smaller fish were caught, more species declined in number and size.

Heavy rains during 1961–64 raised the lake level by two meters. The higher water levels killed much of the shoreline papyrus swamp that provided important habitat for fish in the lake. A healthy fish population can withstand natural anomalies such as unusually heavy rains and temporarily higher lake levels. But populations depressed by excessive fishing were less able to recover from the loss of papyrus swamps. The simultaneous invasion of water hyacinths introduced from South America further depressed the recovery of the native plant communities along the lake margins.

The coup de grâce for many of the native fish in Lake Victoria came with the introduction of the Nile perch (*Lates niloticus*) during the

1960s. Nile perch are in some respects superfish. They grow rapidly, eating other fish to fuel their growth, and can reach 180 kilograms. They are fecund, with each female producing up to sixteen million eggs during a breeding cycle. And they are adaptable. They can feed at different levels of the food chain at each stage in their life history and switch to the most abundant prey at any time. Nile perch were introduced from Lakes Turkana and Albert as a food source for fishermen, but their rapid colonization of Lake Victoria and their rapacious appetites helped to decrease the already stressed communities of native cichlid fish. Goldschmidt describes the Nile perch "eating its way through the lake like a giant vacuum cleaner" while the biologists "felt they should continue working feverishly. A massive front of millions of Nile perch was breathing down our necks and everyone was acutely aware that this was the last opportunity for observing the original fauna."

Decreases in the algae-eating cichlids meant that more organic detritus sank to the bottom of the lake without being consumed, using up oxygen as it decayed. The loss of dissolved oxygen in the lake waters further stressed aquatic organisms. By the early 1990s, 50 to 70 percent of the lake was deoxygenated year-round. A giant bubble of deoxygenated water periodically rose from the depths to the lake surface, suffocating thousands of Nile perch and leaving a rotting mass of dead fish. The lake ecosystem is now unstable, and some experts forecast the collapse of all commercial fisheries in the lake. Nile perch now constitute 80 percent of Victoria's fish population, although the lake still has an estimated 288 native fish species. Yet Goldschmidt ends his account on an unexpectedly upbeat note. After years of absence of any cichlids, biologists discovered new cichlid species during the 1990s, suggesting that these resilient fish may be hybridizing rapidly enough to develop life strategies resistant to the voracious Nile perch.

The Victoria Nile

The fish species of the Nile drainage form a large, persistent segment of an ancient African fish fauna once widely distributed north of latitude 10° south. The river itself has 129 species, 26 of them unique to the Nile and none currently threatened. Some species found in the Nile also occur in the Niger and Congo drainages. If the Nile lakes are included, the number of fish species rises above 350, of which 62 percent are unique to the lakes. The overall high percentage of endemic species in the lakes reflects the physical barriers formed by waterfalls and cataracts between each pair of lakes and between adjacent river segments and the relatively recent establishment of through-flow from the Nile headwaters to the Mediterranean Sea some twelve thousand years ago. Unlike the Amazon and the Ob, which descend from mountainous headwaters onto vast flatlands with extensive floodplains, the vertical drops and floodplain wetlands are interspersed along the length of the Nile, where abrupt falls and gorges alternate downstream with lakes and swamps.

John Speke watched the Victoria Nile drain from the northern end of Lake Victoria through an abrupt drop that beat the waters into thundering white foam. Today the falls are submerged, and the lake's outlet is partially regulated by the Owen Falls Dam, built in 1954 when Uganda was still a British colony. The dam was planned as part of the Century Storage Plan that the British conceived to store water during wet years against the inevitable dry times. Evaporation rates much lower than those of the downstream deserts made a storage reservoir here particularly attractive.

Below Owen Falls Dam the waters of the Victoria Nile make the short northwesterly run to Lake Kyoga. The lake's outline is broken into several long, narrow arms fed by swampy streams dense with papyrus. Much of the region surrounding the lake is also seasonal swampland of some type, because the impermeable clay soils of the shallow basin impede drainage. The seasonal swamp grasslands provide valuable grazing for wild and domestic animals. All the grasses are perennial, and many species can be up to three meters tall at the start of the dry season. People living in the area maintain the quality of the fodder with controlled burns at the close of each dry season.

The swamps are very productive ecosystems that efficiently use water and nutrients in a harsh environment. Traditional agricultural practices acknowledge the scarcity of soil nutrients and the critical role of seasonal floods and swamps in maintaining stability and productivity. But as flood control measures, high-yielding economic crops, and expensive fertilizers have been imported from the temperate zones, ecosystem health is increasingly at risk.

The Victoria Nile drains from the northwestern end of Lake Kyoga and then passes over the Karuma rapids. The gradient steepens and the channel narrows between high cliffs as the river approaches the edge of the western arm of the Rift Valley at Kabarega (Murchison) Falls. Here the river is shunted into a gorge only six meters wide before it thunders more than forty meters down to the floor of the rift valley. This is one of the natural barriers to river migrants that the Nile perch and other fish that are indigenous downstream needed human assistance to travel around.

The course of the river grows gentler below the drama of the falls. Crocodiles bask along mud banks formed where the current slows as the river breaks up into numerous distributary channels across a marshy delta at the head of Lake Albert. Ancient Egyptians revered the Nile crocodile in the form of Sobek, the crocodile god. Hunting and the 1902 construction of the first Aswan Dam eliminated crocodiles from most of the Egyptian Nile, although they are still found in Lake Nasser behind the High Aswan Dam. Only in these upper portions of the Nile drainage do crocodiles remain numerous.

The rivers of this portion of eastern Africa once flowed westward into the mighty Congo basin. Then, six million years ago, the great tectonic plates that form Earth's crust shifted and began to pull eastern Africa apart. Huge faults formed along the surface, dropping blocks of crust into deep, trenchlike rift valleys with steep walls. Magma rose from Earth's interior through the thinning crust, raising volcanic mountains such as the Mfumbiros. As they rose across the southern end of the Nile drainage, the Mfumbiro Mountains cut off the headwaters of the river, forcing waters that had flowed north into the Nile to flow south into Lakes Kivu and Tanganyika instead. These lakes, like all the lakes of eastern Africa except Victoria and Kyoga, formed along the rift valley. What the Nile lost in its southernmost regions, it gained on the west: uplift and tilting along the rift diverted the headwaters of the Congo into the Nile drainage.

As tectonic forces rearranged topography along the rift, waters that had drained west to the Congo or east to the Red Sea and the Indian Ocean were gradually directed northward along the troughs formed by the rift. The rearrangement took time. When the action started six million years ago, the Egyptian Nile was not connected with sub-Saharan Africa. A topographically higher region known as the Nubian massif lay between the Egyptian Nile and the water accumulating in basins farther south. The modern Nile's huge bend in the vicinity of the third and fourth cataracts records the diversions forced on the river as it eroded through this obstacle, and the persistence of the cataracts indicates the resistance of the underlying rocks.

Meanwhile, the lower, Egyptian portion of the Nile cut a deep canyon down to the shores of the Mediterranean. As the African Plate began to pull apart six million years ago, the Strait of Gibraltar was uplifted. Elevation of the Strait cut the Mediterranean Sea off from the Atlantic. The Mediterranean became a lake and then, as its water steadily evaporated, a salt desert. As sea level at its outlet fell, the Egyptian Nile incised downward by as much as four kilometers. This early Nile, which scholars call the Eonile, lasted about five hundred thousand years. Once the connection between the Atlantic and the Mediterranean was restored, the rising water backflooded the river's deep canyon, creating a gulf that lasted until about 3.5 million years ago. The next incarnation of the Nile, the Paleonile, filled the Eonile's canyon with a thick accumulation of sediments over the succeeding 1.5 million years.

Then the Nile ceased to exist. From approximately 1.8 million years ago to 800,000 years ago, the river stopped flowing. Drought settled in to stay, and Egypt became a desert. The rivers that today connect the low-lying basins of lakes also ceased to flow. The lakes became disconnected, internal drainages in which fish and other aquatic organisms evolved into distinct species.

The great climatic pendulum swung back again, and water flowed down the rivers and between the lakes. For the first time, the Nile drainage was integrated from sub-Saharan Africa to the Mediterranean. Lakes Tana and Victoria came into existence. The Prenile flowed mightily down from the Ethiopian highlands and the Lake Plateau for four hundred thousand years. This river carried so much coarse sand and gravel that it built a larger floodplain and delta than those of the modern Nile. The outcrops of sand left behind as the river diminished now provide the settlements of Egypt with their building material.

By four hundred thousand years ago the vigorous Prenile had given way to the much lesser Neonile. The Neonile was often tenuous during the four hundred thousand years of its existence. The Mediterranean portion of the river lost its African connection repeatedly during glacial periods of drier climate when rainfall decreased over the Ethiopian highlands and the Lake Plateau. When global ice sheets retreated, atmospheric circulation patterns and ocean temperatures shifted. Rains came once more to the highlands, and the rivers of the Nile drainage basin sprang back to life and reconnected the isolated rift lakes. The rains that came with the latest phase of glacial retreat caused Lakes Victoria and Albert to overflow into the Nile drainage for the first time, triggering a series of tremendous floods in the Sudd and Egypt about twelve thousand years ago. These floods inaugurated the reign of the modern Nile, a river lesser than its ancestral Prenile, but capable of integrating sub-Saharan Africa with the Mediterranean and supporting civilizations of great wealth and duration.

This latest incarnation of the Great Rift drainage ties the peoples of Africa to the peoples of the Middle East. In the upper basin live Hutu and Tutsi, Twa pygmoid hunters, Bantu cultivators, and Hima pastoralists. Downstream dwell the Nilotes: Nuba, Ingessna, Hamaj, Shankalla, Nuer, Dinka, Shilluk, and Arabs. Along a length of more than seven thousand kilometers, all these people depend on the waters of the Nile drainage, and their fates are closely intertwined.

The Sudd

The outlet from Lake Albert lies north of the entrance of the Victoria Nile. The river flowing out of the lake is briefly called the Albert Nile until it reaches the border of Uganda and Sudan. Here the river makes a hard turn to the west and plunges down into a steep, narrow rocky valley. The river becomes, appropriately, the Bahr al-Jabal, "mountain river."

For nearly two hundred kilometers the Bahr al-Jabal flows closely confined between rock walls. As the river continues northward it gradually leaves the broken and precipitous terrain of the Great Rift Valley. Steep canyon walls give way to sandy banks beyond which lie hills separated by wide, flat plains. At the Bedden rapids the river flows across the last rocks it will encounter until far downstream, past the confluence with the Blue Nile.

The course of the river twines among islands. Tributary channels remain dry most of the year, flowing only after rainstorms. The river passes the city of Juba, flowing beneath the only bridge crossing the Nile for nearly fifteen hundred kilometers. Low-lying marshes begin to appear among the forest surrounding the river. Then the river disappears. Streambanks that were well-defined upstream merge imperceptibly into a soggy, swampy plain. Papyrus and water reeds grow more than three meters tall along the watercourses, and the main channel winds back on itself in intricate meanders or divides into lesser channels that in turn wind into lakes and grassy swamps. This is the Sudd, the great swampland that for millennia frustrated the attempts of explorers and merchants to travel upstream along the Nile from Egypt and the Sudan. A cluster of rivers flow north into the Sudd and converge into the vast wetlands, like fingers converging on a palm. The Bahr al-Jabal divides to produce the Bahr al-Zaraf, "giraffe river."

Writers throughout history have bestowed choice epithets on the Sudd. Phrases such as Collins's 1990 description of "a dreary wasteland of desolation" reflect the difficulty of travel through this region of heat and humidity, where landmarks are few, detached masses of rotting

4.3 Near Guli in the Blue Nile state of southeastern Sudan, upstream from Khartoum. Floodplain wetlands such as these provide diverse habitat and abundant resources that support greater density and diversity of plants and animals than adjacent desert habitats. (Photograph courtesy of Martin Williams.)

vegetation clog the channels, and routes are confusingly circuitous. Like other great wetlands of the world, the Sudd is rich with life, but perhaps only the most dedicated naturalists can appreciate this life. The mosquitoes alone include fifty species.

The Sudd swamps occupy the remains of a former lake. Today the swamps spread outward and shrink back toward their core with the seasons and with wet years and dry years. At some times the Sudd covers only fifty-five hundred square kilometers, at others it sprawls over one hundred thousand square kilometers. The rains of July to November fail to soak into the impermeable, clay-rich soils of the depressions that form the swamps, and the rising waters of rivers in flood add to the inundation. Incoming waters bring a load of dissolved elements, and the Sudd becomes a complicated chemistry laboratory. Nutrients such as potassium, phosphorus, and nitrogen taken up by the vegetation are continuously recycled within the swamp, so that waters leaving the Sudd are depleted in these elements.

Aquatic plants living in the Sudd have particularly effective means of spreading to new colonization sites. Strong winds sweeping across the swamps jostle the plants enough to loosen their roots. When water level rises, the plants break free to float on the water, trailing masses of roots and soil that act as ballast to keep the tall reeds upright. Islands of reeds drift across the open waterways and downstream channels until they reach shallower water or snag against a bend in a channel. The plants then take root once more, creating an obstacle that snags other floating islands. The obstacle spreads completely across the stream, and water is forced to flow beneath the mat of vegetation. As Samuel Baker wrote on encountering such a mass of plants during his return journey downstream: "The river had suddenly disappeared: there was apparently an end to the White Nile. The dam was about three-quarters of a mile wide; it was perfectly firm, and was already overgrown with high reeds and grass, thus forming a continuation of the surrounding country."

Stream velocity increases as the passage for the water grows narrower beneath the raft of reeds and new floating islands are sucked in until it is completely blocked. At this point the pressure from upstream flow can pile up a barrier of vegetation well above the normal water level. Oncoming flow spills around the blockage, and a new channel is cut, sometimes by a small outburst flood.

The waters of the Sudd provide important breeding grounds and nurseries for fish. Water level fluctuates very little in the portions that are permanently swamps, and the shallow channels and ponds are floored by a rich growth of plants on which some forty species of fish feed. The plants provide places where fish can safely lay their eggs, and the newly hatched juveniles shelter among the dense stems of this aquatic jungle. Crocodiles, of which the Sudd is estimated to have the largest population in the world, feed on the fish, as do many birds. Along with the rest of the Nile drainage, the Sudd forms one of the major avian migration routes to and from Africa. More than a million and a half glossy ibis alone spend the dry season in the Sudd.

By the end of the wet season only a few ridges remain exposed across the Sudd, and here the permanent villages of the Nilotic tribes are built. As the rains stop and the temporarily flooded portion of the swampland dries out, the people of these villages move their livestock down onto the newly lush plains. Wild herbivores join the migration back onto the plains. Elephants, buffalo, gazelles, several varieties of antelope, and zebras all come to graze. Feline predators follow their prey down into the tall grasses. Hippopotamuses plow deep tracks through the vegetation fringing the stream channels and lakes.

Much of the water leaving the Sudd goes up into the atmosphere rather than flowing downstream along a river. For this reason, rather than for the hindrance it creates to navigation, many people look disapprovingly at the Sudd. So much water wicking away into the air in this land of overpowering dryness is frustrating. Each drop evaporated from the Sudd is regarded as one drop less for the thirsty crops and cities downstream in the Sudan and Egypt.

Sudan is the largest country in Africa. As of 1982, its seventeen million inhabitants were increasing annually by almost 3 percent; the population reached thirty-nine million in 2007. More than 90 percent of these people depend on agriculture, but average yearly rainfall drops to twelve centimeters in the dry northern third of the country. Under these conditions, the Nile provides the main source of water for irrigated agriculture. Egypt is in much the same situation. In 1959 the two countries signed the Nile Waters Agreement. This agreement divided the available water supplies and stipulated that studies would be undertaken to develop projects that could increase the yield of the Nile. Sudan was already using its full share of allotted Nile waters in 1980.

The idea of the Jonglei Canal Project grew out of these desperate conditions in 1976. This project centers on excavation of a canal more than three hundred kilometers long across the Sudd. The canal is expected to carry forty-two million cubic meters of water each day. The Bahr al-Jabal and the Bahr al-Zaraf would also be "improved"—dredged, straightened, and leveed—so that water would move more rapidly through the depression that now forms the Sudd. Additional structures would be built on adjacent rivers and lakes to regulate the flow of water and to store water during wet seasons. The end result of this engineering would be to completely desiccate the swamps of the Sudd.

As of 2010, the Jonglei Canal remains incomplete and controversial. The canal was 70 percent finished when the Sudanese civil war broke out in 1983. Completing the project would now be extremely costly, and in the intervening two decades, international recognition of the importance of the Sudd wetlands has grown dramatically. Set against the expected increase in downstream water supply are the loss of wetlands, fisheries, wildlife habitat, and the seasonal grazing lands that half a million people depend on. Many argue that present understanding of the likely negative impacts of the canal on the natural systems of the Sudd is not sufficient to justify undertaking such a massive reconfiguration of this portion of the Nile drainage.

The White Nile

The waters of the many rivers that flow into the Sudd, seemingly to disappear forever, eventually make their way to shallow Lake No at the northern end of the swamp. Nineteenth-century traders traveling south up the Nile called the lake Mugran al-Buhur, "the meeting of the rivers," for here the Bahr al-Jabal, the Bahr al-Zaraf, and the Bahr al-Ghazal ("gazelle river") meet and mingle.

Lake No sits at the center of the Nilotic plain, the flatlands stretching almost fifteen hundred kilometers from the Nile watershed in the west to the Ethiopian escarpment in the east. The Bahr al-Arab forms the northern boundary of this plain, as well as the frontier between the Africans of southern Sudan and the Muslim Arab peoples to the north. During the dry season the Bahr al-Arab shrinks back to a series of isolated ponds separated by dry streambed. These ponds provide important water sources amid the arid grazing lands, and people in the borderlands often compete violently for their resources.

Lake No drains eastward into the Bahr al-Abyad, the White Nile, a broad river with swampy banks bordered by ponds and smaller streams. Downstream the river grows more defined, with trees lining the high ground along its banks. The adjacent swamps become narrower and finally disappear. At the northeastern edge of the Sudd, the greenish gray waters of the White Nile are joined by the brick red waters draining down from the Ethiopian highlands via the Sobat River. The highlands are the great nourisher of rivers, and the Sobat contributes half of the total average yearly flow of the White Nile. From Lake No the White Nile flows a thousand kilometers to Khartoum and its confluence with the Blue Nile. In all this length of river, the Sobat is the only significant tributary. From this point downstream, there is only loss from evaporation, infiltration, and withdrawals for human use.

Beyond the riverbanks, forest and swamplands give way to scattered groves of trees and grasslands, to shortgrass plains, and finally to sandy and rocky desert. The horizon remains distant and flat, interrupted only sporadically by an isolated hill. Camels and horses come down to drink in the shallows, where water hyacinths and papyrus form patches of green. The slow-flowing waters divide around islands.

Water hyacinths (*Eichhornia crassipes*) probably invaded the Nile drainage during 1957–58, after being introduced to Egypt from the Amazon in 1912 as an ornamental plant. By the 1970s, the upper Nile swamps had become a perennial center of infestation as currents and winds spread clumps of hyacinths downstream. This superplant can grow by 10 to 15 percent of its fresh weight a day. It can propagate by seeds or from plant parts that break off. Each plant sends out shoots that can bind single plants into rafts several hundred square meters in extent. When these huge rafts break up, the new pieces form new centers from which the hyacinths can spread, like the sorcerer's apprentice's broom. Neither mechanical removal nor use of pesticides has effectively controlled their spread.

Below Kosti, which forms the principal river port of the White Nile, a long stretch of the river vanishes into the huge evaporation pan of the Jabal Auliya reservoir. Jabal Auliya dam was built forty-six kilometers south of Khartoum in order to store the floodwaters of the White Nile during the July to December flood on the Blue Nile. Once the Blue Nile flood recedes, water stored behind Jabal Auliya is released to provide Egypt and the northern Sudan with water during January through June. The dam has provided a partial barrier for the spread of water hyacinths along the river, as well as creating a reservoir from which suck numerous "straws"—entrepreneurial pump schemes. But evaporation from the reservoir also constitutes a large and insatiable straw. In the intense desert heat, the shallow reservoir loses more than 2.5 billion cubic meters of water each year to evaporation.

Released from Jabal Auliya, the waters of the White Nile flow only a short distance before merging with those of the Blue Nile. Samuel Baker described the junction as a "vast flat as far as the eye can reach, the White Nile being about two miles broad some distance above the [junction]." A long strip of land resembling an elephant's trunk lies between the rivers as they converge, and the city of Khartoum takes its name, meaning "elephant trunk," from this landform. Khartoum was established in 1825 as the administrative center for the Egyptian empire in the Sudan and has since remained the commercial and political capital of the region.

The Nile proper begins below Khartoum. The river starts out energetically, for Khartoum marks a transition at which it passes from a substrate of sediments to a substrate of sandstone bedrock. Here the average downstream gradient of the river increases and is periodically interrupted by even steeper reaches at cataracts. For 1,800 kilometers the river is marked by a series of cataracts, numbered from the first cataract farthest downstream at Aswan to the sixth cataract 80 kilometers north of Khartoum. The cataracts, which spread along 570 kilometers of the 1,800 kilometer length, form where the Nile passes over harder, crystalline rocks. This region is one of the steeper sections of the Nile's longitudinal profile, resembling a flight of low, broad steps. Each of the treads, referred to along the Nile as "landings," is separated from the steps upstream and downstream by a steeper section of waterfalls and cataracts. Each landing was an independent, disconnected basin be-

4.4 The Blue Nile at Khartoum. Centers of dense population, such as Khartoum and Cairo, are places where floodplain wetlands have been covered with urban and agricultural areas and where a wide variety of contaminants enter the river. (Photograph courtesy of Martin Williams.)

fore the river was integrated as the modern, through-flowing Nile a few thousand years ago.

The last major tributary to enter the Nile reaches the main channel about three hundred kilometers south of Khartoum. Although the Atbara is another of the many rivers fed by the rains of the Ethiopian highlands and during flood can contribute more than 10 percent of the Nile's annual flow, from January to June it dries back to disconnected pools. Only the Nile itself has enough water to flow year-round from this point downstream to the Mediterranean. Minor tributaries enter the Nile for the rest of its journey, but these channels contribute flow only sporadically, after rare desert rains.

The Nile continues below its junction with the Atbara, forming a thin green line amid the red and tan deserts stretching hundreds of kilometers on either side. The river makes an enormous S as it passes sequentially over the fifth, fourth, and third cataracts before coming to a halt in Lake Nasser, the enormous reservoir behind the Aswan Dam.

CHAPTER FOUR

A Modern Pyramid

In many respects the Aswan Dam represents the culmination to date of water engineering in the lower Nile drainage basin. Although for millennia Egyptian civilizations have relied on irrigation from the Nile, the amount of water diverted grew considerably during the later nineteenth century.

Muhammad Ali introduced perennial irrigation into Egypt in the early nineteenth century. He subsequently ordered construction of the Delta Barrages, a series of diversion dams across the Nile at the head of its delta. These dams, completed by 1861, increased the lands under cultivation. By the end of the century, however, the population of Egypt increased beyond the capacity of already irrigated lands to provide food. Partly in consequence of this desperate need to increase irrigation, projects to measure and control Nile flows increased substantially with the British occupation of Egypt in 1882.

The British realized they needed to understand and control the flow of the Nile in order to feed the Egyptians, to remain in power, and to retain strategic control of Cairo and the Suez Canal. British engineers were given the tasks of increasing the water available for irrigation and of defending against any threat to the water supply throughout the vast Nile drainage basin. Sir Colin Scott-Moncrieff recruited hydrologists and engineers from the Indian Irrigation Service to form a group that for several decades led the world in measuring and managing waters in arid lands.

The first great achievement of the Anglo-Egyptian engineers was the construction of the Aswan Dam in 1902. The dam was enlarged twice by 1934, storing ever more water that could be distributed for irrigation. Meanwhile, Egypt's population grew relentlessly. By 1966 the country held approximately thirty million people, ten times the number who had lived there a century and a half earlier. Existing dams held some nine billion cubic meters of water, or about 10 percent of the Nile's average yearly flow. But as water-intensive cash crops such as cotton became popular and population continued to rise, ever more water was needed. Working with the Soviets to build a still larger dam at Aswan, President Gamal Abdel Nasser boasted that "in antiquity, we built pyramids for the dead. Now we will build pyramids for the living" (Waterbury 1979).

The High Aswan Dam, completed in 1970, rises more than a hundred meters above the river and backs water so far up the valley—almost five hundred kilometers—that it can store almost two years' worth of the

Nile's average yearly flow. Like other rivers draining dry regions, the Nile flow fluctuates substantially from year to year. During 1878–79 the great river poured more than 140 billion cubic meters down to the sea. During 1913–14 only 39 billion cubic meters flowed down from the highlands. The dam was designed to protect Egypt from such climatic fluctuations in water supply, as well as from limitations on water imposed by upstream countries. As Nasser phrased it, "After completion of the High Dam Egypt will no longer be the historic hostage of the upper partners to the Nile basin" (Waterbury 1979).

The dam met its intended use when it helped to avert massive famine during droughts in the early 1970s and the mid-1980s. Egypt's irrigated area rose to nearly 3 million hectares by 2000, and Egyptian officials project that by 2020 this will increase to 4.5 million hectares. It may have to: Egypt's population has reached 70 million and is projected to reach 115 million by 2050. But half of the planned water addition from the High Dam is lost each year to evaporation (ten billion cubic meters) and seepage. Egypt continues to import half or more of its annual food requirements.

Meanwhile, environmental deterioration caused by the High Dam began to appear. Many of the lakes associated with water control structures along the Nile have aided the proliferation of introduced species of aquatic vegetation—principally water hyacinths, water ferns (*Salvinia auriculata*), and water lettuce (*Pistia stratiotes*). These plants impede water flow and navigation. By covering the surface, they reduce phytoplankton that fish feed on and then lower dissolved oxygen levels as they decay. A hyacinth-covered water surface loses three and a half times as much water through evapotranspiration as a clear water surface in semitropical areas. In the desert, it can lose up to eight times as much.

Before the closing of the Aswan High Dam, nearly 40 percent of the Nile's total annual flow (approximately thirty billion cubic meters of water) entered the Mediterranean from the Nile delta. This water transported 90 percent of the sediment washed down onto the delta from the Ethiopian highlands—more than a hundred million tons of silt. The river spread the remaining 10 percent of sediment along its floodplain, creating the fertile soils that supported Egyptians for millennia. Once the dam closed and more of the river flow was diverted from the channel and onto croplands, the annual river flow to the ocean dropped to less than two billion cubic meters. Sediment remained trapped upstream from the dam, where it diminishes the reservoir's storage capacity.

Silt carried by the Nile nourishes plankton growing in the seawater offshore from the delta. The plankton in turn support sardines. When the water and sediment supply decreased, the abundant sardine fisheries collapsed. Contributing to this collapse was the pollution of the brackish lakes on the northern delta, which are connected to the sea, and into which most of the delta farmlands currently drain. Water quality in these lakes is now severely impaired by chemicals and sewage. Diminution in the sediment supply has also caused the delta to subside and erode. The former delta village of Borg-el-Borellos is now two kilometers off the coast.

Along the floodplain farmlands below the dam, soil fertility has diminished in the absence of the yearly influx of sediment and nutrients. Egyptian farmers increasingly rely on expensive chemical fertilizers that pollute surface water and groundwater with excess nitrogen. The nearly constant flow level in the river also raises the regional water table (up to three meters a year in some parts of Egypt) and impedes drainage of irrigated lands. Egypt now has a severe problem with waterlogging and soil salinization as salts dissolved in the irrigation water remain behind when the water finally evaporates. Salts also infuse down into porous rock and soil by capillary action, polluting shallow groundwater aquifers that provide drinking water. By the 1970s, 35 percent of Egypt's cultivated surface was affected by salinity and 90 percent by waterlogging. Artificial land drainage systems are expensive and not always successful. As John Waterbury wrote in 1979, "No regime ever built a monument to itself with tile drains, but it is at that level that Egyptian planners must focus their attention" (Waterbury 1979).

The irrigation ditches that bring the life-sustaining water from the dam can also bring disease and pollution. Bilharzia (schistosomiasis) has become endemic along the river's course in Egypt and Sudan. Irrigation ditches filled with stagnant water and weeds host snails that serve as vectors spreading the disease to people who wade in the shallow ditches. Perennial irrigation eliminates the fallow periods during which the snail hosts would be killed as the canals dried out. Perennial irrigation also expands the area of still water and aquatic weeds where the snails live and increases the time farmers spend in the water, which may contain pathogens, heavy metals, oil, and every other kind of waste.

Egypt continues to aggressively expand its irrigation supplies and acreage. President Hosni Mubarak approved the New Valley Project in 1997 to divert Nile water upstream from Lake Nasser and transfer it hundreds of kilometers to irrigate Egypt's southwestern desert. Mubarak opened another canal to shunt Nile water beneath the Suez Canal to irrigate portions of the Sinai Desert. Like highway systems in large cities, water engineering is now carried out in multiple tiers of structures passing over and under one another.

The source for all this irrigation water remains problematic. The flow of the Nile is anything but infinite. Egypt would like to continue its expansion, but neighboring countries that share the Nile's waters are looking to their own growth. Although the Nile basin includes ten countries, the two downstream nations of Egypt and Sudan claimed most of the river's flow in a 1959 treaty that allocates fifty-three billion cubic meters a year to Egypt and eighteen billion to Sudan. Egypt wants to expand beyond this allotment, but Ethiopia, where nearly 90 percent of the Nile's flow originates, is ready to promote its own development after years of civil war and social unrest. Only 5 percent of the potentially irrigable land in Ethiopia is currently under cultivation, and the Ethiopian government is giving renewed attention to a 1964 U.S. Bureau of Reclamation plan that proposed thirty-three irrigation and hydropower projects on the upper Blue Nile. The plan also calls for four major hydroelectric dams on the Blue Nile that together could store nearly a year's average flow of the river. The new reservoirs behind these dams would increase evaporation, as would the croplands under irrigation, thus reducing downstream water supply.

Three hundred million people in Africa, a third of the continent's population, live with water scarcity at the start of the twenty-first century. By 2025 another twelve African countries are predicted to join the thirteen countries already in this condition. Similarly, nine of the fourteen Middle Eastern countries face water scarcity. The population of the Nile basin, which was 245 million in 1990, is projected to reach 859 million by 2025. By that time, Egypt is projected to have one of the lowest per capita availabilities of water in the world at less than 590 cubic meters per year.

Several observers of international affairs predict that the twenty-first century will see wars over water rivaling the wars fought in past centuries over minerals, oil, and other resources. As Egypt continues to bring on line new projects requiring still more water and upstream nations seek to begin their own projects, the Nile basin seems poised to become a site of international water wars. The Nile thus unfortunately exemplifies the issues of water scarcity and associated environmental degradation that characterize many of the great rivers of the world's drylands.

The Nile Delta

As it leaves Aswan, the Nile is less than a hundred meters above sea level, but it flows twelve hundred kilometers through a labyrinth of irrigation works and canals before it reaches the Mediterranean. The lower river hosts odd juxtapositions of technology and tradition. Without transition, the tall buildings, traffic jams, high-tension lines, and twenty million people of Cairo abut small villages where farmers follow water buffalo dragging wooden plows across irrigated fields. As elsewhere along the lower river, where there is water everything is vividly green. Papyrus line the riverbanks, new crops sprout in the flooded fields, and date palms cast a scanty shade over the dirt roads. Along the linear oasis of the river, simple wooden boats with graceful triangular white sails move quietly beside fishermen casting hand nets across the water. Immediately beyond the reach of the water, there is no apparent life. The orange-red sand and rock of the desert stretch toward a horizon lost in haze.

Seen in satellite images, the thin dark line of the Nile meanders within a slender jade green band that crosses a corrugated landscape of tan, brown, and yellow desert mountains and plains. Just below Cairo the green band expands abruptly into the broad delta that the river has built over millennia. This delta constitutes Lower Egypt. Historical documents and ancient maps depict seven major distributary channels branching across the delta. Sediment carried down by the river has filled five of these branches, but the Rosetta and Damietta branches still run.

Sand, salt marshes, lakes, and swamps intermingle across the twenty-two thousand square kilometers of the delta, but the region still comprises 63 percent of Egypt's fertile land. The sediments of the delta form an enormous freshwater aquifer that, along with the river's annual flood, retains water in the wetlands. Among the wetlands live forty-five species of birds, more than half of the seventy-three species found in all of Egypt. The delta is also the gateway for birds migrating between Africa and Europe. The birds follow the green line of the Nile upstream to their winter homes and downstream to their summer habitat.

British hydrologist H. E. Hurst put together a long-term record of Nile flows based on historical information such as nilometer readings and systematic measurements undertaken by the British. Hurst's statistical analyses of these data, published in 1951, revealed periodicities that eventually became known among hydrologists as "Hurst phenomena": Nile flows tend to be grouped in sequences of higher and lower flows that are longer and more severe than normal statistical theory would predict. But even with Hurst's probability models, the low flows of 1984 to 1987 caught the Egyptians by surprise. It is only in the longer context preserved in the delta sediments, and in the sediments of lakes in the upper Nile drainage, that these low flows seem "normal."

By taking vertical cores through the Nile delta, geologists have been able to estimate changes in Nile discharge as a function of the volume of sediment deposited in the delta. These records indicate dramatic fluctuations in discharge since the Nile was reconnected from the lakes to the Mediterranean twelve thousand years ago. As the Intertropical Convergence Zone migrates poleward into the summer hemisphere each year, it brings abundant rainfall. When global circulation patterns cause the convergence zone to move farther north, the Ethiopian highlands are drenched with rain. If the convergence zone remains closer to the equator, the highlands receive less rainfall and the Nile receives less flow. Sediment volumes deposited in the Nile delta indicate the greatest floods during periods 8,400 to 8,100 years ago, 7,500 years ago, and approximately 7,000 years ago. This corresponds to geologic evidence of stronger monsoons throughout Africa between 10,000 and 5,000 years ago. Climatologists believe that stronger heating over the continent in summer, caused by slight oscillations in Earth's orientation with respect to the sun, creates a steeper ocean-to-continent temperature and pressure gradient, resulting in more rainfall.

By correlating long-term records of Nile flows with independent records of climatic conditions, hydrologists are beginning to understand how global climatic fluctuations translate into changes in water supply in the Nile drainage basin. Nile flows have become more variable since AD 1720, with high- and low-flow sequences that are much more acute and persistent than sequences earlier in history. The twentieth century in particular was characterized by exceptional droughts that climatologists relate to differential increases in ocean temperatures between the Northern and Southern hemispheres. Over-year storage in Lake Nasser and exceptionally high levels in Lake Victoria have buffered the downstream effects of these droughts, but past droughts do not bode well for future reliability of water supplies in the Nile drainage.

The Nile exemplifies environmental alteration along rivers draining arid regions in which rapidly growing human populations demand ever more resources from the river ecosystem. Deforestation and resulting increases in sediment load, overfishing, water withdrawal and soil salinization, and dams and diversions that fundamentally alter the river's natural flood pulse have all changed river form and function. Resource use has supported enormous increases in human population within the river basin, but the situation appears to be unsustainable.

Etymologists believe that the name Nile may be derived from the Semitic root *nahal*, for valley or stream. As population and water demand grow in the Nile basin and the international scientific community attempts to predict the climatic effects of global atmospheric warming, it is sobering to realize that most of the nahals of the Middle East are ephemeral drainages that flow only briefly after rare rainstorms. The Nile already rates a critical red color on a worldwide compilation of flow regulation along major river basins because its waters are among the most highly regulated and utilized of any major river basin. Urban growth in the thirty cities within the Nile drainage basin averaged 4 percent in 1998. As human population soars in the Nile drainage, the margin of adaptability and survival during periods of water shortage decreases for humans and every other form of life dependent on the river. What do we lose when the Nile resembles an irrigation canal more than a river ecosystem? We do not fully know, for there is not a single major dryland river left that has not been heavily altered by humans.

The historical details of the involvement of Europeans and Americans in the Nile drainage amply illustrate the stunning arrogance and aggression behind nineteenth- and twentieth-century colonialism. Contemporary uncontrolled population growth, and its implicit assumption that somehow the Nile will provide, seems equally arrogant and aggressive. It is only a matter of decades before we see this episode of colonization reach its crisis point.

Interlude

Having fallen as July rains on the Ethiopian highlands that feed the Blue Nile, the water droplet flows down the Nile and arrives at the eastern portion of the Mediterranean Sea in December. The Nile flowing into the Mediterranean is in many respects the opposite of the Amazon entering the Atlantic. The predominantly dry climate of the Nile basin did not historically produce immense flows from the river's mouth, and only in recent decades has withdrawal of water for irrigation and other consumption reduced the Nile's flow. And water in the Mediterranean does not rush anywhere at the speeds attained by the water in the North Brazil Current or the Guiana Current.

The Nile did play an important part in the life of the Mediterranean, however, bringing into the sea large quantities of dissolved nutrients and organic matter attached to suspended sediment. Starting with the construction of the Aswan High Dam, which trapped much of this material, and continuing with water withdrawals, the reduction in Nile flows has caused significant changes in the Mediterranean water masses around the Nile delta. One of the most obvious effects of these changes has been the decline in shrimp reproduction and fisheries, but all levels of the Mediterranean food web have been affected by the reduction in nutrients and freshwater and resulting increases in salinity.

Circulation patterns in the Mediterranean Sea are highly complex. A vertical circulation pattern driven by contrasting salinity levels that influence water density operates across the basin's entire 2.5 million square kilometers.

Density contrasts interacting with wind and with coastal and submarine topography also drive progressively smaller-scale circulation patterns at the surface. Density contrasts are pronounced in the Mediterranean because the basin is nearly closed except for small connections such as that with the Atlantic through the Strait of Gibraltar. The strait is only fourteen kilometers wide in places and three hundred meters deep. Through this narrow opening comes a continuous inflow of surface water from the Atlantic. Although this flow is the major source of replenishment and renewal for water in the Mediterranean, it takes more than a hundred years to completely renew Mediterranean waters.

The Mediterranean itself is shallow, reaching a depth of 5,150 meters off the southern coast of Greece but otherwise less than about 2,000 meters deep. All the rivers emptying into the Mediterranean replace only about a third of the water the sea loses through evaporation. The small inflow and high rates of evaporation keep the Mediterranean much saltier than the neighboring Atlantic water. Salty water, combined with low concentrations of critical nutrients such as phosphates and nitrates, limits the quantity of marine life present in the Mediterranean. Many of the species present are so uniquely adapted to the warm, salty water that they are endemic. The greatest threats to marine life there come from fishing, ship traffic, coastal urbanization, water pollution, and invasive exotic species.

The water droplet from the Nile does not remain long in the Mediterranean. Despite the cold rains of winter, the droplet is evaporated from the sea's surface into the Ferrel Cell. There it meets a low-pressure trough coming from the North Atlantic. Swept up briefly into the atmosphere, the Nile droplet falls as snow in the German Alps.

In the past, a snowflake falling in the Alps was likely to have metamorphosed into glacial ice as some of each winter's snowfall persisted through the succeeding summer and was buried beneath subsequent snowfalls. This scenario was a more likely fate for the Nile water droplet even a few decades ago. Glaciers around the world have been retreating for the past two to three hundred years, with dramatic increases in rates of retreat during the past few decades. In today's world, the Nile water droplet spends only a brief winter in the Alpine snowpack before melting and flowing downstream to join the waters feeding the Danube River.

FIVE

The Danube: Remnants of Beauty

The Danube is neither the largest nor the longest river in Europe, but it is the river that most effectively defines and integrates Europe. The Danube collects waters from the territories of eighteen nations and forms the international boundaries for eight of them. Water and sediment from some of Europe's highest peaks and flattest basins flow into the Danube through thousands of tributaries. The river's largely eastward course has served as a corridor for migration and trade and a boundary to be fiercely guarded with castles perched on steep rock walls. Prehistoric peoples hunted the region's forests and fished its waters. Nearly every great power in Europe, the Mediterranean, and western Asia fought over the Danube's rich valley bottoms and clear springs at some point in history. Poets of a dozen languages have celebrated the river's beauty and majesty, and it inspired Johann Strauss's "On the Beautiful Blue Danube," one of the most evocative, lyrical waltzes ever written. Known in English as the Danube, in its own lands the river's name becomes progressively Donau, Dunaj, Duna, Dunav, Dunărea, and Dunay. These variants may stem from the Celtic words *don* and *na,* meaning "two rivers," because the Celts could not agree on the source of the Danube. Or the contemporary names may derive from German words for thunder, fir, or clay. In this river basin with millennia of conflict, even the river's name evokes controversy.

Political cartoonists of the mid-nineteenth century portrayed the Ottoman Empire, which included the southeastern portion of the Danube drainage, as the "sick man of Europe" over whom physicians representing England and France held anxious consultations as they watched death approaching in the form of the Russian Empire. Today the Danube is the sick river of Europe, its once wandering channel straitjacketed into narrow navigational routes, its floods captured by dozens of dams, and its waters heavily polluted with sewage, metals, hydrocarbons, and synthetic chemicals. Some of the eighty million people living within the Danube drainage are working hard to be the physicians who will bring the river back to health, developing a systematic database of the river's current condition, identifying the most pressing problems, and crafting international agreements on water quality and habitat restoration. If the would-be healers can bring back this river of many nations and cultures and a long history of human alteration, then there remains hope for any river on Earth.

Engineering the Mountain Danube

The gentle mountains of the Schwarzwald, Germany's Black Forest, are credited with being the headwaters of the Danube. Here the Danube formally begins at the confluence of the Brigach and Breg rivers. From its confluence the river drops approximately 680 meters vertically to its mouth in the Black Sea. Tributary channels descend the green eastern slope of the Schwarzwald, flowing clear and cold over the crystalline rocks that form the mountains, slip around the southern end of the Schwabian Alps, and then flow northeastward along the margin of the flat Bavarian basin. Cities begin to appear along the Danube in this basin: more than a hundred thousand people each live in Ulm, Ingolstadt, and Regensburg, where the Danube turns southeast as it skirts the mountains of the Bohemian Forest immediately to the north and the German and Austrian Alps farther south. From Regensburg, the Danube flows along an increasingly narrow valley until it is pinched into a gorge between the Austrian cities of Linz and Vienna. Wherever people live along the Danube, there is river engineering. Six major dams arrest the flow of the Danube between its source and Ulm, and the river passes through another seventeen dams between Ulm and the city of Kelheim, just upstream from Regensburg.

Although the Schwarzwald has the honor of being the Danube's origin, the tributaries that enter lower along the river's course begin in

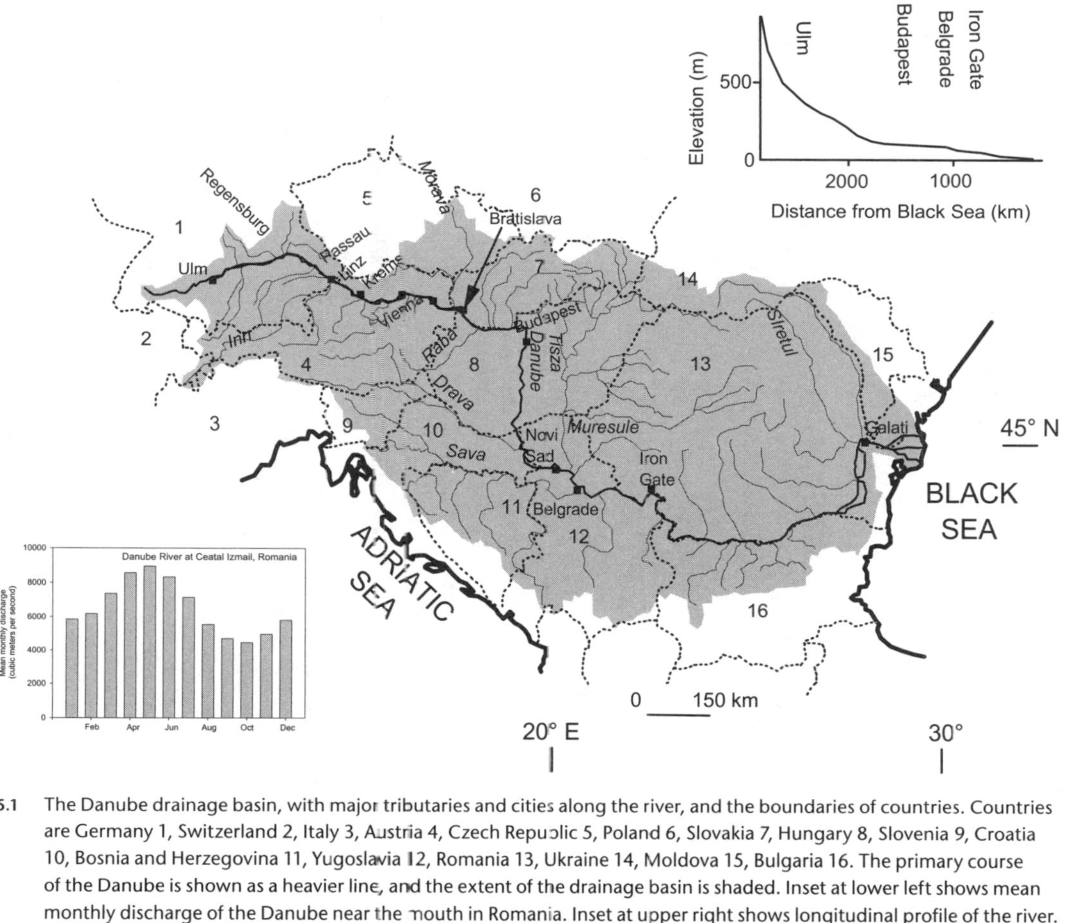

5.1 The Danube drainage basin, with major tributaries and cities along the river, and the boundaries of countries. Countries are Germany 1, Switzerland 2, Italy 3, Austria 4, Czech Republic 5, Poland 6, Slovakia 7, Hungary 8, Slovenia 9, Croatia 10, Bosnia and Herzegovina 11, Yugoslavia 12, Romania 13, Ukraine 14, Moldova 15, Bulgaria 16. The primary course of the Danube is shown as a heavier line, and the extent of the drainage basin is shaded. Inset at lower left shows mean monthly discharge of the Danube near the mouth in Romania. Inset at upper right shows longitudinal profile of the river.

higher, steeper headlands. From the south comes the Inn River, which descends from four thousand meters in Switzerland. Northern and eastern tributaries of the Danube drain heights of more than twenty-four hundred meters in the Carpathian Mountains. It is the higher, steeper mountains, topped with small alpine glaciers and mantled with winter snows, that produce much of the sediment carried by the Danube. Along most of its length the river is a turbid pale green, its waters murky with suspended sediment despite the settling ponds formed by the many dams, and despite the best efforts of the people living in the mountainous headwaters to keep the sediment on the hill slopes.

The little towns of the Austrian Alps exemplify the attempts of humans to arrest the processes of landscape evolution. Stands of pine forest still cover the slopes in places, but much of the region has been cleared for pastures grazed by cattle. White houses clustered about churches in the valley bottoms seem to form a quintessential Alpine scene, a landscape where people have lived productively and securely for centuries, developing an architecture and a culture attuned to the surroundings.

But closer scrutiny reveals pale gray lines arrowing down the steep slopes and bulging into debris fans along the margins of the valley bottom. These are the loose threads in the weave of this landscape, the places where it is in fact unraveling in debris flows and avalanches that bring tons of gravel and boulders down to the thin line of the stream. And the stream is only a thin line because people work very hard to keep it that way. Left to its own devices, this stream into which the hill slopes regularly dump sediment would form a broad, braided channel with numerous shallow side arms and a tendency to periodically migrate from side to side across the valley, or to simply stretch out and encompass the entire valley bottom, as happened during the floods of 1965–66.

Mountainous slopes still funnel masses of sediment into the unconfined channels of the sparsely populated and economically poorer headwater countries along the Amazon, the Ob, and the Nile. In contrast, the Alpine dwellers along the headwaters of the Danube work mightily and at great expense to keep the landscape from unraveling along both the debris torrents and the stream channels. The steepest mountain channels are temporarily shored up with check dams designed to trap sediment. The channel of the Hopfgartnergraben, for example, drains two square kilometers of mountainside above the village of Hopfgarten. Mountain peaks rise two thousand meters above the village of forty-six hundred people, and along the flanks of those peaks

5.2 A channelized stream in the village of St. Jakob in Deferregen, Austria. The channel here is narrowly confined between stone walls. Timber harvest and grazing have replaced much of the forest on the steep slopes above the channel, and the lines trending across the upper slopes in this photo are structures designed to retain snow and sediment on the slopes.

5.2 (*continued*) Filled check dam made of iron "baskets" along the Hopfgartnergraben, Austria. A series of these dams line the short, steep channel to retain sediment on the slopes above a small town.

lie weathered rock and soil that in places reach almost twelve meters thick. During intense rainstorms, infiltrating water builds up pressure that can cause masses of sediment to slide down the hillsides or into the small channel. The rains of 1965–66 destabilized thirty thousand cubic meters of sediment.

A chronology of mass movements of sediment, and engineering responses, in the Hopfgartnergraben during the twentieth century makes the hazards abundantly clear: debris flows in 1917 and 1921; construction of twenty-three check dams in the upper catchment in 1930–33; debris flow in 1945; more protective structures built the same year, as well as in 1956 and 1958; largest sediment disaster in 1965; further protection measures in 1966–1971, 1972–74, 1978, 1981, 1991, and 2000. Picking out just one year of protection measures, 1981—neither the most nor the least expensive—required an estimated 1.75 million euros to protect this one small catchment in a region thick with similarly active mountainsides.

What can be done if the mountainside tends to slide downward and a town is built below? Walls in the form of check dams across stream channels can be built so that each wall collects its own wedge of sediment until the dam is full. Hill slopes can be drained to prevent water from building pressure and lubricating unconsolidated materials on the slope. Vegetation can be preserved, and afforestation of already cleared areas can be encouraged. Training walls can be built that direct avalanches and debris flows away from houses and roads. Retention basins that catch and trap debris near the base of the slope can be built and then periodically emptied as they fill with sediment. Or houses and roads can be moved away from hazardous zones. Only the last option seems to be neglected in the Austrian headwaters of the Danube.

Hill slope and channel protection measures date back to the early sixteenth century in the Austrian Alps, but systematic government programs for hazard zoning and associated hazard reduction really began during the 1950s to 1970s. As population densities and land use in the Alpine regions increased during the later twentieth century, mainly in association with tourism, expenditures on engineering structures designed to decrease natural hazards rose into the billions of euros. At the same time that efforts to keep sediment on hillsides have increased, river scientists are gradually increasing government and public awareness that the lack of sediment, floods, and channel complexity—the hallmarks of a dynamic, natural channel—impoverishes river ecosystems to the point that restoration measures are now needed.

River channels perform a constant balancing act. The amount of sediment that can be carried through any segment of the river depends on the energy of the river's flow, which in turn reflects the discharge and the velocity at which that discharge moves. Water moving through a broad, shallow channel tends to flow more slowly because it encounters more flow resistance along the channel boundaries. The same volume of water moving through a deep, narrow channel is likely to flow more rapidly. River engineers try to design a stable channel with a cross section and velocity distribution just sufficient to move sediment supplied to the river segment. If the engineers overestimate sediment supply, the river flow has energy not being used to carry sediment and will likely erode the channel boundaries. If they underestimate sediment supply, the river will not have sufficient energy to transport the sediment, and it will accumulate along the bed and banks. Part of the challenge is that sediment supplied to natural channels constantly fluctuates over periods of days to decades. A large flood, for example, can unexpectedly supply much more sediment than the channel is capable of transporting. The sedimentation basins built along headwater streams of the Danube have the desired effect of detaining sediment, but they work so well that downstream portions of the rivers, deprived of sediment, are now eroding. Widening an eroding section of the river can help to control the erosion, and this is the latest strategy in the ongoing attempts to fine-tune the rivers of the Alps.

In addition to keeping sediment on hill slopes, people of the Alpine valleys have tried for centuries to constrain their unruly braided rivers within progressively narrower channels. River valleys tend to be the most agriculturally fertile portions of the landscape, and river corridors aid travel through mountainous regions, but floods can disrupt these functions if the channel is not confined to a limited width and predictable location. The Drau River, an Austrian tributary of the Danube, illustrates some of the challenges of this form of river engineering.

Narrowing and bank stabilization along the Drau River date to the 1880s. The initial measures were enhanced following large floods in 1930 and 1965–66. These measures largely eliminated the broad tangle of channels that branched and rejoined across much of the valley bottom, storing and transmitting large quantities of gravel-sized and cobble-sized bed sediment. The control measures worked so well that today the lack of sediment creates problems as the Drau erodes its bed below bridge footings and protective structures along the channel banks. Since 1991, scientists have restored about ten of the seventy most impacted kilometers of the Drau by widening the channel, allowing bars and islands to form, and letting the river occupy former side channels that had been blocked for decades. Results thus far indicate that riverbed erosion has stopped in the restored segments.

Only Connect

River habitat has also been restored along the Drau and other rivers where connectivity has been reestablished between the main channel and secondary channels. To survive and grow larger, smaller fish need refuges such as small side arms and backwaters. The slow water, shallow depth, warmer temperature, finer streambed sediment, and dense aquatic vegetation of these refuges are particularly important for the larvae and young fish of many species, but it is just these refuges that are lost when complex river channels are simplified to straight, uniform canals. Bank stabilization and narrowing reduce the connectivity of river side arms and narrow the main channel, eliminating islands and backwaters. Also lost is the ability of the river channel to provide small refuges that buffer the effect of rapid changes in water level from either natural floods or the flow regime associated with hydroelectric power generation.

The diversity of habitats associated with the swift, narrow portions of the Danube and the broad and diverse floodplains helps to support the richest fish fauna of any European river. River lamprey (*Lampetra fluviatilis*), salmon (*Salmo salar*), sea trout (*Salmo trutta trutta*), barbel (*Barbus barbus*), dace (*Leuciscus leuciscus*), and chub (*Leuciscus cephalus*) prefer the swifter current and deeper flow of the main channel. Burbot (*Lota lota*) spend some stages of their life history in tributaries or backwater channels connected to the main flow. Smelt (*Osmerus eperlanus*) and flounder (*Platichthys flesus*) spend a portion of their lives in the slowly flowing brackish water of poorly connected secondary channels. Brown bullhead (*Ictalurus nebulosus*) and crucian carp (*Carassius carassius*) spend their entire lives in still waters that have rooted aquatic vegetation. The greater the diversity of habitat and food sources along a river, the greater the diversity of invertebrate, fish, and plant species.

One hundred species of fish spend at least part of their lives in the waters of the Danube drainage basin, and thirty of these species are commercially important. Approximately thirty thousand tons of fish are caught yearly by commercial and sport fishermen, and records of Danube fisheries date back to 335 BC, when Greek traders commercialized the fish of the Lower Danube. The Danube is also rich in fish because of its orientation and glacial history. The predominantly east-west alignment of the drainage basin made it a corridor of migration and recolonization both before and after the ice ages as individual fish moved between the Ponto-Caspic and Central Asian fish communities to the east and the Alpine regions to the west. The main stem of the Danube was unglaciated and served as a refuge during periods when the continental ice sheets advanced. As the ice retreated, fish expanded from this refuge to the rest of Europe.

As habitat diversity declined in connection with river engineering along the Danube, the distribution of several fish species was drastically reduced, and some of these species are now close to extinction. The Upper Danube is dominated by salmonids, the lower portions of the river by cyprinids such as nonnative carp originally introduced to Europe from China many centuries ago. Five species of salmonid once existed in the Danube drainage, but now they are mostly either extinct like one type of lake trout (*Salmo schiefermuelleri*), critically endangered like other lake trout (*Salmo labrax*) and Danube salmon (*Hucho hucho*), or endangered like the Alpine char (*Salvelinus umbla*). Austria is the center of the natural distribution of the Danube salmon, for example, yet the fish today occupies only 10 percent of its former distribution along forty-five hundred kilometers of the river basin. Scientists attribute the decline of the salmon, which can live more than fifteen years and attain a weight of sixty kilograms, primarily to channelization and hydropower development. Because the salmon is currently classified as threatened with extinction in Austria, a rehabilitation program aims to reestablish fish passage and habitat complexity along the Pielach and the Melk, two rivers tributary to the Danube that currently host the largest remaining salmon populations.

Most of the fish species historically present along the Danube in Austria are still present today, except for the sturgeon that used to migrate more than twenty-five hundred kilometers upstream from the Black Sea. This suggests that restoring habitat could substantially increase fish abundance. The return of fish to restored zones is limited, however, by the presence of hydroelectric power dams. These dams create physical obstacles to upstream-downstream migratory movements by fish and eliminate the flood peaks that maintain the secondary channels. Operation of the dams creates strange flow regimes, with surges of discharge during periods of power generation—mostly three times a day—and extremely low-water conditions in between. This is so stressful to river organisms that scientists have recorded 95 percent decreases in abundance of organisms downstream from power dams. From the

5.3 *Above*, downstream side of the dam Donaukraftwerk Jochenstein, a hydroelectric facility built along the Danube between Passau and Engelhart during 1953-56. Dams like this alter the river flow, store sediment, and impede fish movement along the length of the Danube. *Below*, an "ecohydroelectric" power plant along the Drau River, tributary to the Danube, in Austria. The water in the foreground flowing through the main channel is pale because it carries fine sediment in suspension. The water in the background is clear and dark as it flows back into the main channel from a side arm along which the power plant is built.

German headwaters down to the massive Djerdap dams in Romania, dams built in confined, formerly swift-flowing portions of the river create ponded water that is then taken over by reed beds and other rooted aquatic plants more characteristic of broad, slower-flowing stretches of river. Dams also alter spawning habitat and reduce the seasonal peak flows many fish rely on for spawning cues.

All along the Middle Danube, fish numbers rise and fall with the extent of flooding. During a thirty-eight-year period in the twentieth century, annual fish yields dropped to less than five hundred tons when the annual flood lasted twenty days or less and rose to more than fifteen hundred tons when it lasted two hundred days. As on the Amazon and the Ob, fish productivity along the Danube is closely tied to annual flooding.

It is trends like these that have spurred the restoration of side arms and floodplains along the Danube. These restoration measures are at present very limited in spatial extent, at least in part because much of the historical floodplain and secondary channels along the Danube and its tributaries is now occupied by towns or agricultural lands. Changes in flow regime associated with dams and hydropower generation also limit the efficacy of the restoration measures. Some scientists have proposed replacing large hydropower facilities with smaller dams that generate only up to ten megawatts of electricity. The argument in favor of these "ecohydro" dams is that because they are smaller, they backflood smaller areas upstream and disrupt the natural flow regime of shorter downstream segments. Some ecologists, however, argue that there is no basis for the ten megawatt limit, and that numerous ecohydro dams can be just as disruptive as the larger dams.

The Upper Danube

The Upper Danube extends from the headwaters in Germany approximately eleven hundred kilometers downstream to the border of Austria and Slovakia. This is the steepest portion of the Danube, falling an average of forty centimeters vertically for each kilometer traveled. At this slope the waters can transport large volumes of the sand, gravel, and cobbles that travel in contact with the streambed, as well as silt and clay suspended in the river's flow. The river is generally less than 240 meters wide in this upper portion, but it flows alternately through canyons with only narrow floodplains and broad basins across which the river spreads out and splits into multiple channels separated by side arms, backwaters, and forested floodplain.

Cities have developed in the broader basins interspersed along the Upper Danube. Most of the cities have a history going back centuries, to a Roman or Celtic military camp. Among these cities is Regensburg, which grew from a nucleus that remains closely walled along the south bank of the river to the contemporary urban area that spreads broadly along both banks of the river. Just as the history of the city is inseparably linked to the river, so the condition of the river is closely intertwined with the activities of people living in the city.

Like the old city, the Danube is walled by stone where it passes through Regensburg. By this point, 370 kilometers below the confluence of the Brigach and the Breg, the river has passed through numerous storage reservoirs. The Danube here has up to four times the sodium, potassium, and chlorine of rivers with little human effect on water quality, up to eight times the calcium, magnesium, and sulfate, and ten to sixteen times the nitrogen and phosphorus compounds, mainly as a result of agricultural, municipal, and industrial sources of pollution. Regensburg itself forms another large point source of contaminants, although the water quality of municipal releases in Germany and Austria has improved substantially during the past two decades.

Forested hills crowd the north bank of the river downstream from Regensburg, but the south bank spreads broadly across an alluvial plain. The Danube is riprapped at every point visible from a bridge, but a small side road leads to a wilder portion of the river. A swiftly flowing main channel splits around forested islands and into secondary channels where great blue herons perch in the trees and ducks paddle in midstream. Vegetation grows thickly along the banks: reeds at the water's edge, shrubby willows farther away, and tall deciduous trees on the sloping banks covered only at high water. The river corridor looks green and messy, but in this apparent messiness is the diversity of ecological health. In the other direction spread kilometers of former floodplain now neatly parceled into croplands and villages. Only seventy kilometers of the great river remain less regulated here. The main channel has riprap and levees. Dams upstream and downstream limit the natural flow present here. But the levees are farther apart, the dams are spaced more widely along the main channel, and secondary channels and wetlands are present.

Beyond Passau, the Danube drops more than 150 meters vertically as it flows approximately 330 kilometers through Austria. This portion

5.4 Side arm of the Danube downstream from Regensburg, Germany. Although flow is regulated here and much of the former floodplain is isolated from the channel by levees, this portion of the upper Danube retains some of the diversity once present along the entire river.

of the river tended to be relatively steep and swift-flowing, with gravel and sand forming the streambed, before it was arrested by ten hydroelectric dams. Downstream the river alternates repeatedly between narrow, steep sections where it flows through the foothills of low mountain ranges and wide valleys where it crosses basins. In these valleys the river historically dropped the sediment it had carried from the Alps, creating large alluvial fans as it divided into a braided channel with a broad floodplain.

Severe regulation of the Danube in Austria began in 1850 and continues today. During the past fifty years more than 90 percent of the length of the Danube and its major tributaries in Germany and Austria was dammed for hydropower. Nineteenth-century channelization and twentieth-century dams have reduced most of the formerly braided reaches to a single, much narrower channel that engineers dredge to maintain minimum depths for navigation. Surface water area declined by 50 percent in the Linz basin and 55 percent in the Vienna basin, for example, as a result of all this engineering. Throughout Austria, an estimated 80 percent of former floodplain areas, backwaters, and side arms along the Danube have been filled in or disconnected from the river, with a corresponding loss of hydrological connectivity, shoreline habitat, and ability to retain fine sediments and nutrients.

Amphibians provide another example of organisms whose ability to survive is strongly dependent on the characteristics of floods and on the complexity of channel margins. Floods allow fish that prey on newts, toads, and frogs to colonize breeding ponds normally not connected with the river, but they also leave behind numerous small water bodies with nutrient-enriched water that favors the growth of amphibian larvae. Loss of small water bodies and the floods that periodically enrich them thus results in loss of both amphibians and fish.

Today only two relatively free-flowing sections of the Danube remain in Austria: the Wachau section between Melk and Krems, and the portion of river between Vienna and the Slovakian border. It is downstream from Vienna that remnants of the river's formerly extensive floodplain can still be found, and it is on these remnants in the wider valleys that contemporary rehabilitation measures focus.

Before the Danube reaches these less impacted reaches, however, it passes through Vienna, former capital of the Austro-Hungarian Empire. This city of 1.6 million people is one of the great romantic capitals of the Danube, thanks in part to the association with Strauss and the Blue Danube waltz. For centuries the citizens of strategically located Vienna walled themselves in against the Turks and other would-be masters of the Danube, and against the river itself. The great flood of 1501 poured an estimated fourteen thousand cubic meters per second of water through and around the city, more than five times the average flow.

Today the metropolitan area covers more than four hundred square kilometers. The various portions of the Danube run from northwest to southeast through Vienna, dividing the city into long, narrow slices. The Danube has various portions in Vienna thanks to engineers and flood control. Curving remnants of the original river channel, now known as the Old Danube, lie along the northeastern margins of the river course. During 1870–75 a long, straight channel was created that would contain floodwaters and speed them through the city. This channel is now called the Danube. A century later this artificial channel was supplemented by another parallel channel for flood control, known as the New Danube. Finally, there is the Danube Canal, which swings south and then east in a long arc through the city.

The Old Danube is the healthiest-looking of the multiple channels within Vienna; the worst of them, the Vienna River, deteriorates to a concrete trench. But all the river courses have some problem with water quality. Even the flowing portions of the river experience eutrophication associated with excess nutrients and algal blooms. Six hydropower plants were built along the Austrian section of the Danube during the last three decades of the twentieth century. These dams decreased average flow velocities by nearly two-thirds and promoted sedimentation. Discharge of cooling waters further raised water temperatures in the river, and these changes promoted algal blooms. Mollusks living in the river channels have elevated concentrations of cadmium, copper, and zinc, especially at sites along the New Danube where traffic emissions add to metal loading of the river.

The situation has improved, however. The Romans built the first underground pipes to rapidly direct wastewater and storm runoff to the nearest river channel when Vienna was a Roman military camp during the first century AD. The Roman system was periodically enlarged and repaired during the succeeding centuries, until it was replaced by brick pipes during the late Middle Ages. By the early eighteenth century the growing city's drainage system had been substantially expanded, but wastes remained untreated before being dumped into rivers. City engineers responded to a severe cholera outbreak in 1831 by intercepting wastewaters at local outfalls and directing the untreated water into the Danube Canal and thence downstream from the city. Continued growth of the city eventually caused an increase in the number of points where untreated waste was discharged directly into the Danube River. It was not until 1969–80 that a series of intercepting sewers and a central treatment plant were constructed. These facilities were further upgraded between 1986 and 2000, although the sewer system still occasionally overflows during heavy rainstorms.

Despite the impaired water quality in the rivers crossing through Vienna, many organisms can survive if provided with some habitat. In 1997 engineers restructured a portion of shoreline along the Danube by creating artificial side channels, coves, gravel banks, and pools. Over the next two years, twenty-eight species of dragonflies colonized these sites, along with twelve species of amphibians and thirty-three species of fish. Dragonflies in particular are used as indicator species all along the Danube in Austria. The abundance and species diversity of these mobile insects reflect the presence of diverse small habitats that dragonflies need to survive, which range from water currents over floating leaves, required for territorial males, to sets of water plants that females use for egg laying. A 2004 study of dragonflies at fourteen sites along the Danube throughout Austria ranked ten of these sites in good ecological condition.

One of the more environmentally healthy portions of the Danube in Austria lies about twenty-five kilometers downstream from Vienna and extends to the Slovakian border. This stretch of river was designated the Alluvial Zone National Park in 1996. The floodplain within the national park supports very diverse species of river organisms, from rooted aquatic plants through mollusks, dragonflies, amphibians, and fish. Even here the Danube has a long history of engineering, but current restoration efforts seek to revitalize some of the river's natural function.

An 1812 map of this portion of the Danube's valley resembles the tracks left behind by several skiers slaloming down a snowy slope. The Danube splits into multiple branches that rejoin seemingly at random farther downstream before splitting once more. Historical records going back to the sixteenth century indicate that floods repeatedly deposited large amounts of wood on the streambed, causing jams that trapped sediment, formed bars and islands, and encouraged channel branching. When more water came down the Danube during periods of high flow, the water spread among all these channels, allowing aquatic organisms to migrate between different habitats during an estimated ninety days each year. Nearly 60 percent of the active channels and floodplain was inundated at moderate flow volumes.

The generous spread of water among many channels rendered navigation difficult during periods of low flow and created extensive flooding during high flows. Attempts to straighten and narrow the river's course began during 1826–59 with placing a guide wall made of bundles of wood and stone across the upstream end of some of the side channels. Construction of the Ybbs-Persenbeug hydropower plant twenty-three kilometers downstream in 1957 and the Wallsee-Mitterkirchen plant just upstream in 1968 further altered flow into side channels and across the floodplain by reducing seasonal flood pulses. With the addition of artificial levees and embankments in 1991, the historical flood pulse decreased by 50 percent. A 1991 map of the same area resembles a highway more than the tracks of skiers; the channel forms a single straight line across the former floodplain that averages 350 meters wide.

Large portions of the former floodplain are now drained or disconnected from the river as a result of channelization and flow regulation. Restoration efforts currently focus on the five square kilometers of the Regelsbrunn floodplain downstream from Vienna. Here a former river channel about ten kilometers long was artificially cut off from the main channel at its upstream end more than a century ago. This side channel was subsequently divided by weirs into partially isolated water bodies that are now connected to the main channel at the surface for less than half the year during times of high water. The porous sands and gravels forming the floodplain aid subsurface flow between the side arm and main channel, however, and water levels in the side arm follow those in the main channel after a short time lag. Scientists now seek to raise the water level across the floodplain in order to restore surface connectivity with the main channel. A 1997 restoration project at the artificially isolated side channel lowered the upstream barrier between the side and main channels. To be successful, these restorations require a thorough understanding of river dynamics. A similar restoration project on the Danube tributary the Morava River along the Austrian-Slovakian border merely led to sediment accumulation in reopened side arms.

Ecologists correlate the degree of surface connectivity, or how much flow secondary channels and other floodplain areas receive directly from the main channel, with richness of surface-living species and invertebrates that live in groundwater and also with the complexity of bacterial and planktonic communities. Bacterial complexity is not something most of us think about when looking at a river, but the unseen microbes control the productivity of the entire river system.

River water transports organic nutrients such as carbon both in solution and as fine bits of organic detritus in suspension in the flowing water. Bacteria, which are the most important consumers of this carbon, provide food for higher organisms, so that the rate and manner in which bacteria make carbon biologically available governs the rest of the aquatic food web. A single relatively uniform channel along the Danube represents complete connectivity within the channel, which decreases bacterial productivity in the river because the food is moving by too fast. A single channel also represents complete disconnectivity from the floodplain, which decreases bacterial productivity on the floodplain because there is limited nutrient input. The ideal situation is one in which river waters periodically inundate the floodplain and then slowly drain, providing nutrients that bacteria can consume over days to weeks.

The Middle Danube

Downstream from the Vienna basin, the Danube valley again narrows as the river cuts through a spur of the Carpathian Mountains in a short gorge known as the Porta Hungarica before crossing the Slovakian border and passing through the city of Bratislava, home to approximately 450,000 people.

Historical maps dating back to 1712 record changes in the Danube at Bratislava that mirror those described for the Alluvial Zone National Park. Originally the Danube branched and meandered across a large alluvial fan. Increased population growth resulted in channelization of the entire river course during 1886–96, but flooding continued, exacerbated by changes in the upstream drainage basin. Soil erosion intensified as new crops—potatoes, maize, and tobacco—reached Europe from the Americas. Completion of numerous dams along the Danube drainage in Germany and Austria during the later twentieth century decreased sediment load on the main river to about one-quarter of historical levels, worsening bed and bank erosion downstream. The economic and environmental costs of channel engineering at Bratislava continue to rise as the Danube continues to respond to changes in flow, sediment supply, and channel configuration.

The flood-prone site of Bratislava also marks the start of the Middle Danube, which continues for eight hundred kilometers downstream to the Iron Gate where the Danube passes through the Carpathian Mountains in Romania. The Middle Danube drops six centimeters per kilometer, compared with the forty centimeters per kilometer of the Upper Danube. Consequently, the river's ability to carry sand and gravel declines in this section, although it still transports large amounts of finer sediment in suspension. The main channel grows to widths of 180 to 300 meters at the start of the middle portion and expands further to 600 to 900 meters approaching the Iron Gate.

A few kilometers beyond Bratislava, the Danube crosses into Hungary and meanders broadly across another wide basin as it moves gradually southeast. There is little direct evidence of the country's turbulent history in the appearance of the landscape, which features endless kilometers of corn and pumpkin fields where clusters of large modern windmills churn slowly as they convert the breezes to electric power. Occasionally the flat lands give way to low, rolling hills partially covered in deciduous woodlands.

In addition to crossing a political border into Hungary, the river crosses a border of human perceptions and management. The upstream countries have changed direction in their interactions with the Danube. There, construction of dams and channelization are now largely complete, and they focus on improving water quality and restoring river habitat. The downstream countries are trying to build capitalist economies after the ruinous world wars and decades of communism.

5.5 *Above*, the Danube at Szentendre, Hungary, a wide and relatively natural portion of the river upstream from Budapest. *Below*, the right bank of the Danube in Budapest is completely covered in concrete, preventing infiltration of river water through the bank, as well as adjustments by the river to altered discharge or sediment supply.

They seek to build dams and channelize the river, and they do little to protect water quality.

Gabčíkovo, between the cities of Bratislava and Győr, Hungary, exemplifies the difficulties associated with hydroelectric power generation in this region. In September 1977 Hungary and Slovakia agreed to jointly construct two hydroelectric power plants on the Danube where the river forms their common border. The plan was to divert most of the river's flow at the dam into a bypass canal on Slovak territory. Flow would be used by a hydroelectric plant halfway down the bypass canal at Gabčíkovo, and the water would eventually be returned to the Danube's channel a few tens of kilometers downstream. A second hydroelectric dam would be built another 115 kilometers downstream where the Danube enters Hungary near Nagymoros. The second dam would generate power and attenuate the peak flows coming from the upstream power station so as not to hinder navigation.

Although some of the potential environmental impacts of this project were known in 1977, their extent and intensity were more fully realized during the late 1980s, just as environmental degradation was becoming a source of political protest in former communist countries of eastern Europe. The Hungarian government reassessed the project's environmental implications in 1989, when construction of the upstream portion of the project was well under way. Sources of concern in Hungary were decreases in groundwater levels that would damage agricultural and drinking water supplies and loss of floodplain wetlands along the nearly sixty kilometers of artificially dried Danube. Slovakia protested Hungary's hesitation in completing its portion of the project; joint meetings suggested that Hungary was going to withdraw; and the Slovakian engineers went to plan B, which was to build a dam upstream, where the Danube is solely in Slovakia.

The Gabčíkovo power plant began operation in 1992, with most of the river's water diverted to the bypass canal. The two governments submitted the dispute to the International Court in June 1993. Four years later the court ruled that both parties had breached their obligations and must renegotiate the agreement. Engineers and ecologists from Hungary and Slovakia now work together to devise a flow management plan that allows both power generation and the maintenance of floodplain wetlands along the largely abandoned, multibranched portion of the Danube downstream from the bypass.

Meanwhile, at least some of the predicted environmental effects have materialized in the few years since project completion. The water table dropped in the diversion zone along the Danube. The abundance of fish of all ages decreased in the affected floodplain section, although three new fish species colonized the area as quiet-water fish replaced those that prefer flowing water. The mean annual fish catch dropped by 87 percent compared with the decade before construction of Gabčíkovo. Aquatic plants characteristic of lakes have begun to colonize the floodplain side channels.

The Middle Danube continues to flow alternately through confined and broad reaches on its path toward Romania. Cities along the wider portions create some of the greatest challenges to water quality, as exemplified by the Hungarian capital of Budapest. Buda on the west bank of the river and Pest on the east bank together form a metropolis of approximately 1.8 million people that is known as the Queen of the Danube. By the time it reaches this point, the river's flow is altered by twenty-eight storage reservoirs. Flow regulation limits the river's ability to dilute and process wastes from Budapest. Scientists sampling barbel fish and their intestinal parasites in the Danube at Budapest in 2004 found elevated concentrations of a wide variety of toxic heavy metals, including platinum group metals that derive from chemotherapy drugs, dental treatments, and catalytic converters in automobiles.

Despite the contamination of water and sediment in the Danube, Budapest draws a large portion of its drinking water from the river. Throughout Hungary, approximately 40 percent of the water delivered by public water companies is drawn from shallow groundwater deposits along the Danube and its primary tributaries. This water comes from the rivers, having first filtered through the silt, sand, and gravel of the riverbed and riverbanks. Its quality depends on the quality of water in the river, but also on how quickly the river water filters through the bed and bank sediments, on how well the sediments remove impurities, and on other sources of contamination in the area. Filtration can trap the fine particles of silt and clay some contaminants are attached to and provide a site for the microbial reactions that reduce the concentrations of nitrogen, phosphorus, and other excess nutrients.

In recent decades, however, Hungarian scientists have grown concerned about the effects of river regulation on filtered water quality. Riprap and other forms of bank stabilization can reduce infiltration into riverbanks, and infiltrating waters can dissolve iron or manganese contained in the stabilization materials. Dredging the riverbed can remove the biologically active filter layer of microbes and aquatic insects, which does not recover quickly. Reduced infiltration rates can also allow shallow groundwater from other sources, potentially contaminated, to flow into the aquifer around the river. The town of Vác, for example, obtains bank-filtered water from the Danube but had to shut down some of its wells in 1981 as a result of complex interactions between river regulation and local industry. A dike constructed upstream from Vác's well field partially closed off a secondary channel of the Danube, creating stagnant water during low flow. This reduced filtration into the riverbank, allowing a plume of polluted groundwater to spread toward the river from a nearby pharmaceutical factory.

From Budapest the Danube again flows south through agricultural lands and small villages of Hungary and then Yugoslavia. During prehistory the Great Hungarian Plain that the river crosses was covered in oak woodlands. Thousands of years of land use caused deforestation, soil erosion, and salinization, resulting in today's farm fields and grasslands with only scattered patches of trees. Farming has a long history in the basins surrounded by the Carpathian Mountains. Archaeologists find evidence of systematic agriculture along the Danube's floodplain in the ninth century. By the Middle Ages Hungarian farmers had devised a system of small drainage canals and artificial inlets to direct water through the Danube's natural levees and onto the floodplain. Alteration of the Danube for navigation and flood control spread in earnest through the region during the nineteenth century. Today the Danube is strictly confined between embankments as it crosses this wide alluvial basin, and the width of contemporary flooding seldom exceeds five hundred meters, with the two exceptions of the Gemenc and Béda-Karapancsa floodplains. These relatively small remnants of functional floodplains encompass a diverse mix of wetlands and bottomland forest.

Interspersed among the farm fields along the Danube's course lie cities such as Novi Sad, once known as the Serbian Athens. Situated on the Danube approximately seventy kilometers upstream from Belgrade, Novi Sad today has three hundred thousand inhabitants and is the second largest city in Serbia. As with other major commercial and industrial centers along the Danube, scientists have analyzed the quality of river water and sediments here. Potential sources of contamination include an oil refinery. Oil refineries are known for polluting the region around them even in the course of normal operations, but the refinery at Novi Sad was bombed by NATO forces in 1999. Destruction of the refinery released more than 73,500 tons of crude oil. Most of this was incinerated, but an estimated 560 tons reached the Danube. Scientists analyzing soils in the area during the year 2000 found that concentrations of oil decreased dramatically with increasing depth, but that the oil was slowly infiltrating toward the groundwater. About half of Novi Sad's drinking water comes from the aquifer the oil is moving toward, which is predominantly fed by infiltration from the Danube. In October 2000, the Danube itself had an oil content about 20 percent of that measured immediately after the bombing. The river also contained concentrations of mutagenic and carcinogenic polyaromatic hydrocarbons of up to ten thousand times the background levels common in the United States. In the absence of containment and remediation measures, which are not very likely in the depressed economy of contemporary Yugoslavia, these contaminants will move inexorably down into the local drinking water supply and downstream along the Danube.

Early investigations of water quality along the 580 kilometers of the Danube's length in Yugoslavia indicated deteriorating quality between 1975 and 1988. As water quality declined, the number and diversity of aquatic organisms also decreased. Decline of economically important stocks of Danubian carp (*Cyprinus carpio*) and Black Sea migrant sturgeon species (Acipenseridae) became particularly noticeable during the 1980s, presumably reflecting the combined effects of water pollution, destruction of floodplain spawning grounds through channelization, construction of the Djerdap dams downstream that blocked migration routes, and extensive fluctuations in water levels associated with flow regulation, which cause mass deaths of very young fish.

A study published in 1990 noted that for many samples from the Danube, permitted concentrations of oils, grease, and phenols were exceeded; dissolved oxygen levels were below stipulated standards; and there was heavy metal pollution of sediments in the main channel and tributary mouths. Contaminants also show up in elevated concentrations in living organisms. Aquatic plants can accumulate heavy metals at concentrations up to a million times those found in adjacent water. Studies far upstream on the Danube in Germany indicate that it can take almost fifteen years for the number and diversity of aquatic plants to reflect an improvement in water quality. Some of the highest concentrations documented in a 2002 study of heavy metals along the Yugoslavian portion of the Danube came from plants in the 960 kilometer Danube-Tisza-Danube canal system near Novi Sad. Sources for these metals include wastewater that enters the canal system from a fertilizer factory and an oil refinery. Fish also show an accumulation of heavy metals.

As with the results from the canal study, many of the worst examples of water and sediment pollution in Yugoslavia come from specific point sources. The river Sava, which enters the Danube at Belgrade, receives wastewater from pulp and paper mills. These mills have water treatment plants, but the plants do not always work. When the plants are not operating, and when low summer water levels reduce dilution, dissolved oxygen levels drop and high concentrations develop of such toxic substances as phenols, oils and grease, and dyes. Increasing industrial and municipal development along the Sava contributes its share of pollution as well. The river drains more than ninety-four thousand square kilometers, being fringed by some of the largest floodplains in Europe, and the area has a population of 8.5 million spread among small towns and large cities such as Zagreb. A large number of these people rely on drinking water from shallow aquifers supplied by the river, yet most cities and industries release untreated or only partially treated wastewater into the river. One sign of an overburdened river is the massive fish kills that have become more frequent since 1965. As noted by the authors of a 1990 study on water quality below the pulp and paper mills, the toxins constantly accumulate, creating more and more severe pollution, regardless of water levels and various water treatment plants that remove only some or none of the contaminants.

Other examples of water contamination in the lower half of the Danube drainage come from mine accidents in 2000 along the Tisza River, which heads in the Carpathian Mountains. Breaches of tailings dams at gold and silver mines in the Carpathians released hundreds of thousands of cubic meters of cyanide, heavy metals, and metal-rich liquid waste. From the nearest creeks, the contaminants moved downstream toward the Danube, quickly killing most living organisms in their path. These accidents gained the attention of the other Danube basin countries, particularly because Romania's Maramures County, where one of the accidents occurred, has 215 tailings dams as a legacy from its long history of mining.

The Iron Gate

The Danube turns east and begins to leave the Hungarian basin at Belgrade. The southern tributaries that drain the Alps contribute the bulk of the water in the German and Austrian sections of the Danube above Budapest, but the single largest increases in the river's flow come from the Carpathians via the Tisza and from the mountains of Bosnia and Herzegovina via the Sava. Together the contributions of these tributaries more than double the Danube's total flow, which increases only incrementally from there downstream.

At the border between Yugoslavia and Romania the river is once again tightly confined between mountains. Metamorphic and intrusive igneous rocks of the Transylvanian Alps abut the Danube on the north, and northward-curving arms of the Balkan Mountains confine the valley to the south. The large-scale course of the Danube grows sinuous, bending first south, then north, then southeast before nearly doubling back on itself to the northwest. Within this confined section of abrupt bends lies the Iron Gate, or Djerdap, where the river has responded to the continuing rise of the surrounding mountains during the past two million years by incising a deep, narrow channel. As the name implies, five gorges stretching a hundred kilometers along the Danube create a gate that many ships could not pass.

Today the Iron Gate is famous for the two hydroelectric power plants Romania built there. Djerdap I, built in 1970, remains Romania's most important hydropower plant. The complex is equipped with six hydropower units of 175 megawatts each. The dam creates a backwater 270 kilometers long up the Danube. Djerdap I was augmented by 279 megawatt Djerdap II, built eighty kilometers downstream in 1984. These power plants undoubtedly help to reduce air pollution that would be generated by coal- or oil-burning power plants, as well as radioactive contamination from nuclear power plants, but they come with their own costs.

The dams create two reservoirs with a combined area of more than twenty-five thousand hectares. They also block migration routes for fish species that move between the Black Sea and upstream parts of the Danube during their life cycle. Fish catches of the anadromous species beluga (*Huso huso*), Russian sturgeon (*Acipenser gueldenstaedtii*), and stellate sturgeon (*Acipenser stellatus*) decreased after the dams were built. Numbers of sterlet (*Acipenser ruthenus*), which also migrate and prefer faster and more abundant flows, also decreased significantly.

The impassable barriers created by the Djerdap dams are only the most recent blow to the sturgeon of the Danube. Six species of sturgeon historically lived throughout much of the river. All these species are now extinct in the Upper and Middle Danube, and only four species still reproduce in the Lower Danube. Sturgeon are endangered throughout the world and have been included in the Convention on International Trade in Endangered Species since 1998. Sturgeon belong to the order Acipenseriformes. Fish of this order originated in Europe about two hundred million years ago, then diversified as they spread to Asia.

Sturgeon have long been prized for caviar and meat. Fishermen along the Danube severely depressed sturgeon numbers as early as the Middle Ages when they built thousands of fishing weirs, sometimes across the entire channel, to catch the migrating fish.

Some fish could always get by the weirs, but the inauguration of dam building signaled the end of sturgeon migrations for reproduction and feeding. The 1927 construction of the Kachlet power plant just upstream from Passau blocked fish migration to the upper and middle portions of the Danube, but the dam included a fish passage. Fisheries scientists recorded abundant use of the passage by species such as nase (*Chondrostoma nasus*) and barbel, among others, but this was the last fish passage built with a dam for more than sixty years as another twenty-nine dams went up between Ulm and Vienna.

Although the loss of migration routes to dams occurred simultaneously with many other changes detrimental to fish populations, it is difficult to overestimate the effect of these structures. As on the Amazon, the Ob, the Nile, and other great rivers, most species of fish in the Danube undergo true migration or at least exhibit substantial movement patterns during various stages of their life history. Beluga, Russian sturgeon, stellate sturgeon, Pontic shad (*Alosa pontica*), and Caspian shad (*Alosa caspia*) are among those that travel more than three hundred kilometers in one direction within a year. Barbel, nase, and Danube salmon travel anywhere from thirty to three hundred kilometers in one direction within a year, and other species undertake short migrations of less than thirty kilometers. Fish undertake energy-intensive and dangerous annual migrations because they gain more in reproductive potential and food than they expend during the migration. Both individuals and species can be hard-pressed to survive when migration corridors are blocked. Combined with other stresses such as overfishing, pollution, and habitat loss caused by channel engineering, the cumulative effect of the hundreds of dams along the rivers of the Danube basin is inexorably pushing many fish toward extinction.

Beyond the constriction of the Iron Gate, where channel width seldom exceeds 200 meters, the Lower Danube spreads out into a channel 790 to 1,200 meters wide. The river flows for 1,080 kilometers through Romania, creating along its course one of the largest and most complex river margins in Europe. The primary floodplain is 780 kilometers long and varies from 6 to 30 kilometers in width, for a total area of more than 5,830 square kilometers. Only 20 percent of this once dynamic environment retains a natural flow regime, unregulated channel movement, and native plant and animal species, but these remaining natural areas help preserve the health of the delta that lies immediately downstream.

Over its last 940 kilometers the river drops less than four centimeters per kilometer traveled. As the river's downstream slope declines, more of the sediment it has carried in suspension is deposited across the broad valleys along its course. Beyond the Iron Gate, the Danube enters the Wallachian Plain, the last of the vast basins that characterize its course. The river flows eastward along the southern margin of the plain, at the base of the Danubian Plateau, before making one last hard turn north around the Dobruja Hills and then east at Galaţi into its delta on the Black Sea. The northern edge of the delta forms the border between Romania and Ukraine, but scientists from all over the world come here to study the history of the delta and the effects of all the channelization, flow regulation, and pollution in the upstream drainage basin.

The Danube Delta

As it enters the Black Sea, the Danube splits into three main branches. Wetlands across portions of the delta have been drained since the mid-1700s, and one of the branch channels that is dredged and used for shipping has been engineered since 1857. The distributary channels carry an average fifty million tons of suspended solids each year and encompass a surface area of 4,150 square kilometers. The three main distributary channels carried about 97 percent of the river's flow before 1980, with the remaining 3 percent making its way across the delta plain. Human creation of new channels for wetland drainage reduced flow in the three main distributaries to 90 percent after 1980.

Like many deltas, the Danube delta is a region of water as much as of land. Most of the delta plain is flooded from April to June, and water dominates 70 percent of the delta throughout the year. The water flows through the three main channels and lesser distributaries, and it pools in hundreds of shallow lakes, wetlands, and secondary channels. Between the water bodies lie levees, vegetated islands, sand dunes, forests of oak, alder, willow, and poplar, and pastures. This diverse and dynamic landscape supports 1,688 plant species and 3,800 animal species, including 300 species of birds and 75 of fish. The majority of the world populations of red-breasted goose (*Branta ruficollis*) and pygmy cormorant (*Phalacrocorax pygmeus*) spend part of each year on the delta. Up to thirty-five hundred pairs of great white pelicans (*Pelecanus onocro-*

talus) form the largest breeding colony of this endangered species that is sensitive to disturbance. The delta provides one of the last refuges for European mink (*Mustela lutreola*), as well as habitat for wild cats (*Felis sylvestris*), otters (*Lutra lutra*), wild boars (*Sus scrofa*), red deer (*Cervus elaphus*), and jackals (*Canis aureus*). Not all species have been able to thrive unmolested in the delta, but it retains a wealth of biological diversity.

The Danube delta is also the home of the carp, a much-maligned introduced fish in many other parts of the world, but also an important source of protein for millions of people around the world. Common carp (*Cyprinus carpio*) was likely the first domesticated fish. These fish were abundant in the inland portion of the Danube delta two thousand years ago, with large schools that reproduced on the river's extensive floodplains. The Romans kept ponds for raising carp, and after the Romans left, their successors kept raising them. All the domesticated European forms of carp present today originated as wild carp from the Danube.

The history of carp highlights how many resources the delta provides in the form of fish, livestock pasture, wood, and *Phragmites* reeds (from which cellulose and sugar are extracted) to the people who live within the delta and along its margins. Although the constantly shifting landscape of the delta is not an ideal place to preserve archaeological records, artifacts as ancient as a fifth-century BC Greek vase and scattered Roman settlements suggest the long history of human occupation in the delta. The delta also buffers water and sediment entering the Black Sea and draws tourists from around the world. International commissions have recognized its importance by designating it as a UNESCO Biosphere Reserve in 1990, a World Heritage Site in 1991, and in 1995 a Wetland of International Importance under the Ramsar Convention on Wetlands.

The Danube delta is under increasing threat from developments both locally and upstream. Scientists assess the condition of particular subenvironments within the delta as indicators of its ecological health. One of the subenvironments these inquiries focus on is the submerged aquatic plants that until very recently dominated most shallow lakes in the delta. These plants require clear water for photosynthesis at least part of the year. Because water flowing into the delta tends to be turbid with suspended sediments, growth of submerged plants in water bodies that are highly connected to the Danube distributary channels is suppressed immediately following the spring flood. As the suspended sediment gradually settles out of the standing water bodies on the delta, submerged vegetation grows during the summer. These interactions have been changing, however, as water quality and connectivity change.

Studies between 1980 and 2000 revealed that eutrophication in the delta lakes resulted in nearly a 50 percent reduction in species richness and surface area of submerged aquatic plants. Eutrophication usually occurs where nutrient levels in water increase to the point that microscopic blue-green algae populations rise dramatically. The resulting algal blooms reduce dissolved oxygen in the water, making it difficult for other species such as fish to survive. Algal blooms directly release compounds that are toxic to many animal species and can also effectively block the penetration of sunlight to lower levels of a lake and thus limit photosynthesis by submerged plants.

Eutrophication-induced stresses to submerged aquatic plants in the delta affected plant populations already stressed by reduced habitat. An estimated 80 percent of the former Lower Danube floodplain and 15 percent of the delta were drained during the final decades of the twentieth century. As former wetlands and channels were lost, the sediment transport capacity of the delta as a whole decreased to about 45 percent of historical levels. The delta's ability to attenuate incoming floods and spread the floodwaters across broad areas also decreased. These alterations, accompanied by changes in water and sediment chemistry resulting mainly from increasing eutrophication and contamination by heavy metals and pesticides, caused replacement of submerged aquatic vegetation by phytoplankton. Regular algal blooms began in 1981.

As might be expected, decline in wetland surface area and the biomass of aquatic plants further decreased water quality and altered invertebrate and fish communities. Oligochaete worms make up as much as 36 percent of the total mass of bottom-dwelling animals in the delta lakes, and the worms carry energy from organic carbon to bottom-feeding fish species. Eutrophication has reduced the numbers of these worms and consequently reduced the energy flow through these communities. During the past forty years, fish communities in the lakes have shifted from clearwater species toward species adapted for turbid water.

The Danube delta supports one of the most important inland fisheries in Europe, although fish catches declined during the last four decades of the twentieth century, from fifteen thousand tons of fish each year to less than five thousand. The species caught during this period also shifted from high-value fish such as pike (*Esox lucius*) to less valuable species such as Prussian carp (*Carassius auratus*). An estimated fifteen thousand people living within the delta and sixteen thousand adjacent to the delta still depend on this fishery for their livelihood.

The Danube Delta Biosphere Reserve Authority, established in 1993, struggles to maintain sustainable fisheries in the delta, largely because of the continuing effects of historical changes. Romanian dictator Nicolae Ceauşescu attempted to transform large portions of the delta wetlands into farmland during the 1980s. Some of this area has recovered since his execution in 1989, but dams built between 1972 and 1989 continue to reduce spawning zones. The dams built during the 1970s and 1980s particularly affected valuable species such as the migratory sturgeon (*Huso huso, Acipenser guldenstaedtii,* and *Acipenser stellatus*). The commercial catch of these species declined from one thousand tons a year at the start of the twentieth century to ten tons a year in 1990. The drastic declines in fisheries translate all the way downstream to the Black Sea, where catches of small fish such as sprat (*Sprattus sprattus*) and anchovy (*Engraulis encrasicolus*) have decreased from ten thousand to less than one thousand tons a year.

The dynamics of the Danube delta influence the Black Sea in many ways. Compared with the early 1960s, nitrogen loads entering the Black Sea via the Danube have increased fivefold, phosphorus loads have doubled, and silica has decreased by about two-thirds. The net result has been to take eutrophication all the way down to the Black Sea. Scientists propose at least two reasons for the increase in nitrogen and phosphorus coming out of the Danube: loss of delta lakes and flooded surfaces, and upstream land use. Delta lakes and other flooded surfaces provide efficient sinks for nutrients during the growing season because all those submerged plants, algae, aquatic insects, and fish need nitrogen and phosphorus. Loss of habitat for these organisms means much less nitrogen and phosphorus are extracted from the Danube's flow and retained in the delta. A delta composed of primarily through-flowing channels simply passes incoming nutrients downstream. Increasing intensity and extent of upstream land use in the form of commercial agriculture, and to a lesser extent industrial and municipal areas, also means that surface runoff and groundwater carry more nitrogen and phosphorus fertilizers and other compounds down into the Danube and its tributaries.

Decreased silica loads during recent decades are a little more difficult to explain. The most widely accepted idea is that the drop in silica reflects the tremendous potential of upstream reservoirs for trapping sediment. Although increased erosion of downstream riverbed and riverbanks has partially compensated for these losses, sediment yields to the delta now remain 70 percent lower than before the later twentieth-century episode of dam building. Declines in silica entering the Black

Sea have caused a shift in the sea's phytoplankton communities, from diatoms that use silica to construct their microscopic shells to coccolithophores and flagellates that do not require as much silica. These changes have stimulated algal blooms that have destabilized the Black Sea ecosystem.

In addition to trapping silica, the dams also trap contaminants that move attached to suspended sediment. Of the highly toxic mercury transported by the Danube, for example, 70 to 80 percent travels attached to suspended sediment. Plenty of contaminants, however, still make it through to the delta. Mussels in the delta have elevated concentrations of heavy metals. The river deposits nearly thirty tons of mercury on the delta each year, as well as carrying another thirty to forty tons into the Black Sea. Scientists estimate that 75 percent of the pollutants entering the Black Sea come from the Danube, which contributes particularly large amounts of fresh oil and polyaromatic hydrocarbons and lead.

To Cooperate and Restore

People in the Danube drainage basin have come together at various times during the past two centuries to coordinate engineering of the river. Many of these coordinated efforts achieved their original aims. Barges can now travel hundreds of kilometers on the Danube throughout the year, thanks to channelization, flow regulation, and dredging. The countries the river flows through derive drinking water, irrigation water, and water for industrial uses from the river, as well as hydroelectric power that does not generate air or soil pollution.

Intensive engineering of the river also caused many unforeseen or unintended side effects. The progressive loss of channel diversity and floodplains all along the river's course resulted in loss of habitat and of the diverse plant and animal species that depend on the river corridor. As human communities encroach on the artificially narrowed and straightened rivers, loss of life and property damage rise during the large floods that inevitably break through or overtop the engineered channel. Nutrients and contaminants that used to be filtered from the river water and stored in floodplain wetlands now continue downstream to reach damaging levels in the ecologically rich delta and Black Sea, even as the concentrations of nutrients and contaminants entering the Danube and its tributaries increase steadily in response to changing land use throughout the drainage. There are few problems with water quantity in the water-rich Danube basin, which receives seventy to one hundred centimeters of precipitation each year, but as with the Ob, many of the eighty-three million people who live in the Danube basin do not have access to drinking water clean enough not to endanger their health.

The most pressing problems vary with location in the drainage. The upstream Danube countries of western Europe have relatively clean water flowing through highly engineered channels. The downstream countries of eastern Europe have polluted water flowing through more natural channels. Industrial chemicals and fertilizers create the single largest manufacturing discharge to the Danube, with the bulk of this discharge coming from Romania and the Slovak Republic. The percentage of the national population linked to sewerage as of 2005 ranges from unknown levels in Ukraine, to 10 percent in Moldavia, 50 percent in Romania, and 93 percent in Germany. By comparison, in 1960 these values were 5 percent for Moldavia, 15 percent in Romania, and 40 percent in Germany.

The greater naturalness of channel segments in the lower portions of the Danube basin is only relative to the upper basin. At the start of the twenty-first century, 95 percent of the original floodplain had been lost in the Upper Danube, 75 percent in the Middle Danube, 72 percent in the Lower Danube, and 30 percent in the delta.

The good news is that problems associated with channel destruction and pollution can be decreased. The inputs of many heavily polluted tributaries in the lower Danube are diluted by the river's relatively large discharge, keeping the lower river only moderately polluted except downstream from local point sources. The Danube thus has a certain natural potential for cleansing. Water quality in the upstream basin improved during the latter decades of the twentieth century, when it became a public and government priority to install water treatment plants and alter land use practices. Mean nutrient concentrations in the German and Austrian portions of the Danube, for example, fell by at least half between 1980 and 2000. Water quality also improved slightly in the downstream basin during the last few years of the twentieth century because of regional economic problems and reduced agriculture and industrial outputs. Regional economies in the Lower Danube basin will presumably improve during the first decades of the twenty-first century, and strengthening economies can be used to justify greater effort and expenditure on maintaining clean water. The prospects for improving water quality appear brighter throughout the Danube drainage basin than in the Ob drainage.

Improvements in water quality have also translated into local recovery of species. The Danube streber (*Zingel streber*), a small fish that feeds on bottom-dwelling animals along gravel portions of the river, reappeared at the confluence of the Danube tributaries the Morava and Dyje rivers in the Czech Republic during the first years of the twenty-first century after being absent almost a century. Scientists attribute the streber's return to the marked improvement in water quality at the site after 1990, as well as the lack of barriers between the confluence and the Danube.

The lesser levels of channel engineering and floodplain loss in the Lower Danube basin also mean that downstream countries will have to do less of the restoration now being locally undertaken in the upper basin. This restoration becomes expensive once humans encroach on the former floodplain, because people and their infrastructure must be either physically moved or altered to coexist with channel movement and inundation. Restoration undertaken to date has produced at least some local improvements in species abundance and diversity.

Some skeptics dismiss the current attempts in upstream Danube countries to reconnect side channels and floodplains to the main channel as "gardening" because these projects do not address the flow regulation that will ultimately limit habitat viability and colonization of the reconnected areas by plants, insects, and fish. The presence in the Danube drainage of more than five hundred larger dams and reservoirs with a capacity of over five million cubic meters of water indicates that changes in dam operation must ultimately occur to restore natural function along the Danube. But existing restoration projects represent an important start that can help to demonstrate the positive benefits of river restoration and thus build public support for more widespread and substantial changes in river form and function.

A potential difficulty to restoring the Danube arises from the international nature of the river basin. The Gabčíkovo hydropower project illustrates some of the issues that come into play where adjoining countries have different states of economic development and associated public opinion regarding environmental protection. In 1994 the Danube River basin countries and the European Union signed a convention stipulating cooperation in measures to maintain and improve the current environmental and water quality conditions in the basin. Gradually strengthening economies in the former communist countries and eventual incorporation into the European Union are likely to be crucial in fostering international cooperation within the Danube drainage.

The Danube thus stands poised for change between a past of steadily deteriorating water quality and loss of complexity and ecosystem sustainability along the river corridor and a future that can include cleaner water; more natural and functional rivers, floodplains, and delta; greater number and diversity of species; and a generally healthier environment. With luck and perseverance, the epithet "beautiful" can once again be inseparably linked with the great river of Europe.

Interlude

Having flowed through the Danube drainage in just over a month, the water droplet enters the Black Sea in late June. Navigators have ventured across the Black Sea since the ancient Greeks and Romans, but it was the Turks who named it Karadeniz, "black sea," because of its dark water and violent storms.

The Black Sea, although now connected to the Atlantic via the Mediterranean, averages only about half the salinity of the Atlantic. Inflows of the rivers emptying into the Black Sea—the Danube, Dniester, Dnieper, Don, Halys, and others—help to keep the sea's salinity low and greatly influence the sea's productivity. The Danube alone contributes nearly 80 percent of the total river flow entering the Black Sea.

The northern part of the Black Sea, where the Danube and Dnieper enter, is a region of intense recycling of the organic matter carried in by the rivers. River sediment settling onto the seafloor is also an efficient sink that traps about half of the inorganic nitrogen brought in by the rivers. The multiple dams in the Danube basin store so much sediment that now the river brings only 30 to 40 percent of its historical load to the Black Sea. These changes, along with changes in nutrient input from the various rivers, have overwhelmed the sea's ecosystem in recent decades.

The Black Sea ecosystem was near collapse during the 1980s to the 1990s from a combination of excessive nutrient enrichment from domestic and agricultural wastes entering via the rivers, other forms of pollution, overfishing, and invasion by comb jellies that reached the Black Sea in ballast water dumped by cargo ships in 1982. By eating zooplankton that normally feed anchovies, the comb jellies nearly caused the Black Sea ecosystem to collapse.

Biologists were pleasantly surprised when the Black Sea ecosystem began to recover. Nutrient inputs from most rivers stabilized or decreased after the mid-1980s, in part because of water pollution control programs and in part because of economic collapse in the former Soviet Union. As a result, the biomass of phytoplankton decreased and so did fish mortality from algal blooms, although the Black Sea still faces severe threats from pollution and overfishing.

The Danube water droplet evaporates from the surface of the Black Sea and is carried east-southeast at low levels across the Middle East, the Arabian Sea, and India. Moving at elevations of only six hundred to nine hundred meters above the surface, the water droplet is joined by millions of particles of silt and clay that the winds have picked up from the deserts of northern Africa and the Middle East.

The air mass in which the Middle Eastern dust and the Danube water droplet are traveling crosses India and moves out above the Bay of Bengal before reversing direction off the coast near Calcutta. The air mass moves back toward the northwest as part of the summer monsoonal circulation, rising as it approaches the Himalaya. Although the Himalayan peaks reach the highest elevations in the world, they are not high enough to keep airborne contaminants from passing over them. Samples of water chemistry at three stations in the Ganges River basin south of the Himalaya—Delhi, Sharanpur, and Jagmanpur—showed a peak in the radioactive isotope cesium-137 after the Chernobyl nuclear accident in the former Soviet Union. The Danube water droplet mingles with some of this air coming over the Himalaya from Siberia and falls in July on the lower elevations of the Nepalese Himalaya. There the droplet joins the rush of uncountable masses of water droplets flowing into the Ganges River.

SIX

The Ganges: Eternally Pure?

The Ganges is a river of stories. Scholars have collected many tales about this river that springs from a dozen sources on the roof of the world. Each of the stories shares the theme that Ganga, daughter of the Himalaya, is eventually persuaded to shed her purifying waters on the sinful earth and thus bring salvation to humanity.

The Aryan invaders of India named the Ganges Sursari, "river of the gods." Today Indians bestow more than a hundred epithets on this river and print them in books so they can be chanted devotionally. Ganga is the melodious, the fortunate, the cow that gives much milk, the eternally pure, the delightful, the body that is full of fish, affords delight to the eye, and leaps over mountains in sport, the being that bestows water and happiness, and the friend or benefactor of all that lives or moves. To drink the Ganges, bathe in it, and carry it to distant places is to perform a pilgrimage. To be cremated on its banks and float down its eternal stream is to reach salvation. Even to speak the words "Ganga, ganga" can atone for sins committed in past lives. This is a river of stories by which people regulate their lives.

The Ganges is also the great collector of Himalayan snows. Its asymmetrical network of channels reflects the enormous forces at work in Earth. The Ganges lies at an immense suture, where the Indian tectonic plate has smashed northward into the southern margin of Asia. The meeting of the continents creates the highest region on the planet, folding and faulting the rocks up into the Himalayan Mountains and the Tibetan Plateau. The many

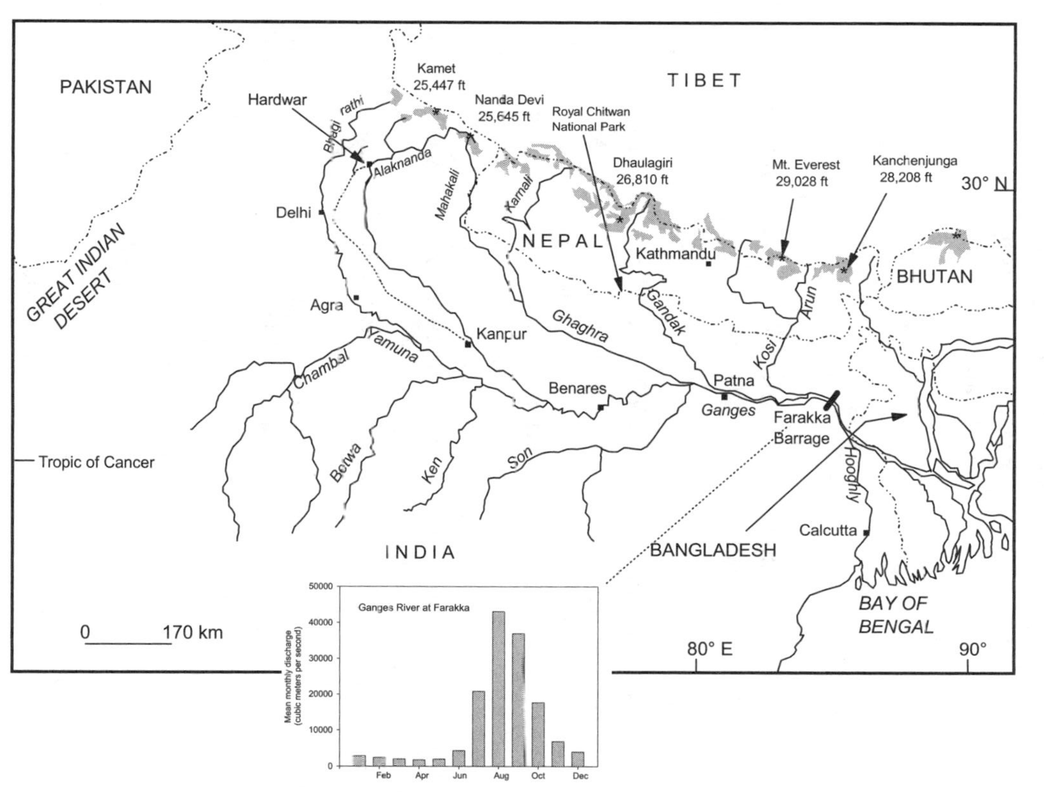

6.1 Schematic map of the Ganges River drainage basin, showing the principal rivers and cities mentioned in the text. Principal Himalayan glaciers are shown as gray shaded areas, and the highest Himalayan peaks in the drainage are labeled. The Upper Ganges Canal is the dashed line from Hardwar to Kanpur.

sources of the Ganges flow south and east from these highlands, collecting into the great trunk stream that flows due east before bending slightly southward into the Bay of Bengal. No water comes from the Thar and Indian deserts at the western end of the basin, and relatively little flows from the south, where a narrow band of mountains divides the Ganges drainage from that of the Narmada River.

Each of the four primary Himalayan sources of the Ganges flows from the melting ice of a glacier. The Gangotri Glacier is usually accorded first status as the source of the Ganges. The uppermost portions of this glacier lie at nearly six thousand meters. Two thousand meters below, the glacial meltwaters collect sufficient volume to give rise to the Bhagirathi River. Approximately two hundred kilometers downstream, the Bhagirathi joins the glacier-borne Alaknanda River to form the Ganges. Twenty-eight hundred kilometers from the base of the Gangotri Glacier, the Ganges enters the sea.

The Gangotri and other Himalayan glaciers depend on monsoonal precipitation during July and August for about 80 percent of their snowfall. The word monsoon comes from the Arabic word *mausim*, "season." A monsoon refers to a seasonal shift in the wind direction that usually causes a dramatic change in temperature and precipitation. Anywhere from 100 to 250 centimeters of precipitation fall on different portions of the Ganges headwaters during the monsoon.

Although the monsoonal circulation occurs consistently every year, its strength varies. And when the air grows warmer, the summer precipitation falls as rain rather than snow, accelerating glacial melting. Summers have been unusually warm since the 1980s, with air temperatures rising on average by more than 1.5°C each year. As a result, the glaciers are melting rapidly. At least two things happen when glaciers melt more quickly; the meltwater fuels larger and more variable river flows, and the sediment frozen into the glacial ice is released to form moraines. Moraines are loosely consolidated piles and ridges of sediments that form beside, beneath, and downvalley from retreating glacial ice. Moraines downvalley from the retreating ice can temporarily pond meltwater, creating lakes that are apt to fail catastrophically and produce outburst floods when the meltwater either overtops the moraine or seeps through the moraine sediments.

Increased river discharges below melting glaciers might provide a temporary boon to the thirsty regions downstream if they do not produce large floods. But after the glacier melts completely, the loss of dependable, perennial headwater flows will be catastrophic. The Himalaya contain the largest volume of ice outside the polar regions, and

6.2 *Above*, headwater tributary of the Ganges in the High Himalaya of Nepal. The steep, bouldery channel with abundant rockfall and debris flows along the side slopes is typical of rivers in the upper portion of the Ganges River system. *Below*, people winnowing grain in terraced fields of the Middle Himalaya, Nepal. These subsistence farmers rely on wood for heating and cooking, and these terraces were once covered with forest.

as it slowly melts this ice creates the Ganges, Brahmaputra, and Indus rivers. These rivers in turn support more than 10 percent of the world's population in India, Pakistan, Bhutan, Nepal, and Bangladesh. Without the estimated ten thousand billion cubic meters of water stored in the Himalayan glaciers, these people face an uncertain water supply.

It is difficult to imagine the disappearance of the Himalayan glaciers. Everything on the roof of the world is on a giant scale, seemingly invulnerable to human effects. Pale blue glaciers lie like a necklace of ice at the base of abrupt rock horns. The sharp black lines of arêtes rise in knives of rock between individual valley glaciers. Smaller valleys hang above the deeper main valley where tributary glaciers have receded. Waterfalls twist precipitously down the cliffs. The glaciers and the broad, braided rivers they give rise to have carved deep, steep-sided valleys. Outburst floods from the glaciers roar down these valleys at unpredictable intervals. People cluster their villages atop the huge terraces of sediment left behind as the rivers cut steadily downward.

The Himalayan ranges have some of the highest rates of uplift in the world, but the mighty glacier-fed rivers cut down equally fast. Geologists estimate erosion rates of a quarter of a centimeter per year in the High Himalaya, with slightly lower rates of a tenth of a centimeter on the southern edge of the Tibetan Plateau and in the foothills south of the high mountains. These numbers might not sound like much, but they represent average rates of lowering across the entire landscape, from rockfalls and landslides on the steep peaks to the river floods that carry sediment downstream from the adjacent hill slopes. Much of the core of the Himalaya consists of the crystalline igneous and metamorphic rocks of the High Himalaya that were intruded by molten magma, locally melted, then folded and faulted during uplift of the mountains. Sedimentary rocks underlying the regions to the north and south are more easily weathered and eroded, but it is the central crystalline rocks that supply about 80 percent of the sand, silt, and clay that ultimately travels all the way downstream to the Bay of Bengal by the great conveyor belts of the Ganges and Brahmaputra rivers.

To Dam Nepal

People living along the northernmost fringes of the Ganges drainage, in the highest inhabited portions of the Himalaya, are largely Buddhists from Tibet. Moving south and lower in elevation within the Himalaya, the shrubby slopes give way to pine forests and eventually to rhododendrons and oaks. This is also the transition zone between the Tibetan Buddhists and Hindu Aryan peoples from India. From Hindu culture comes the most commonly used name for the mountains, Himalaya, derived from the Sanskrit words *hima* and *alaya*, the abode of snow.

Below the abode of snow, the monsoons bring only rain, and the landscape is richly green. Rice and corn supplant potatoes and barley, and yaks give way to water buffalo. Beans, oranges, lentils, bananas, and vegetables grow in small terraced plots that farmers have carved out of the steep hillsides.

The rivers that flow through these steep green valleys, making their way down toward the Ganges, reflect the power of the landscape. They are big churning rivers, pale and milky with the rock flour of fine sediment released into them by melting glaciers. Steep banks of giant boulders line the channels, and vegetation crowds thickly down to the scour line left by the last high flow. For long stretches the rivers are more or less continuous rapids that provide a roaring background for every other sound in the valley. Boulders crash together underwater during the high flows of the monsoon. Tributaries come charging down steep valleys filled with equally huge boulders, but the main streams are so powerful that even the flash floods and debris flows that carry the tributary boulders into the main channel leave only temporary alluvial fans at the tributary junction. Where the slope of the stream channel grows gentler, the main rivers meander across beds of cobbles, gravel, and patches of bedrock. The valleys grow broader downstream, and the wide bars exposed at low flow have silt and sand as well as cobbles, but the channels are still mainly riffles and swift runs, with few quiet pools.

As the people of the Himalaya go steadily about their physically strenuous lives, the rest of the world debates whether they are creating an environmental catastrophe. Much of this debate focuses on Nepal, by any measure one of the poorest and least developed countries in the world. Ninety-five percent of the Nepalese are farmers who require land cleared of native forest cover to grow crops. The farmers also rely on wood for fuel. The introduction of modern medicine and health care, along with malaria suppression, supported enormous population growth after 1950. Growth rates reached 2.6 percent a year during 1971–81, which translates to doubling the country's population every twenty-seven years. This indigenous growth is exacerbated by large-scale illegal migration from India into the Terai, the lowland plains at the base of the Himalaya.

The authors of a 1978 study estimated that Nepalese in the Himalaya used more than five hundred kilograms of firewood per person each year, whereas the sustainable yield from the forest was less than eighty kilograms. As firewood becomes harder to obtain, farmers turn to dried animal dung for fuel, robbing the fields of fertilizer and lowering crop yields.

The next step in the debate involves extrapolating deforestation in the Himalaya to flooding and siltation downstream, as well as enhanced population pressures as people leave the increasingly marginal highlands to settle in the lowlands. Other studies from the late 1970s noted that by exacerbating flooding the 40 million people living in the Hindu Kush likely were adversely affecting the lives of the 350 million people inhabiting the adjacent large river basins and plains. As an example these studies cite the 1979 floods that caused two billion dollars in property damage in India. Some scientists, however, argue that even forested watersheds lose much of their buffering effect during the type of extreme rains that occurred during 1978–79. These scientists assert that the argument that halting Himalayan deforestation will end flooding in the plains is simplistic, and they question whether a Himalayan ecocrisis is in fact translating all the way to the middle and lower portions of the Ganges drainage.

Everyone agrees that recent changes in the Himalaya are stressing the forest environment severely. Half of the forest reserves of Nepal were lost within thirty years. But it remains very difficult to isolate the effects on the lower Ganges basin of these upstream changes from the effects of concomitant downstream changes in population, land use, and river engineering.

Given that headwaters deforestation causes myriad environmental problems, from localized erosion and loss of soil fertility to potential exacerbation of downstream flooding and siltation, the Indians and Nepalese and seemingly everyone paying attention to them are advocating using hydroelectric power to reduce reliance on wood as a fuel. Scientists and engineers repeatedly cite the enormous hydroelectric potential of Nepal, which currently generates about 2 percent of its national energy requirements this way. Estimated potential hydropower generation is about 83,000 megawatts, of which 244 megawatts has been developed. Twelve percent of the Nepalese had access to electricity in 1996. Developing water resources for irrigation is also frequently spoken of as a means to raise the standard of living in this country of farmers.

On the one hand, landlocked Nepal is rich in water, with more than four times the world's average water availability per square kilometer.

On the other hand, only 60 percent of the country's people have access to safe water. The 8 percent of the population living in the mountains are more likely to have safe, uncontaminated water supplies than people in the hills and the Terai. People at lower elevations have not only the contaminants introduced by people upstream, but also strongly seasonal precipitation and river flows that limit water availability outside of the June through September monsoon. The rivers of Nepal contribute more than 40 percent of the total flow of the Ganges and over 70 percent of its dry season flow. Much of this comes from primary tributaries of the Ganges that originate among the snowfields and glaciers of the High Himalaya, which flow year-round. Rivers originating in the middle mountains and lower hills have very low or no flow during the dry season, and people living along them accordingly suffer water shortages and increased contamination.

India has also exerted an enormous influence on water development in Nepal. India is by far the larger country in terms of land area, population, wealth, and international political power, but most of the Ganges basin that lies within India ultimately depends on water flows from Nepal. Water agreements between the two countries date to 1920, when India built the Sarada Barrage on the Mahakali River following a land exchange between the countries. A barrage is a gated dam built across a river to regulate water discharge. The Sarada Barrage lies in India and stores water for irrigation, some of which is supplied to Nepal. Development continued with the Kosi Barrage in 1954 and the Gandak Barrage in 1959, both constructed on or close to the international border in response to India's initiative and needs. All these projects were financed by India and mostly benefit Indians, with the result that many Nepalese have come to resent such development as a sellout of their water resources. Part of this resentment also stems from the fact that, to provide convenient proximity to the Indian border, the barrages were constructed not in deep, narrow mountain valleys, but on broad alluvial plains where the ponded waters drown more of the already limited arable land of Nepal. Now both India and Bangladesh are urging Nepal to construct large storage dams in the mountain gorges to enhance low-season flows on the Ganges. Such dams would also generate hydroelectric power that can be sold to India, which has an estimated power deficit of fifty thousand megawatts in the northern part of the country.

In the early twenty-first century, expert commentary on the development of water resources in the Ganges basin repeatedly emphasizes that hydropower is clean, cheap, endlessly renewable, and environmentally friendly. Some commentators even stress that the enormous energy concentrated in river networks, which currently contributes to erosion and sedimentation and is otherwise "released," could instead be converted to "a cleaner and usable form of energy." This sounds like a self-evident prescription for better living through engineering, but it is worth examining the records of the existing water development projects.

Three Cautionary Tales

Three projects illustrate the problems with water engineering in the Ganges basin: the Tehri Dam in the Garhwal Himalaya of India, the Arun III Dam in Nepal, and the Farakka Barrage in India.

Tehri

The Tehri Dam on the Bhagirathi River was officially approved in 1972. Within a few years, an antidam movement formed around the issues of displacement, compensation, and rehabilitation. The movement incorporated environmental considerations after a massive landslide during the 1978 monsoon temporarily dammed the river and then burst, killing hundreds of people.

As a result of opposition to the dam, an environmental impact assessment was begun in 1980. As the assessment continued for years, new issues gained greater visibility: earthquake hazards, given that Nepal has had fifteen major earthquakes since 1866; the inability of Indian engineers to accurately estimate siltation rates for the dam impoundment in this region that has some of the highest sediment yields in the world; the forced displacement of eighty thousand people in a country that has a poor record of resettlement and compensation; and the sacrilegious disruption of the journey of pilgrims to the headwaters of the Ganges. A committee of the Indian government recommended against environmental approval of the dam project as designed. Nonetheless, the dam—more than 150 meters tall and 456 meters long—generated its first electricity in 2006 and now creates a reservoir 25 kilometers long.

Arun

The proposed Arun III Dam in Nepal has a long history of controversy that exemplifies the trade-offs among flood hazards, environmental damage, and massive social change on the one hand, and improved standard of living on the other.

The Arun River drains the Himalaya east of Kathmandu. Approximately 450,000 indigenous people from ten ethnic groups live in small villages widely dispersed among the steep ridges and deep valleys of the region, which also shelter more than a hundred species of rare and endangered plants and animals. People in this region live a subsistence lifestyle, relying on the crops they can grow on tiny plots carved out of the steep slopes. As in other parts of the Himalaya, their life is physically strenuous in a way that is unimaginable to most people in developed countries. Nutrition is limited, professional medical care requires days of walking, and upper respiratory infections are widespread as a result of years of inhaling wood smoke in small, poorly ventilated houses. In this nearly vertical landscape, everything must be carried: water from the nearest creek or spring, food, building supplies. The small thatched houses have no electricity, no running water or plumbing, and nothing that the inhabitants do not build themselves or carry in on their backs.

In this setting the World Bank proposed to build the Arun III Dam, a 201 megawatt run-of-the-river hydroelectric project and the first in a series of three dams to be built in the valley. A five-day walk from the nearest road, the project would require construction of a 120 kilometer access road. Up to ten thousand construction workers and their families would move into the valley for the project. Proponents cited the increased standard of living that the project would bring to the region: roads, electricity, and all the medical, educational, and material benefits of increased exchange with the outside world. Opponents cited the massive social and environmental disruption that would ensue from the sudden influx of people and from damming this free-flowing river.

In 1994 a coalition of Nepali nongovernmental organizations filed the first claim against the World Bank and proposed alternatives that were less expensive, were less environmentally and socially damaging, and would build domestic industrial capacity and develop hydropower more evenly throughout Nepal. They were joined by local hydrologic experts, who promoted alternatives to Arun III that focused on small- and medium-scale projects of up to 100 megawatts that would easily meet Nepal's growing demand for electricity.

The list of those opposing Arun III grew rapidly as deliberate or accidental oversights in the numbers and arguments used to justify the project became obvious. Experts noted that to build a huge dam in this region of landslides, earthquakes, and glacier outburst floods was to invite the potential for a massive flood if the dam failed or was overtopped. The World Bank dropped the project in 1995, and private financing was alternately promised and withdrawn during the succeeding decade, but as of 2008 the Nepalese government had awarded a private company a concession to build the dam.

Farakka

After nearly twenty years of negotiation between India and Pakistan, India reached an agreement with newly independent Bangladesh and commissioned a barrage on the Ganges at Farakka in 1975, eighteen kilometers from the western border of Bangladesh. At least thirty-four barrages and other structures alter the flow of the Ganges and its tributaries in India and Nepal, and as many as four hundred irrigation structures divert somewhere between six hundred and eleven hundred cubic meters per second from the Ganges before the river reaches Farakka.

As it approaches the ocean, the Ganges breaks up into numerous distributary channels. One of these, the Hooghly River, was once the primary channel. This led the British East India Company to found Calcutta as a major port along the river in 1790. Unfortunately for Calcutta, distributary channels on an actively building delta like that of the Ganges are constantly shifting and conveying more or less flow as sediment deposition alters their characteristics. Within a hundred years of Calcutta's founding, the main flow of the Ganges began to shift east toward the Brahmaputra. By the time of India's independence in 1947, a silt deposit effectively blocked flow into the Hooghly except during the largest monsoon discharges. Meanwhile, saline intrusions during high tides were starting to affect water supplies for Calcutta, which remained a major industrial city and India's busiest port in the early 1950s. Once called the City of Palaces, other names come to mind for this metropolis of more than eleven million people that now discharges 398 million tons of municipal sewage and domestic effluents each year. Because of these discharges, Calcutta needs flowing water for dilution.

The Indian government finally stepped in and built the Farakka Barrage to divert 1,100 cubic meters per second from the Ganges into the Hooghly. This is up to 58 percent of the dry season flow on the Ganges. The diversion has successfully maintained flow and channel dimensions in the Hooghly, but it has created havoc elsewhere.

Landings of the commercially important fish hilsa (*Tenualosa toli*) were reduced by more than 99 percent upstream from the barrage after completion, because the barrage blocked the fish's migration route. These fish are born in large rivers within Bangladesh, enter coastal waters by the time they are one year old, then return to freshwater once they reach sexual maturity.

July through October monsoon flows have increased on the Ganges downstream from the barrage, but November through May dry season flows have decreased considerably. Increased monsoon flows have exacerbated flooding in Bangladesh, but lower dry season flows have reduced downstream fish habitat. The lower flows have also accelerated sedimentation in the Gorai River, another downstream distributary of the Ganges that provides much of the freshwater for the southwest region of Bangladesh. Sedimentation in the channel means that when high flows do occur they are more likely to spill overbank and cause damaging floods, as occurred in August 1998.

Decreased Gorai flows also translate to further inland penetration of saline water from the ocean in this region with very low elevation. Increasing salinity negatively affects agriculture, forestry, and industrial and drinking water supplies in southwest Bangladesh. The rice varieties grown in this region are very sensitive to increases in salinity, as are the coastal mangrove forests that protect low-lying Bangladesh from storm surges. Bangladesh draws 90 percent of its drinking water from the subsurface. Because saline water can move upstream along the bed of the Gorai River during periods of low discharge, the saline waters can also infiltrate and contaminate the groundwater. Sampling of groundwater wells indicates that this process has occurred since diversions for the Farakka Barrage.

Loss of the overbank floods that ameliorate local climate has resulted in rising temperatures during summer and declining temperatures during winter downstream from the barrage. The number of low rainfall events has increased, but the number of large rainfall events has been halved since the barrage became operational. Although climate is notoriously difficult to understand because so many factors influence it, some scientists believe that the reduction in surface water and drop in the groundwater table have altered the thermal properties of landforms in the affected area, providing a mechanism that explains the observed climatic changes.

These unforeseen side effects of the Farakka Barrage stimulated India and Bangladesh to sign a thirty-year treaty in December 1996 stipulating rules for sharing the waters of the Ganges, although both sides still claim more water than the river provides. Poverty-stricken Bangladesh continues to struggle with the consequences of the Farakka Barrage.

In each of these three examples, large-scale water engineering on the Ganges causes massive side effects that range from the dislocation of human communities to climate change, loss of resources such as freshwater and fishery stocks, and increased hazards from flooding. Proponents of megaprojects commonly claim that the projects represent a necessary and inevitable phase of development and that to oppose them is to oppose raising the standard of living for people in the Ganges basin. Development in the Ganges, however, is not a case of megaprojects or nothing. Numerous smaller-scale technologies that are more sustainable and much less environmentally disruptive exist. These technologies can be implemented to provide electricity and drinking water to people living along the Ganges, but they are less likely to make manufacturers and governments wealthy. Those who stand to profit handsomely from the megaprojects are often the most vocal proponents of this development path.

The Meeting of Mountains and Plains

As they leave the steep topography of the High and Middle Himalaya, the rivers of the Ganges basin enter the hill country to the south. Elevations remain high here relative to much of the world—Kathmandu is at 1,370 meters, for example—but topography grows more subdued, and the rivers no long rush downstream as torrents of whitewater.

Kathmandu is now the political and urban center of Nepal and, like many urban centers in the Ganges basin, a site of rapid population growth as a result of rural migration and high birthrate. Rapid population growth also produces various forms of pollution in the Bishnumati River, a Ganges tributary that flows through the city.

The Bishnumati also represents a transition from the narrowly confined mountain rivers churned into whitewater as they flow over steep, bouldery beds to the broad, more gently sloping rivers of the Gangetic Plains. The point at which the mountain rivers give way to a broad network of braided channels on the plains marks the farthest upstream range of many unique inhabitants of the Ganges drainage. Three species of river otters range upstream on the plains rivers. Several species of hard- and soft-shell turtles apportion the diverse river habitats among themselves. Gharial (*Gavialis gangeticus*) and mugger (*Crocodylus palustris*) crocodiles range upstream along the Ganges as far as the base of the mountains. Both species of crocodiles are precipitously declining in number as a result of habitat loss, hunting for skins and body

6.3 River in Kathmandu. A variety of wastes are dumped directly into the river, and the smoke comes from cremation of animal and human bodies.

parts believed to have medicinal value, and incidental catches in fishing nets.

Detailed studies of the Karnali and Narayani rivers in Nepal found 121 and 135 fish species, respectively, below and within the Himalayan foothills. This high diversity reflects the transition between cool Himalayan waters and the warm, slower-moving waters of the Gangetic Plains. As the rivers change form across this transition, they provide an assortment of habitats that fish such as golden mahseer (*Tor putitora*) can use. Despite the mahseer's diverse food and feeding habitats, and the reverence in which it is held by Hindus, the fish is now endangered throughout its distribution. Fishing pressures have contributed to the mahseer's decline, as has habitat destruction from removing wood along the riverbanks that provides cover and feeding spots and from sand and gravel mining for road building, which alters the composition of the streambed and thus both spawning habitat and the habitat used by aquatic insects the fish feed on. The fish also suffer from dams that block their migration routes.

Mrigal carp (*Cirrhinus mrigala*) and sucker catfish (*Bagarius bagarius*) provide other examples of species sought by local fishermen that are also declining in number. The Ganges basin as a whole supports 265 fish species, many of them sought by commercial and subsistence fishermen. Hooks and lines, drag nets, gill nets, cast nets, trap nets, dip nets, purse seines, traps, and in some cases dynamite are all used to catch fish. Fish farms, mainly devoted to carp species, also provide much of the fish eaten. Although fish stocks and habitat in the Ganges can be protected under international conventions to which Nepal, India, or Bangladesh is a signatory as well as by national legislation, all three countries have given very little attention to fish conservation. Scientists suggest that unregulated fishing and spawn collection for fish farms, as well as water pollution, habitat destruction, and dams that impede fish migration, all harm fish populations, although very few systematic surveys of fish have been undertaken to provide data for assessing these effects.

During the faunal survey of the Karnali and Narayani rivers in 1995, one elderly man told the scientists conducting the survey that "until fifteen years ago, the land bordering the river was completely covered with dense forest and aquatic animals could be seen in high abundance. Now more than fifty percent of the forest has completely disappeared and, although he occasionally sees aquatic wildlife and big fish, he is skeptical that they will remain for much longer" (Smith et al. 1996).

Perhaps the most charismatic river animal that ranges as far as the base of the foothills is the Ganges river dolphin (*Platanista gangetica*), or susu. Susus are freshwater dolphins that live throughout the Ganges and Brahmaputra rivers in Nepal, India, and Bangladesh. Their total numbers are estimated to be less than five thousand and declining, and the International Union for Conservation of Nature regards the species as vulnerable. Like other big river dolphins, the susu has evolved distinctive adaptations to life in the big rivers draining the Himalaya. The animal's eye is only slightly larger than a pinhole and lacks a crystalline lens, presumably because it is so hard to see in the turbid waters of these rivers. Instead, the susu relies on echolocation to get around and find its prey. The dolphins most commonly spend their time downstream from shallow areas or tributary junctions where convergent streams create an eddy countercurrent. These countercurrents attract the fish the susus feed on, but they also attract humans seeking to catch fish, who often catch susus in their nets by accident.

Susus living in the upper portions of the species' range are relatively uncontaminated. Susus in the lower portions of the rivers now carry a heavy burden of toxins in their bodies. Scientists analyzing tissue from dolphins and from their prey fish throughout the Ganges in India and

Bangladesh consistently find DDT and PCBs, as well as lesser amounts of other organochlorine pesticides such as chlordane and dieldrin, and heavy metals. As of 1989, farmers applied an estimated 2,500 tons of pesticides and 1.2 million tons of fertilizers within the Ganges basin. Perhaps most disturbing, contaminant concentrations increased between sampling in 1988–92 and 1993–96. Scientists do not know enough about these increasingly rare animals to understand the effects of such contaminants, except by extrapolation from other organisms, which suffer reproductive and developmental disruption, as well as cancers and death.

The use of DDT and PCBs continues in India, decades after these compounds were banned in Europe and the United States. DDT is used to control mosquitoes that spread malaria and sandflies that spread kala-azar disease. Both of these diseases occur near the Ganges River. PCBs come primarily from industrial discharges or from fluids leaching from discarded electrical equipment such as capacitors and transformers. DDT and PCBs reach the rivers of the Ganges network through both aerial deposition and attachment to silt and clay particles entering the river.

In addition to toxic contamination, the dolphins are threatened by incidental and intentional killing for meat and oil and by habitat destruction associated with channelization, flow regulation, and dams. Channelization methods such as dredging and artificially cutting off meander bends largely eliminate the eddy countercurrents that form prime susu habitat. Barrage dams block the migratory routes of prey fish and isolate small populations of dolphins. Above barrages on Ganges tributaries such as the Mahakali, Sapta Kosi, and Narayani rivers of Nepal too few susus remain to constitute viable populations, and they are on their way to extinction.

Among the habitat alterations that threaten fish and susus are those caused by canals' diverting water. Diversions from the Ganges have a long history. The Western Jamuna Canal was originally built in 1355 and remodeled four hundred years later during the Mughal Empire. British engineers continued to build and repair canals, renovating the western and eastern Yamuna canals in 1820 and 1830. When the Upper Ganges Canal was completed in 1854, it initiated irrigation development across half a million hectares supplied by 3,830 kilometers of main canals and distributaries. The resulting agricultural growth not only ended serious famines in the region but produced surpluses that could be exported, a development enthusiastically taken up elsewhere in the Ganges basin. A Lower Ganges Canal was completed in 1878.

The Great Plains

South from the hill country, the rivers of Nepal enter the Terai, and the rivers of India flow onto the vast Gangetic Plains. Both regions are low-lying flatlands, once covered with jungle and rivers that migrated freely back and forth. Hunter-gatherer Tharu people occupied the Terai historically, but increasing population pressures in the highlands to the north, combined with the control of malaria during the 1950s by the U.S. Agency for International Development, drove agriculturalists into the Terai. As regional population shot up, the jungle was cut down. Now most of the Terai is an endless flat plain where water buffalo graze among fields of rice and corn.

The original Terai ecosystem is preserved in only a few places such as Royal Chitwan National Park. Here the larger animals including elephants, one-horned rhinoceroses (*Rhinoceros unicornis*), Royal Bengal tigers (*Panthera tigris*), sloth bears (*Melursus ursinus*), and sambar deer (*Cervus unicolor*) still maintain small populations. The tall deciduous trees of the forest shelter epiphytes and bromeliads. Vines twisted like strands of braided rope hang down from the canopy or surround tree trunks in a smothering embrace. Massive *Ficus* trees flare into buttress

6.4 Early morning along the Rapti River, Royal Chitwan National Park, Nepal. The Rapti River flows over cobbles and gravels and spreads across a broad floodplain during the wet season.

roots that are lost in the thick understory. Sal trees (*Shorea robusta*) and enormous pale-barked kapok trees (*Ceiba pentandra*) protrude above the main canopy. Broad floodplains cut through the forest, but the rivers recede to shallow streams over cobble and gravel beds during the dry season. Grasses six meters tall line the riverbanks, sheltering the birds that concentrate along the river corridor.

The Terai and the Gangetic Plains lie between the fault that defines the Himalaya to the north and the ancient rocks of the Indian Shield exposed in the peninsular hill ranges to the south. Within these bounds, a wedge of sediments more than seven hundred meters thick has been deposited as rivers have carried sediments weathered and eroded from the Himalaya down to the adjacent lowlands during the past two million years. As might be expected, the sediment of the plains thins toward the south, away from the Himalayan source. These thick deposits of gravel, sand, silt, and clay store plenty of groundwater, forming one of the largest aquifers in India. Irrigated agriculture in the region draws heavily on this groundwater, with thousands of wells dug during the past three decades.

Each well is like a giant straw, sucking the groundwater table downward more rapidly than rainfall and infiltration from floods fills it back up. In the worst case scenario, groundwater can be drawn down so fast that the surrounding sediment collapses into the pores that formerly held water particles and the surface subsides. Subsidence now affects the city of Calcutta at the downstream end of the Ganges basin, where damaged buildings tilt as their foundations sink irregularly. Across the Gangetic Plains, subsidence rates of a fraction of a centimeter per year do not yet pose a hazard. They are low in part because of subsurface layers of clay that resist subsidence and in part because of high rates of groundwater recharge from the Ganges. But if flows in the Ganges are further altered by upstream flow storage or diversion, this effective bolstering of the groundwater aquifer will decrease, and subsidence on the plains may become a problem.

Until very recently, the Ganges and its tributaries migrated freely across the Gangetic Plains, leaving abandoned river channels that record two prominent humid climatic phases and an intervening drier phase during the last glacial maximum eighteen thousand years ago. Meandering rivers became braided during the drier phase, as water flow decreased and sediment load increased. The rivers began to meander once again sometime after four thousand years ago.

Today the Gangetic Plains are an all-or-nothing environment in terms of water. The cold season of November to February gives way to the hot season of March to June. The monsoon season of July to October follows the hot season, and 80 percent of the annual rainfall occurs during July, August, and September. The rivers have only low, sluggish flow during the rest of the year, but the during the monsoons the rivers surge up and out of their banks, flooding the flat plains and creating a landscape that is more water than land. The plains receive an estimated million cubic meters of water per square kilometer each year.

For centuries human ingenuity has been devoted to storing the abundance of the monsoon season for use during the leaner times of the dry seasons in order to make agriculture feasible on the plains. Agricultural productivity has in turn made the central Ganges drainage the historical seat of Indian civilization.

No Longer Pure

Approximately halfway through its journey across the Gangetic Plains, the Ganges is joined by the Yamuna. The Yamuna is the only major tributary that enters from the south. Fatepuhr Sikri, Mughal emperor Akbar's capital from 1571 to 1585; Agra, site of the Taj Mahal and Agra Fort; and Delhi, home to nearly nine million people, lie along the Yamuna. The millions of tourists who visit these sites each year are greeted with scenes of incredible filth and poverty as well as architectural beauty and past grandeur.

Foreign visitors to India have commented on the filth of the Ganges water for as long as they have been writing their impressions. Indians historically believed the Ganges was physically as well as spiritually pure and thus had no trouble bathing in and drinking water contaminated with partially cremated corpses. Nonetheless, 80 percent of the health problems in contemporary India come from waterborne diseases. Raghubir Singh noted that no one in India spoke of the Ganges as polluted until the late 1970s, by which time large stretches of the river—over six hundred kilometers—were effectively ecologically dead. National attitudes have now changed dramatically. Grassroots environmental concern about water pollution and government attempts to control pollution are growing.

The sources of organic pollution remain much the same as described by foreign travelers for centuries. At Benares alone, an estimated forty thousand bodies are cremated each year. Many are dumped into the river only half-burned. These obvious sources of contamination are not nearly as dangerous and persistent, however, as the often-invisible chemical pollution. Hundreds of factories along the Ganges, from an antibiotics factory upstream from Hardwar to a shoe factory below Calcutta, release untreated wastes into the Ganges. A single industrial facility can be a substantial source of persistent contaminants. One coal-based power plant on the bank of the Pandu River, a Ganges tributary near the city of Kanpur, burns 600,000 tons of coal each year and produces 210,000 tons of fly ash. The ash is dumped into ponds from which a slurry is filtered, mixed with domestic wastewater, and then released into the Pandu River. The fly ash is full of highly toxic heavy metals such as lead and copper; one study found that it can release up to seventy parts per million of copper, one of the most toxic of metals to freshwater organisms. Uncontaminated rivers commonly have copper concentration levels of one to seven parts per billion, a thousandfold lower than that released by the fly ash in the Pandu. When the river water reaches Kanpur, it encounters sixty-six more industrial units, including many tanneries that dump their untreated effluent directly into the river.

As of 2004, half a billion people, almost one-tenth of the world's population, lived within the Ganges River basin. Nearly all the sewage from fifty-two cities, forty-eight towns, and thousands of villages goes directly into the Ganges and its tributaries, an estimated 18.5 billion liters each day. The rivers also receive 3,730 million liters of industrial waste and the runoff from the nine thousand tons of pesticides and six million tons of fertilizers applied in the basin each year.

The Yamuna picks up a hefty load of contaminants as it flows through Delhi and then Agra. These cities occupy just over 2 percent of the Yamuna's thousand-kilometer length but contain 78 percent of the population in the river's drainage. Industrial, municipal, and domestic wastes flow directly into the Yamuna through several open drainage systems such as the tributary Delhi River. A portion of the Delhi River receives an estimated 240 billion liters of urban effluents each day from seventeen drains throughout the city of Delhi. As much as 1,630 million liters of water is drawn from the Yamuna daily to meet Delhi's water requirements. About 80 percent of the water supplied to the city is returned to the Yamuna as wastewater, most of which is only partially treated and contaminated. Yamuna riverbed sediments contain exceptionally high levels of chromium, nickel, copper, zinc, lead, and cadmium derived from Delhi and Agra.

High levels of phosphorus, chemical fertilizers, and pesticides occur elsewhere along the Yamuna's length. A study published in 2003 noted that mercury emissions in India increased significantly from

1990 to 1995 (Subramanian et al. 2003). Highly toxic mercury enters rivers through atmospheric deposition from thermal power plants and chlor-alkali plants, as well as by moving attached to particles of silt and clay. Transport and deposition of mercury have increased downstream along the Yamuna River, particularly during the past fifty years. Some of this sediment-mercury combination continues downstream to the Yamuna's junction with the Ganges at Allahabad and eventually to the Bay of Bengal.

Studies of groundwater quality in the central Ganges basin also indicate various forms of contamination. Concentrations of toxic metals such as iron, manganese, cadmium, lead, and chromium are above permissible limits for the safety of human health. Deep waters remain relatively uncontaminated, but shallow groundwater receives contaminants from sewage effluents and excessive use of fertilizers and pesticides in agriculture.

Ultimately, of course, these contaminants show up in the tissues of every living creature that comes in contact with the river or the groundwater, or that eats river plants or animals. A 2001 study of contaminants in tissue from humans, domestic animals, wildlife, and fish throughout India found that compounds such as polychlorinated dibenzo-p-dioxins (PCDDs), polychlorinated dibenzofurans (PCDFs), and polychlorinated biphenyls (PCBs) are ubiquitous (Kumar et al. 2001). These compounds persist in the environment and in the tissues of living creatures, where they accumulate with time and reach higher concentrations than the level being ingested by the organism. Sources of the compounds include chlorinated pesticides such as 2,4-D that are still used in India; pulp and paper mills that use chlorine bleaching and then discharge the effluent onto agricultural land for use in crop irrigation; and chlor-alkali plants. Once they get into living tissue, the PCDDs, PCDFs, and PCBs wreak havoc with hormones, disrupting reproduction and development, and creating hormone-dependent cancers. The 2001 study found some of the highest national concentrations of contaminants in fish from the Ganges River at Patna, which the authors of the study attributed to chlorine bleaching at pulp and paper mills. Carnivorous fish species had higher concentrations than other types of fish, being at the top of the fish food web. Ganges River dolphins had even higher contaminant concentrations in their livers than the fish did. Fish well upstream at Hardwar had the highest concentrations of DDT, partly because of widespread continuing use of DDT, and partly because of waste disposal from several DDT manufacturing units along the Ganges.

The movement of sediment and associated contaminants in the Ganges drainage is highly seasonal. Individual channels can have discharges of water and sediment that increase by a factor of fifty to one hundred during the monsoon, when unchannelized flow spilling over the channel banks and across the flat Gangetic Plains also carries sediment and contaminants into shallow ponds or depressions. These depressions can act as sinks that concentrate contaminants, as shown by the elevated levels of arsenic, chromium, copper, nickel, lead, and zinc found in some of them. At sufficiently high concentrations these metals can leach into the groundwater, creating a plume of subsurface contamination that is very difficult to contain or treat. Contaminants can travel attached to silt and clay or in association with fine organic matter. Changes in organic transport and the chemistry of river water during the monsoon also affect contaminant dispersal.

River pollution is in some ways a more difficult problem for people to grapple with than, for example, deforestation. A would-be activist cannot literally embrace a portion of a river the way a protester can hug a tree. Because a river integrates an entire drainage basin, the contaminants present at any point along the river represent everything entering the main channel upstream, as well as tributaries, surface runoff, and subsurface flow. Unless the contaminants produce a highly visible result such as a massive fish kill below an industrial point source, the effects of water pollution are usually subtle and slow-acting.

As public awareness of water pollution in the Ganges has increased, the Indian government has gradually instituted water quality standards and protective legislation. A first step was taken when the Ganga Action Plan, or GAP, was launched by the national government in February 1985. GAP activities include interception and diversion of sewage, construction of sewage treatment plants, low-cost sanitation projects, and riverfront development projects, the costs for all of them met by the central government and local governments. Current environmental regulations stipulate effluent standards for water-polluting industries, which must cover the cost of meeting these standards, but the standards are poorly enforced. As a result of GAP activities thus far, certain parameters such as dissolved oxygen levels or phosphate and nitrate concentrations have improved locally. In other portions of the Ganges drainage, however, these parameters continue to deteriorate, albeit perhaps more slowly than they would have without GAP. Other contaminants that are more difficult to control, such as pesticides in agricultural runoff, remain largely unchanged because they are not treated in wastewater plants. A 1994 study found that residues of the insecticide aldrin, for example, often exceeded the guidelines for drinking water set by the World Health Organization (Agnihotri et al. 1994).

Engineering the River

Contaminants flowing downstream are only one of the international management issues within the Ganges drainage that Nepal, India, and Bangladesh must somehow resolve. Negative effects in Bangladesh from India's Farakka Barrage clearly illustrate the magnitude and diversity of potential conflicts that can arise as population densities, urbanization, and water use all grow within the Ganges basin, which as of 2005 was rated only "moderately impacted" by flow regulation compared with other large rivers around the world. Several books and many articles have been written about international water issues in the Ganges, and from them emerge consistent themes.

The first of these themes is that various proposed hydroelectric projects, whether in the headwaters or in the lower basin, will likely cause a great deal of environmental, social, and economic disruption. The hazards associated with Arun III and other headwater projects now receive international attention. Less publicized are the problems with large reservoirs in the lower, flatter portion of the Ganges basin. Bangladesh's only very large hydroelectric project to date, the Kaptai project, has "proven terribly disruptive," in the words of one author. Created in 1964, the new Kaptai Lake inundated virtually all of the prime agricultural lowlands of the Karnafuli Valley. Rehabilitation efforts aimed at the eighty thousand tribal people evicted from the site were unsuccessful, and many of these displaced people resorted to slash-and-burn agriculture on steep hill slopes inside regional forest reserves. This resulted in premature siltation of Kaptai Lake, as well as continuing armed resistance by the displaced tribes. In return, the project yielded only 120 megawatts of hydropower (subsequently expanded to 230 megawatts).

A second theme is the need for equitable water sharing among countries and regions, particularly during the low flows of the dry season. From the perspective of Nepal and Bangladesh, India has thrown its weight around in designing and implementing past projects such as the Farakka Barrage. The smaller countries now insist on more mutually beneficial designs. Agriculture contributes approximately a third of Bangladesh's gross domestic product, and the Ganges basin contributes a quarter of the GDP. Freshwater fisheries contribute almost 90 percent of the total fish production in southwestern Bangladesh, where fish furnishes the great bulk of animal protein in human diets and ranks second only to agriculture in economic importance. As low flows below the Farakka Barrage have reduced the area of wetlands and impeded fish migrations, fish production has dropped substantially. The livelihood for a third of the country's people is at risk if flow in the Ganges basin is decreased, so anything that affects the availability and quality of land and water in the basin is viewed by Bangladeshis as a threat to national security.

India strongly favors constructing a link canal across Bangladesh to divert water from the Brahmaputra to the Ganges. Bangladesh opposes this idea, asserting that the canal would only compound the damage caused by the Farakka Barrage. Bangladesh favors storage dams in Nepal that would dampen seasonal flow fluctuations and generate hydroelectric power. Nepal is wary of having its resources used to generate power that will go primarily to other countries. It generally favors large dams in the Himalaya, such as the Arun III project or a dam in the Chisapani Gorge on the Karnali River. Financing has been problematic, however, thanks to international opposition to the social and environmental hazards associated with large dams and the consequent political pressure on the World Bank and other potential financers of such projects. A more stable and achievable approach may be to have small, decentralized hydroelectric projects to meet local needs, medium-scale projects for national needs, and perhaps large projects for transborder regional demands.

A third theme is the need to somehow reduce flood damage along the Ganges. Extensive overbank floods are an integral part of life in the middle and lower basin, but during 1987 and 1988 60 percent of Bangladesh was flooded and twelve hundred people died. The total flood-prone area of the lower basin countries is just under 5 percent of India (sixteen million hectares) and 15 percent of Bangladesh (two million hectares).

Another theme is the necessity of wisely using the seasonally abundant waters of this region to alleviate the extreme poverty and related health hazards that threaten millions of people in the countries of the Ganges basin. From waterborne diseases associated with sewage pollution to organochlorine residues in food, salinization of water supplies, and arsenic contamination of groundwater, perhaps the ultimate irony of the water-rich Ganges basin is that many people cannot get clean water to drink.

In many parts of the world, groundwater provides a more reliable and less expensive water source for the urban and rural poor than does treated surface water. Reliance on groundwater becomes problematic, however, if the water table drops, necessitating expensive pumping from deeper levels or long waits for recharge, or if the groundwater becomes contaminated. Groundwater contamination can result from humans applying fertilizers, pesticides, heavy metals, or other toxic materials that infiltrate into the subsurface water or else from naturally occurring concentrations of salts or metals. Arsenic, which is highly toxic to living organisms, contaminates portions of the groundwater reservoirs on five continents. Some of the worst contamination occurs on the Bengal delta plain of the lower Ganges basin. Contamination in this region exposes an estimated seventy-five to one hundred million people to the daily risk of chronic arsenic poisoning. Most of these people are village farmers growing rice, making an average income of about a dollar a day. They are exposed to contamination directly through the water they drink and indirectly through their food, which is grown using the same contaminated water.

People living on the delta plain have relied for centuries on pools and on shallow, hand-dug wells to supply water during the dry season. These surface or near-surface sites of water storage are vulnerable to bacterial contamination. Diseases such as cholera periodically broke out across the delta plain historically. In the early 1970s, international aid agencies began to supply drilled tube wells that people could pump by hand to draw water from deeper levels. About four million tube wells are now in place, sealed with tight concrete tubes that reach down twenty meters or more, past the surface contamination. Drawing water from tube wells dramatically reduced diarrheal diseases, but by 1983 people noticed widespread occurrence of strange symptoms such as skin disorders, and nearly two hundred thousand people in West Bengal were diagnosed with arsenicosis. It took ten years, and repeated warnings by a Bengali chemist in Calcutta, before the danger was traced specifically to high concentrations of arsenic in the water being drawn from the tube wells. By 1998 the number of arsenicosis victims rose to nearly three hundred thousand.

Arsenic is the most insidious of poisons. It is invisible, tasteless, and odorless. Symptoms of arsenic poisoning may not manifest themselves until after two years or twenty years. The World Health Organization set a limit of ten parts per billion in drinking water. At fifty parts per billion, arsenic makes drinking water marginally unsafe. At one hundred parts per billion, as noted by writers Philip and Phylis Morrison, "Those who drink such water imbibe only enough arsenic to fill the volume of a grain of rice in five years." Yet a couple of grains of rice worth of arsenic produces skin disorders, and double that amount greatly increases the risk of skin cancer. A lifetime of daily intake produces fatal cancers of several internal organs. People with poor nutrition suffer more from arsenic poisoning than better-nourished individuals whose bodies are more able to withstand the onslaughts of the arsenic.

Scientists have scrambled to understand the controls on the naturally occurring arsenic that poisons the deeper wells on the delta plain, in part because the distribution of contaminated wells is very patchy. The most likely explanation is that oxygen-free conditions associated with buried peat deposits facilitate chemical reactions that release arsenic to the surrounding groundwater. Another twist in the story comes from the possible effects of the Farakka Barrage. By reducing stream flow to this portion of the delta plain, the barrage indirectly reduces overbank flows across the floodplains and associated groundwater recharge. This in turn alters water and oxygen levels in the vicinity of the buried peat deposits and may promote release of arsenic.

The Delta and Its Beautiful Forest

Discussions about international water issues in the Ganges have heightened the realization that environmental degradation caused by incautious water management will ultimately destroy the region's ability to support not only diverse species of plants and animals, but also humans. The lower portions of the Ganges in Bangladesh provide a clear warning.

The Ganges breaks up into distributary channels that meander broadly across extensive flat plains as they near the ocean. Eventually the Ganges channels merge laterally with those of the mighty Brahmaputra River, which also drains the Himalaya. Together they create one of the largest deltas in the world, covering an area of approximately two hundred thousand square kilometers within India and Bangladesh.

As the continental ice sheets began to melt after the last major glaciation, rising sea level intersected the gently sloping area that is now the continental shelf beyond the delta. The rising ocean waters trapped sediment being carried down the Ganges and Brahmaputra, initiating growth of the delta. The large sediment loads in these rivers allowed the delta to grow even though sea level was rising rapidly, at just over a centimeter a year. The rise in sea level slowed by seven thousand years ago. The enlarging delta both migrated seaward and forced the incoming rivers to deposit sediment farther upstream, creating lateral river migration and widespread sediment deposition. The present configuration of the delta developed by about three thousand years ago.

The Ganges and Brahmaputra continue to rearrange the smaller details of the delta, depositing sediment here and eroding it there. Each year the Ganges in Bangladesh transports an average 316 million tons of sediment and the Brahmaputra an average 721 million tons. During the high monsoon flows, the Ganges brings more sediment than the Brahmaputra. The situation reverses during the season of low flows. About half of the sediment the rivers carry suspended in their flows is deposited on the riverbeds and on the deltaic plain. The rest goes out to sea. But the sediment deposited on the delta is not enlarging the delta with time; the delta front has been stable over the past two hundred years. Instead, the sediment seems to be balancing the subsidence that is occurring within the delta as previously deposited sediment gradually compacts under the weight of new material.

Siltation in both the Ganges and the Brahmaputra has increased about fourfold during the past fifty years, creating bars in the channels that hinder the downstream passage of floods and increase overbank flooding. Siltation also hinders navigation and the movement of fish and reduces the productivity of the river ecosystem. In some cases siltation forces the river to move laterally until a new channel is created. Siltation rates are much higher along the Ganges than along the Brahmaputra, despite the Brahmaputra's higher sediment load, as the Ganges flow is decreased by irrigation withdrawals.

Natural levees along the meander bends form the high points in the delta landscape. The delta rises only twenty meters above sea level at its northwestern end and only a couple of meters at the southern end. Among the levees and channels lie floodplains, abandoned channels, and oxbow lakes, back swamps between levees, and tidal marshy lowlands. All of these areas of seasonally ponded water provide critical habitat for fish species in the Ganges River basin. Draining and farming of these low-lying areas, along with reduced connection with the main channel because of water withdrawals, have substantially decreased this type of habitat. This is one of the primary causes of reduced fish abundance and diversity in the Ganges basin. A few species, such as the freshwater Ganges shark (*Glyphis gangeticus*), which lived in the lower reaches of the Ganges-Hooghly river system, are already gone.

The fertile soils and abundant rains of the delta facilitate agriculture. The humidity is always greater than 65 percent and the air is always hot, varying between 15°C and 43°C throughout the year. Each year 120 to 400 centimeters of rain falls across the delta. As a result, the delta supports one of the highest population densities in the world: anywhere from four hundred to twelve hundred people per square kilometer. This is nonetheless a challenging place for humans to live. The big rivers that have created the landscape can flood massively, turning the low-lying landscape into a world of water. Or the ocean can cause floods with surges driven by tropical storms. Each year an average of 30,000 square kilometers are flooded, although the area flooded has fluctuated from 3,140 square kilometers in 1982 to 89,920 square kilometers in 1988. This translates to anywhere from 18 to 40 percent of Bangladesh's landmass being flooded each year. Almost three hundred thousand people were killed in one storm during 1970. Fifteen to twenty million people are displaced by flood-induced erosion each year, and the results can be grim. Having lost their farmland, these individuals are often compelled to live on the periphery of urban life in squatter settlements along railway lines, abandoned brickyards, or flood-protection embankments. Climate change is an unknown, but potentially major, component of flood hazards in Bangladesh. Predicted rises in global sea level will increase inundation during coastal storm surges, and these may be exacerbated by predicted increases of up to 45 percent in seasonal floods on the Ganges River because of a shift from snow to rain in the headwaters.

Among the many environmental challenges facing Bangladesh is the crisis developing in the Sundarbans, a name that means "beautiful forest." The Sundarbans are littoral mangrove forests that cover more than four hundred thousand hectares in the western Ganges-Brahmaputra delta. The forests extend about 80 kilometers north of the Bay of Bengal and 130 kilometers west from the Baleshwar River into India. Sundri trees (*Heritiera fomes*) and Gewa trees (*Excocaria agallocha*) predominate, but many other species are present in this ecologically rich area. Intertidal mangrove forest makes up 44 percent of the Sundarbans, and 18 percent is aquatic. The rest is cleared for settlement and agriculture. The portion still in forest cover is the largest single continuous tract of diverse mangrove forest in the world and includes the world's only mangrove tiger preserve. An estimated 350 Royal Bengal tigers live here. The tigers have adapted to semiaquatic lives in which they swim from island to island, eating crabs and fish as well as their more usual diet of deer and wild pigs. Forty-one other mammal species also live in the Sundarbans, along with 350 species of birds, 50 species of reptiles and amphibians, 177 fish species, and 130 plant species. Other species such as the Javan rhinoceros and the water buffalo disappeared during the twentieth century.

Like most of the delta, the Sundarbans are completely flat. A complex network of streams dissects the land into clumps of elongated islands. From this maze of land and water come nutrients that increase the productivity of the rivers and the coastal waters. The Sundarbans have been managed for more than 120 years as a commercially exploited reserved forest and are now listed as a World Heritage Site. Fishermen take more than 120 fish species from the area. An estimated 180,000 kilograms of honey and 45,000 kilograms of wax are harvested each year from the western part of the forest.

The problem is that now the Sundarbans are showing signs of stress. Salinity increases from east to west across the area and toward the south. These salinity gradients structure species distribution, growth, and productivity of the forest; vegetation tends to be more luxuriant at lower salinity levels. Freshwater diversions upstream, including the Farakka Barrage, have reduced downstream flows to the Sundarbans and thus increased salinity levels. Along with siltation in the northern part of the forest, these changes appear to be stressing the mangrove trees and lowering productivity. Top dying, in which trees die from the top downward, now affects nearly 20 percent of the Sundri trees. Biodiversity of the plant communities is declining sharply, and average plant height is decreasing. Local overexploitation in the form of indiscriminate cutting of mangrove plants, intensive prawn culture along the streambanks, excavation of tidal mudflats for brick making, and wastes from riverside cremation grounds only exacerbate the problems of depleted water from upstream. As the Sundarbans mangrove forests disappear, the 110 million people living in the lowest-elevation portions of the delta are increasingly exposed to the dangers of storms and cyclonic surges.

Changes in the flow of surface water and groundwater through the Sundarbans and other farthest downstream portions of the delta also have the potential to alter the chemistry and productivity of sea water in the Bay of Bengal. Scientists recently discovered that substantial groundwater flows from the delta to coastal waters and that this groundwater carries significant levels of elements such as strontium, radium, and barium. Reducing this groundwater flux, or altering its chemistry by adding contaminants such as excess arsenic, fertilizers, or pesticides, may alter the nearshore ecosystem.

The massive delta formed by the Ganges and Brahmaputra supplies sediments to the Bay of Bengal and from there to the world's largest submarine fan, the Bengal Fan. This fan, which is nearly seventeen kilometers thick at its head and approximately one thousand kilometers wide and four thousand kilometers long, dwarfs the fans of the Mississippi and Amazon rivers combined. An estimated thirty million cubic kilometers of sediment are currently stored in the fan. Rivers have been transporting sediment to the Bengal Fan since the Himalaya began to rise twenty million years ago. So much sediment piles up in the shallow zone just offshore that periodically the sediment slumps, forming enormous submarine debris flows that follow the deep canyon known as the "Swatch of No Ground" cut into the continental shelf. This canyon channels the sediment out to the Bengal Fan, a huge submarine depositional zone that lies nine hundred to two thousand meters below sea level. Of the estimated one billion tons of sediment that the Ganges and Brahmaputra bring downstream each year, between 30 and 39 percent is deposited on inland floodplains. Another 21 percent is deposited on the delta, and the remaining 20 to 29 percent reaches the Swatch of No Ground and the fan. Geochemical properties of sediment cores from the fan indicate that times of greater sediment transport energy in the Bay of Bengal correspond to climatic fluctuations recorded as distantly as ice cores from the Greenland ice sheet.

The Ganges in the Future

Just as the sediments of the fan record past changes in the Ganges drainage basin, so they will undoubtedly record what occurs in the future. The sediments will likely show that flows reaching the mouth of the Ganges decreased during the later twentieth century. Varying levels of contaminants will be buried with the fan sediments; recent sampling shows detectable levels of DDT and other synthetic chemicals in the surface sediments of the Ganges estuary. Farmers who irrigate crops use on average twice as much fertilizer and pesticides as those who depend on rainwater, so the fan sediments will record the increase in irrigated agriculture in the Ganges basin as well as pollutants coming from the Bangladeshi industries that manufacture leather, paper and pulp, fertilizer, pharmaceuticals, sugar, jute, textiles, and petrochemicals. All these industries discharge their untreated wastes directly into the nearest river. The fan sediments will include organic compounds left by the profuse blooms of marine algae of species of *Valonia* that now occur on the lower Ganges during winter months. In a variety of ways, the sediments of the fan will record the ever-growing number of people living upstream in the basin and all the changes they create in the rivers.

Average population density across the Ganges drainage basin was more than 280 people per square kilometer at the start of the twenty-first century. This average hides the differences between the relatively sparsely populated Himalayan headwaters and the densely populated middle and lower basin. Bangladesh alone has 535 million people, whom one writer described as the largest concentration of the world's poorest. It is difficult to be optimistic about the chances of preserving diverse, functional riverine ecosystems under conditions where so many of the actions people take to improve their situation worsen their plight in other ways. In flood-prone Bangladesh, constructing embankments along the channels to decrease overbank flooding also restricts the access of fish to their floodplain habitat, which is critical for reproduction and growth of 60 percent of the riverine fish species in the country. And in Bangladesh, which depends on fish for protein in human diets, a study on the effects of embankments on the Megna River near Chandpur found that fish production declined by 35 percent and several economically important fish species disappeared entirely within two years of construction.

As in the Ob and, to a lesser extent, the Danube, clean water is now a limiting resource across the Ganges drainage despite the relatively wet climate of most of the region. Water and channel engineering have created the same types of negative side effects seen in the Ob, Nile, and Danube drainages, and the rapidly increasing population and severe poverty of the countries drained by the Ganges create challenges to river protection and restoration similar to those faced by the countries of the Nile drainage.

What is clear throughout the Ganges basin is that technologies developed in more industrialized countries, which have often wreaked environmental havoc there, cannot be exported to Nepal, India, and Bangladesh without causing even worse environmental and social devastation in these densely populated, economically vulnerable countries. Smaller scale run-of-river hydroelectric projects that do not impede fish migration or sediment transport provide a more environmentally and economically sound strategy for supplying electricity to communities in the Himalaya than do grandiose megadams. Agricultural technologies and crop varieties that reduce farmers' reliance on intensive irrigation, fertilizers, and pesticides will be more economically and environmentally sustainable than the massive engineering projects and diversions that destroy downstream rivers, flood arable lands, and result in a continual drain of toxins into surface and groundwaters. Enforcement of water quality standards and legislation through such technologically simple procedures as functional wastewater treatment plants and treatment of industrial effluents will avert the looming crisis of widespread toxic contamination of waters, sediments, and the bodies of living organisms from invertebrates to people.

Even the sacred Ganges cannot withstand the onslaught of modern practices and remain eternally pure. The recognition that people can, and must, make a difference in the quality of the region where they live has the potential not only to save the rivers of the Ganges drainage, but to empower the people living along the banks of these rivers and relying on river water for life itself.

Interlude

Having fallen on the Ganges drainage as part of the monsoon rains, the water droplet moves rapidly through the basin and reaches the Bay of Bengal in the Indian Ocean late in the summer. There the Ganges droplet joins a tongue of less saline water two hundred to three hundred kilometers wide that migrates southward parallel to the coast of India.

The Ganges water droplet moves some two thousand kilometers south, from the Bay of Bengal into the Indian Ocean. There it joins a huge cyclonic circulation system operating at the surface of the Indian Ocean. First the North Equatorial Current carries the water droplet another twenty-eight hundred kilometers, west along the equator and then south. Then the droplet reverses direction and flows fifty-four hundred kilometers east toward Indonesia as part of the Equatorial Countercurrent. At the Indonesian archipelago, some of the flow turns north and west once more into the North Equatorial Current, while another portion continues south and east around Indonesia to flow through the Torres Strait between Australia and New Guinea or among the islands of Indonesia and the Philippines.

Near Indonesia the Ganges water droplet is caught up once more in the shallow surface limb of the Great Ocean Conveyor. Carried south and then west around Africa, the droplet moves northward into the Atlantic.

In its long journeys through the ocean, the water droplet absorbs some of the contaminants released into the marine environment each year by shipping. A great many of the things that we wear, use, and eat every day now reach us by ship. Seaborne trade increased by about 45 percent during the decade 1986–95. Shipping now accounts for 80 percent of the volume of goods traded around the world, and the tremendous number of ships crisscrossing the oceans each year leave a very large wake of trash and poisons.

As it moves across the Atlantic toward the Gulf of Mexico, the Ganges water droplet once again passes among the coral reefs of the Caribbean. Like coral reefs around the world, these reefs are showing the strain of recent changes from increasing sea surface temperatures and other stressors. Reef decline in the Caribbean may be reflecting increased nutrients coming down the Mississippi as well as the Orinoco and Amazon rivers. Approaching the nutrient bath coming from the Gulf of Mexico via the Mississippi, the water droplet also encounters contaminants associated with the offshore oil industry.

A more benign flow moves through the Caribbean region at the same time as the Ganges water droplet. No less miraculous for the fact that it occurs twice each year, the great flow of migratory birds moving from Central and South America to their summering grounds in North America uses Cuba and other Caribbean islands as staging posts on the journey. Millions of birds pass this way each year, drawn in part by the nutrients carried in on the Guiana Current that in turn support invertebrates and fish. The birds of course drink during their migrations, so some of the water accompanying the droplet from the Ganges gets diverted and carried across the northern regions in the bodies of birds. The Ganges droplet keeps going in the ocean.

As it moves past the east coast of Florida, the water droplet gets caught up in some of the extremely complex secondary currents associated with the Gulf Stream. Scientists have recorded the paths of these secondary currents for twenty years using freely drifting floats. A map of their paths resembles an impossibly tangled mass of string. Counterclockwise rings form when meanders in the Gulf Stream pinch off and drift south. Clockwise rings form when meanders detach and drift north. Both types of rings remain at the surface for months or sometimes years as a closed segment of warm water revolving around a cold core.

Caught in these complex motions at the ocean's surface, nearly three years pass from the time the Ganges water droplet enters the Bay of Bengal until it is evaporated near the Gulf of Mexico and carried northwest across the United States to fall as April rain on the northern headwaters of the Mississippi River.

SEVEN

The Mississippi: Once and Future River

The Mississippi River basin encompasses the tremendous natural diversity of the continental United States. The headwaters of the various subbasins spread from the Appalachian Mountains of the east across the forests of the central-northern United States and the grasslands of the western Great Plains to the Rocky Mountains. This huge span includes humid continental, semiarid steppe, and wet subtropical climates. The river is the center of this drainage basin that encompasses 3.5 million square kilometers and 41 percent of the contiguous forty-eight states.

Peak annual discharge averages 39,300 cubic meters per second at Vicksburg, Mississippi. The river splits into distributary channels some kilometers farther downstream from this point, and these combined channels discharge an average of 580 cubic kilometers of water a year to the Gulf of Mexico. This is the seventh largest river discharge to the ocean, but the Mississippi is unpredictable. In 1954 the river peaked at only 19,800 cubic meters a second at Vicksburg, whereas in 1927 it peaked at 64,000 cubic meters a second. No single flood has ever encompassed the entire Mississippi basin.

From the western half of the basin comes the sediment that keeps the river a turbid brown. From the northern and eastern portions comes the water. The heaviest precipitation falls from spring thunderstorms triggered by warm, humid air moving northward from the Gulf of Mexico. But winter frontal storms and rains during summer and autumn contribute to totals that vary from 80 centimeters in the northern part of the upper basin to 114 centimeters in southern Illinois and 150 centimeters in parts of Tennessee.

The Upper Mississippi

The snowmelt and rainwater collect perceptibly at Minnesota's Lake Itasca, which is accorded the honor of being the headwaters of North America's largest river. Although the name Itasca sounds vaguely Native American, it was created by Henry Schoolcraft when he traced the Mississippi upstream to the small lake in 1832 as part of an expedition by the U.S. War Department. Itasca comes from the Latin *veritas caput*, "true head." Despite Schoolcraft's assertion, others quibble that the true head of the river is at Little Elk Lake, eight kilometers above Itasca. And then there are those who argue that the headwaters of the Mississippi's tributary, the Missouri, having the longest journey down to the Gulf, should be considered the source.

Average velocity along much of the Mississippi's length is five to seven kilometers an hour, about the pace of a fast walk. Despite this seemingly unimpressive rate of flow, it takes only a little more than thirty days for the average drop of water to travel from Lake Itasca to the Gulf of Mexico. Along this four thousand kilometer trip more than 250 tributary rivers add their waters to the main channel, speeding the water drop on its way.

The placid waters of Lake Itasca overflow across a bouldery sill to begin their downstream journey as the Mississippi River. The narrow stream channel winds slightly as it moves north and then east across a lightly populated landscape covered with bogs and spruce forests. The eight hundred kilometer headwaters portion of the river descends 200 meters from an elevation of 440 meters above sea level at Itasca's outlet to St. Anthony Falls. Natural falls and rapids, and nine glacially formed lakes, punctuate this headwaters portion of the river. Beyond Itasca, the river grows more sinuous as the valley widens. Small rapids separate wetlands dotted with beaver lodges, cranberry bogs, and beds of wild rice from natural lakes and artificial impoundments.

The river follows a circuitous course through greater and lesser lakes—Bemidji, Cass, Winnibigoshish—slowly changing course toward the southeast. The headwaters portion is so sinuous that a third of the river's entire length is in Minnesota. Seen from the sky, the viewpoint of millions of migratory waterfowl, this portion of the river basin is a

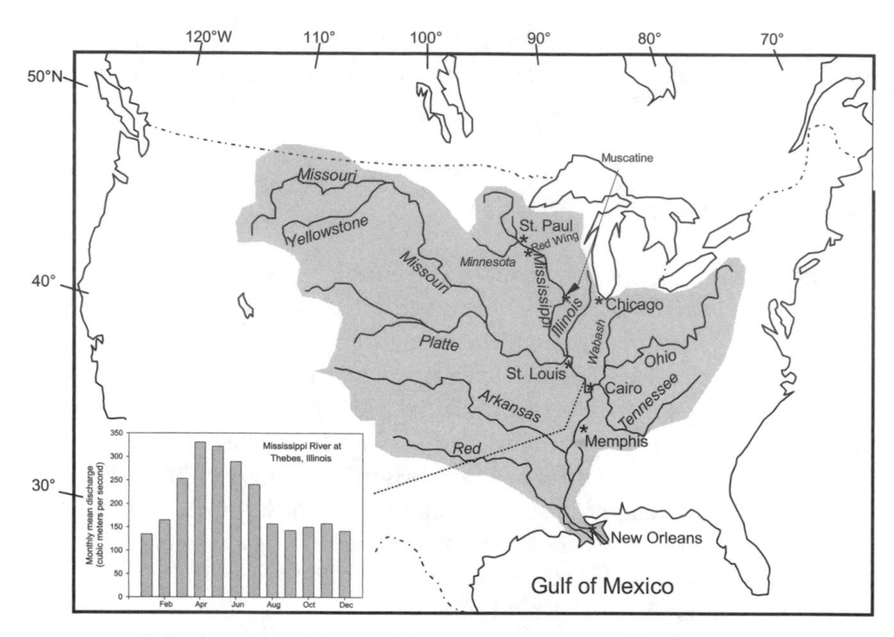

7.1 Location map for the Mississippi River drainage basin (shaded), with principal tributaries and cities labeled. Inset figure shows a sample hydrograph with the average monthly flow at Thebes, Illinois, just upstream from the junction with the Ohio River.

green-and-blue mosaic of wetlands, mixed conifer and deciduous forests, grasslands, and thousands of ponds and small lakes. The blue river moves sinuously across this mosaic, as though aiming to touch every watery depression along its route.

These depressions are an inheritance of the enormous continental ice sheets that advanced and retreated four times during the past two million years. Unlike the headwaters of most great rivers, no mountains create precipitous slopes that rush the Mississippi onward. The last great geologic force shaping this landscape was the final advance of the continental ice sheet, which reached its farthest southward extent about fourteen thousand years ago. Each time an ice sheet nearly two kilometers thick flowed southward, its great weight depressed Earth's crust. Meltwater ponded beneath the massive ice, breaking out periodically in catastrophic floods that sculpted the landscape so thoroughly that their routes can still be traced today. Some of these floods discharged volumes of water comparable to the 1993 flood, the largest flood ever recorded on the Upper Mississippi.

When it retreated, the ice released massive quantities of sediment. Rivers of meltwater flowing beneath the ice sometimes arranged the sediment in neatly sorted, linear mounds called eskers. Braided rivers beyond the ice front deposited sand and gravel across broad plains. Huge chunks of ice left behind as the glacier receded were buried and insulated by this sediment, so that when they finally melted, the overlying sediment collapsed into the resulting depression to create a kettle lake. The history of glacial deposition can be traced in the modern bed of the Mississippi as the river flows alternately over boulders, bedrock, mud, and sand.

Along much of the ice front, the melting ice simply dumped sediment in long lobes called moraines. The depressed Earth rebounded more slowly than the ice sheet melted, so that for a time the land sloped back toward the center of the ice, allowing the moraines to effectively trap enormous volumes of water melting from the ice sheet. The retreat of the most recent ice sheet blocked its own drainage into Hudson Bay, creating Lake Agassiz. This lake, hundreds of kilometers wide and nearly sixteen hundred kilometers long from northwest to southeast, episodically overflowed into the Glacial River Warren. River Warren cut the valleys now occupied by the Minnesota River and the Mississippi below the Minnesota's mouth at the southern end of the Minneapolis–St. Paul metropolitan area. The Lake Agassiz overflows produced floods two to ten times the size of the 1993 flood.

The Mississippi first appeared about five hundred thousand years ago, cutting a valley into sedimentary rocks as it flowed along the edges of an earlier ice sheet. The river system persisted through an interglacial period and then was partially overwhelmed by the most recent ice sheet, which formed Lake Agassiz as it began to retreat. Each overflow from Agassiz into the River Warren sent such a pulse of cold freshwater down to the Gulf of Mexico that isotope contents of marine sediments clearly record the influx. As the ice sheet retreated back into Canada, the Great Lakes, the Hudson River, and the St. Lawrence River were progressively uncovered, and some of the meltwater began to flow into the North Atlantic Ocean through these alternative routes. So much freshwater flowed into the North Atlantic that scientists believe it temporarily shut down the Great Ocean Conveyor Belt that modulates climate in the higher latitudes.

The meltwater of the great ice sheets has long since drained back to the oceans to resume its endless cycling between ocean, land, and atmosphere. But the irregular topography and poorly drained soils left by the ice sheets still store the precipitation that falls west and south of the Great Lakes, releasing the water slowly into springs and rivers that ultimately feed the Mississippi.

The first Europeans to reach this region were French fur trappers seeking to profit from the abundant beavers. Driven by commerce and curiosity, the *coureurs de bois*, "runners of the woods," explored much of the upper watershed and main channel within a period of only thirty years between the 1650s and 1680s. Britons and then Americans subsequently dominated the fur trade in the Upper Mississippi region. By the late 1800s, the original beaver population of the Mississippi River basin, estimated at ten to forty million animals, was nearly gone. Removal of millions of beavers, and the dams they built, likely reduced local overbank flooding and the extent of wetlands in the upper basin, making it that much easier for loggers and farmers to follow the fur trappers.

The fur trade moved west into the Rocky Mountains as the beavers were trapped out of the northern Midwest and the Upper Mississippi basin, and people began to look about for the next resource that could be exploited. Commercial logging began in the 1820s, subsequently aided by rapid expansion of the railroad network. Numerous commercial sawmills competed for the wood coming down the river in great lumber rafts as large as 1.2 to 1.6 hectares. As historian R. D. Tweet wrote, "Down this river, between 1835 and 1915, came virtually every usable white pine log in the states of Minnesota and Wisconsin." Some of the earliest engineering modifications to the Upper Mississippi were undertaken in connection with log floating. Wing dams of piled stone constricted the channel and prevented the logs from spreading out across the floodplains or into secondary channels. Naturally occurring snags and other obstacles were blasted or dredged. Like the removal of beavers, these changes reduced the extent of flooding and wetlands in the valley bottoms.

Farmers quickly moved in, although diversified farming was only marginally successful in the cutover areas of the Upper Mississippi watershed during the nineteenth and early twentieth centuries. Today the primary regional industries continue to be logging and farming, as well as recreation. Farming becomes more widespread downstream, and the river picks up progressively more sediment along its route as erodible croplands replace stable forested lands. These effects were even more dramatic in the recent past. Increasing deforestation and agriculture in the upper portion of the Mississippi drainage during the nineteenth and early twentieth century produced so much excess sediment that rivers and streams throughout the drainage lost some of their ability to convey water downstream as channels clogged up. Combined with decreased infiltration capacity and increased runoff from the denuded slopes, these changes produced larger floods in many of the drainages tributary to the Mississippi. (Knowledge of these changes helped to fuel the argument that deforestation in the Himalaya might exacerbate flooding lower in the Ganges.) Flooding, erosion, and sediment deposition on floodplains have all decreased since the mid-twentieth century in response to better land conservation practices, but the Mississippi today has already taken on its trademark muddy brown as it flows into Minneapolis and St. Paul, Minnesota.

The twin cities lie well up the Mississippi River basin, nearly forty-two hundred kilometers upstream from the river's mouth at the Gulf of Mexico. Yet the river is already impressively large thanks to the consistent rain and snow of the uppermost basin. Here the Mississippi becomes, briefly, an urban river. Concrete and stone riprap lines the rigidly channelized banks. Roads and railways crowd the banks, and storm sewers drain contaminants into the brown water.

The Twin Cities metropolitan area is one of the principal sources of heavy metals such as lead and mercury, insecticides, and polychlorinated biphenyls (PCBs) in the Upper Mississippi. Although improved municipal water treatment and changes in industrial practices have reduced levels of mercury and lead since the 1970s, other contaminants such as PCBs remain widely distributed. PCBs are a very stable group of industrial chemicals that leach into the river primarily from urban and industrial areas. They were used for electrical transformers and capacitors, among other things, until banned in the United States in 1979. When the U.S. Geological Survey studied contaminants in the Mississippi River during 1987–92, they found PCBs in almost every silt sample taken throughout the river basin. Survey scientists found comparable PCB concentrations in the tissues of catfish. Most of the PCBs moving insidiously through the Mississippi River drainage are attached to clay and silt particles. Where clay and silt settle out of suspension in backwaters or on the floodplain, the PCBs settle with them and become concentrated. PCBs entering the Mississippi from St. Paul and Minneapolis ride downstream on silt and clay particles until the sediment is mostly trapped and stored in Lake Pepin. Concentrations of contaminants are generally smaller just downstream from the lake and remain so until the entry point of the next contaminant source.

PCBs are linked to birth defects, reproductive failure, liver damage, tumors, a wasting syndrome, and death, both in humans and in other living creatures. These contaminants bioaccumulate within an organism and biomagnify as organisms pass on their accumulated doses through the food web. Invertebrates that ingest sediment to get at the bits of organic detritus mixed in with the silt and clay ingest PCBs and other contaminants as well, then pass them up the food chain. Because PCBs are very stable and slow to chemically degrade under environmental conditions, they accumulate in the environment.

Pollution-sensitive mayflies disappeared from parts of the Mississippi by 1927, including the stretch downstream from the Twin Cities and Lake Pepin, then reappeared by 1984 as water quality began to improve. In areas that remain too polluted, mayflies are replaced by pollution-tolerant organisms such as midges. In *Immortal River,* Calvin Fremling compares *Hexagenia* mayflies to canaries in a coal mine. Just as miners used the sensitive canaries to detect poisonous, odorless gases, so the distribution of mayflies along rivers reflects the distribution of contaminants. Burrowing mayfly nymphs live in sediments where contaminants can accumulate. The mayflies cannot swim long distances to escape environmental stress from toxins in the sediment or from low oxygen levels, so although their presence does not mean that the river is unpolluted, it does indicate that some minimum level of water quality has been maintained for the mayfly's life span of up to a year.

When they are present, mayflies efficiently convert organic muck on the streambed into their own adult bodies, which provide high-quality food for fish and birds. The bottom-feeding mayfly nymphs unfortunately also concentrate organochlorine compounds and heavy metals that predators ingest with the nymphs. Digging by nymphs can also resuspend contaminants that might otherwise remain segregated in streambed sediments. Mayfly nymphs might not sound like a particularly disruptive force in a river the size of the Mississippi, but the sheer number of insects makes their burrowing significant: as mayflies returned to the waters of Lake Erie, *Hexagenia* adults formed clouds twenty-three kilometers long and over six kilometers wide that were observed on Doppler radar.

Downstream from the twin cities, the Mississippi winds among a maze of loosely woven channels, islands, and wetlands. Low buff cliffs of crumbling sandstone, shale, and limestone represent the great ancient oceans and sedimentary basins characteristic of the upper Midwest. The river has carved a broad, flat valley among the bluffs and ridges of the undulating uplands, and the wetlands and sloughs record the meandering passage of the river through time.

By the time it reaches Red Wing, Minnesota, the river has flowed 80 kilometers southeast and passed through two lock and dam complexes. Today the Upper Mississippi drops only eighty meters along the

3,170 kilometers of its course between St. Anthony Falls and its junction with the Missouri. Like much of the Danube, the Upper Mississippi is a tamed river, moving in measured lockstep between the twenty-nine locks and dams along these 3,170 kilometers. On a global compilation of intensity of flow regulation, the Mississippi basin is colored a vivid red for strongly impacted.

Like many place-names along the river, Red Wing derives from Native American inhabitants. In many cases the language of indigenous peoples as reflected in place-names is all that now remains of their traditional culture. "Mississippi" comes from a native language. Some authors trace the name's origin to a Chippewa word meaning Father of Waters, whereas others trace it to an Ojibwa word meaning Big River. Of the ten states bordering the Mississippi, only Louisiana (named for France's Louis XIV) does not take its name from a Native American word.

Scattered roads, farm fields, and houses lie superimposed on the varied landforms created by the Mississippi across the bottomlands around Red Wing. Like the Amazon today, the Mississippi historically had large tracts of seasonally flooded forest. A wide variety of organisms used these forests: migrating or seasonal birds such as the North American wood duck (*Aix sponsa*), fish such as black bullheads (*Ictalurus melas*) or orangespotted sunfish (*Lepomis gibbosus*), Blanding's turtles (*Emydoidea blandingii*) and other reptiles, amphibians including the amphiuma, a salamander a meter or more long, and invertebrates that scientists may not yet have even identified. Early travelers describe catfish big enough to overturn a canoe and flocks of waterfowl darkening the sky as they passed. Big floods coming through at intervals recontoured the riverbanks, ripping out mature trees and leaving fresh sediment in which seedlings could germinate, creating a continually changing forest with age and species composition structured by the dynamics of the big river. More than 80 percent of the floodplain is still connected to the river along the Upper Mississippi between the headwaters and the junction with the Ohio River, although much of the floodplain forest has been cleared. The riverbanks remain thickly wooded, but in many places the trees form only a fringe, with cleared lands beyond.

Environmental historians debate how extensive the forests of the northeastern quarter of the United States once were. The old adage that a squirrel could travel from the Atlantic coast to the edge of the central prairies without touching the ground has been called into question by research indicating that Native Americans used fire extensively to create more open, parklike woodlands favored by game such as deer. Some

7.2 Riverbank vegetation and wetlands along the upper Mississippi. Here a thin fringe of shrubs and trees separates the main channel from a floodplain wetland, with a denser forest beyond. Although less common today, this mosaic of bottomland forest and wetlands once extended for hundreds of kilometers along the river.

of this research relies on pollen and sediment deposited in lakes, some uses historical accounts. The Englishman Jonathan Carver described intermixed stretches of prairie along the Mississippi between the junction of the Wisconsin River and Lake Pepin during his travels in 1766. Zebulon Pike also described prairies with groves of fire-resistant trees such as bur oak (*Quercus macrocarpa*) during his 1805 journey upstream from St. Louis, as did Henry Schoolcraft in 1820 (Pike 1810; Schoolcraft 1953). Forests of more fire-sensitive trees such as sugar maple (*Acer saccharum*) grew in deep, moist valleys and on north-facing slopes. A mix of fire-resistant and fire-sensitive trees grew along the floodplains. Regardless of the local density of the Upper Mississippi River basin forests, the forests supplied plentiful wood to the river, creating large logjams that enhanced overbank flooding, as well as the partly buried logs that snagged and sometimes sank passing boats.

Navigation on the big river and many of its tributaries was notoriously difficult during the first decades of the nineteenth century. Navigators had to thread their way among numerous side channels and backwaters where the river split around hundreds of natural islands. In summer, low flows and shifting sandbars limited depth even in the main channel. Logs partially anchored in the streambed formed obstacles that could rip the hull out of a large steamboat. Charles Dickens penned a particularly vivid description of these snags during his steamboat trip in the early 1840s. "[The Mississippi] running liquid mud, six miles an hour: its strong and frothy current choked and obstructed everywhere by huge logs and whole forest trees: now twining themselves together in great rafts . . . now rolling past like monstrous bodies, their tangled roots showing like matted hair; now glancing singly by like giant leeches; and now writhing round and round in the vortex of some small whirlpool, like wounded snakes."

Dead trees floating down the river might snag on a sandbar or a bend of the channel, forming a larger obstacle that snagged more wood floating downstream. These logjams grew to formidable size and would quickly reform once they were cleared. Logjams were ubiquitous on the Mississippi's tributaries from the upper Ohio River, where John Bruce began removing snags in 1825, to the Red and Atchafalaya rivers in Louisiana. When first described in 1803, the Atchafalaya was choked with logs that prevented boatmen from following the river's shorter course to the Gulf of Mexico. Repeated efforts to clear the channel starting in 1812 produced only temporary navigation access. One enormous logjam nearly 170 kilometers long on the Red River withstood repeated attempts at clearing, despite a six-year effort during the 1830s by Henry Shreve, a steamboat captain who invented a steam-powered "snag machine" that speeded up the laborious process of removing tangled masses of logs from the river. Shreve's long effort temporarily cleared the jam, which then reformed and persisted until the 1850s. The logs blocking navigation along the Atchafalaya finally were thoroughly removed in 1880.

The big logjams did more than impede navigation. By also impeding the downstream movement of floodwaters, they helped to sustain extensive floodplain wetlands that supported a wealth of plants and animals throughout the Mississippi River drainage. Once the logjams were removed, floodwaters spilled out of the riverbanks less frequently and for shorter periods, and the extent of wetland vegetation shrank. In places the energetic floodwaters enlarged the river channels. Once the jams were removed, animals ranging from fish to wading birds lost some of their quiet-water habitat. The combined effects of beaver trapping, modifying channels for log floating, and removing logjams tended to reduce the complexity and flooding of the bottomlands, thus lessening the amount and diversity of habitat available to a wide variety of species.

Shackles for a Giant

Beyond Red Wing, the current of the Mississippi slows as the river enters Lake Pepin. The lake originally formed about ninety-five hundred years ago when a large delta from the tributary Chippewa River partially blocked the Mississippi valley. Although lock and dam complexes are now present above and below Lake Pepin, the delta of the Chippewa River remains a primary control on lake level. The U.S. Army Corps of Engineers regularly dredges the lake because of large annual inputs of sand from the Chippewa River. Attempts to stabilize or flush sediments from Lake Pepin and backwaters above each lock and dam complex must be undertaken with care, for the sediments stored in pools upstream from every lock and dam of the Upper Mississippi have elevated concentrations of heavy metals.

Lake Pepin is now less noticeable as an interruption of the river than it would have been historically because of the reservoirs created by water ponded upstream from each of the numerous lock and dam structures along the river. The sequence of locks and dams along the upper river, which a 1970s Corps of Engineers promotional film described as shackles for a giant, is designed to maintain a minimum water depth of 2.7 meters (nine feet) in the main channel used for boat navigation. The channelization has been carried much further in the Lower Mississippi, where the floodplains and secondary channels are largely disconnected from the main river. Instead of completely channelizing the upper and middle portions of the Mississippi in a similar manner, decades of engineering have focused on creating and maintaining a central navigation channel.

The Upper Mississippi River historically had a distinct seasonal pulse of high flows during March to May and low flows during August to October. These summer and autumn low flows made navigation difficult, if not impossible, and channel modification quickly followed the first steamboat that forged its way upstream to St. Anthony Falls in 1823. Within a year, the U.S. Army Corps of Engineers began removing snags, dredging sandbars, excavating the rocks that formed rapids, and damming sloughs to keep the water within the main channel. Natural processes would have kept the Corps sufficiently busy maintaining a waterway for the shallow-draft steamboats, but land clearing and

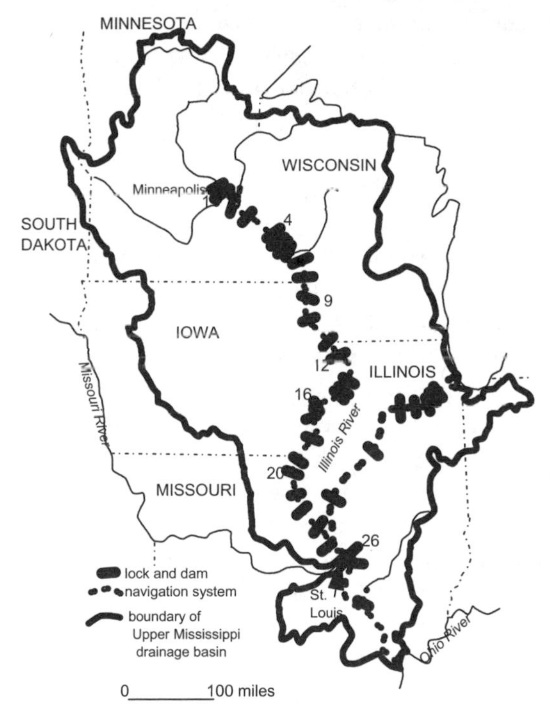

7.3 *Above*, downstream from a lock and dam along the upper Mississippi River. As on the Danube, these navigation dams substantially alter the seasonal fluctuations of river flow, store sediment, impede the movement of fish along the river, and alter habitat available for plants and animals. *Below*, the navigation system of the upper Mississippi River basin, showing lock and dam complexes on the Mississippi and major tributaries. (After Changnon 1996, figure 5-1.)

agriculture accelerated the rush of sediment from the hill slopes into the river.

Steamboat traffic on the upper river reached its peak during the 1850s and 1860s with increasing immigration and the shipment of agricultural products such as wheat to railheads on the river's eastern bank. As more sediment came in and boat drafts grew larger after the 1860s Civil War, the Corps ramped up its efforts to channelize the Mississippi for navigation from the mouth of the Ohio River upstream to Minneapolis. Railroads reduced steamboat commerce during the 1860s to 1890s except for lumber being rafted downstream. Congress nonetheless authorized maintenance of a navigation channel that grew successively deeper, from 1.2 meters in 1866 to 1.4 meters in 1878 and 1.8 meters in 1907. This channel was created through nearly continual dredging combined with use of "closing dams" that blocked the river from entering side channels and wing dams that constricted the flow, increased the current, and thus encouraged "natural" dredging.

Barge traffic on the Upper Mississippi waxed and waned following completion of a 2.7 meter channel in 1940. Subsequent proposals to enlarge the channel have been defeated. The Corps has also considered altering lock and dam operation in a climate of stable or declining grain shipments, competition for midwestern grain from other navigation routes such as the St. Lawrence Seaway, and growing environmental concern about the negative side effects of navigation improvements. Historians and economists question the necessity of spending hundreds of millions of dollars in tax subsidies and navigation projects to sustain a barge industry that exists primarily to ship grain overseas. Although Congress instituted user fees in 1978 for barges taking advantage of federal engineering, these fees do not begin to cover the full cost of maintaining the navigation channel.

The numerous locks and dams along the Upper Mississippi finally transformed the seasonal pulses of the river's flow into a more sluggish current between a series of shallow impoundments that now occupy some of the historical river floodplain. The locks and dams fulfill their purpose of creating more uniform water depths that make navigation easier, but as on the Danube, they also have many unintended effects on the plants and animals living along the river corridor.

The seasonally flooded and exposed wetlands along the 1,420 river kilometers between St. Paul and Cairo, Illinois, historically supported millions of fish, birds, reptiles and amphibians, invertebrates, and mammals. Although the ponded water above each of the dams between St. Paul and Alton, Illinois, now attracts migrating ducks, the water also floods part of the historically seasonal wetlands year-round, creating greater water depth, turbidity, and sedimentation as the slowing current deposits some of the sediment it carries. Increased depth and sedimentation have killed the aquatic plants many aquatic organisms feed on.

To mitigate these side effects, the Nature Conservancy is working with the Corps of Engineers, experimenting with several techniques. These include dredging the lakes formed above dams to reduce sediment thickness and adding islands built from the dredged sediment to create shallow water. Backwater isolation projects involving dike construction, pump installation, and management of water levels are used to produce food or nesting areas for migratory waterfowl or habitat for fish. Flow is introduced to other backwaters to counteract oxygen depletion. Engineers actively replant aquatic plants and create periodic reductions in water level in some pools above dams as a means of compacting and dehydrating sediments and restoring vegetation. The success of local experiments with lower summer water levels above dams is encouraging, but such efforts are constrained by navigation requirements and access to riverside docks for commercial and navigational boaters. Because federal law mandates that the Corps maintain the 2.7 meter deep, 91 meter wide navigation channel from Minneapolis to south of St. Louis, restoring floodplain wetlands through seasonal drawdowns may not be feasible during low-water periods. As on the Danube, the basic requirements of contemporary navigation severely limit the scope of restoration efforts, but even limited efforts to restore some of the form and function present before river engineering can make a difference for declining fish and bird populations.

Fish of the Mississippi

Downstream from Lake Pepin, the width of the channel alternately expands and contracts among numerous islands. More than five hundred islands divide the channel between St. Paul and St. Louis. Beyond the sandy, wooded banks lie the extensive riparian wetlands of the Upper Mississippi National Fish and Wildlife Refuge. The refuge was established in 1924 to protect populations of smallmouth bass but is now managed primarily for migratory waterfowl.

More than three hundred bird species, including 60 percent of the species in North America, use the river corridor. Every spring and fall millions of birds migrate along the Mississippi or remain as year-round residents. The river's north-south orientation and nearly continuous habitat are critical to the life cycle of these birds. The Mississippi flyway draws birds across a broad band from northern Alaska to Baffin Island and funnels them down the river valley, creating a winged river above the liquid one.

As exemplified in the creation of the Upper Mississippi National Fish and Wildlife Refuge, initial attempts to protect fish populations along the Mississippi focused on game fish such as bass. Attention has gradually shifted toward native species as scientists realize that many of the turbid river's miraculously adapted fish are now at a disadvantage in a world of regulated flows, limited floodplain access, and locks and dams. Among these are paddlefish.

Paddlefish (*Polyodon spathula*) in the Mississippi River historically grew to at least 1.8 meters and ninety kilograms. Part of their size derived from longevity. The fish can live at least thirty years, feeding almost constantly and continuing to grow as long as they live. They are fish to match the big river. They swim eight hundred kilometers to spawn and can travel at five kilometers an hour against a current running at the same rate. Yet, like most whales, these big fish live on very small food organisms. Up to a third of a paddlefish's body length consists of a broad, rigid flattened snout containing tens of thousands of minute electroreceptors. Paddlefish adapted to the murky waters of the Mississippi by evolving these sensitive receptors to detect the tiny electrical signals generated by zooplankton measuring only a fraction of a centimeter. A paddlefish swims along with its mouth open, using sievelike gill rakers within the mouth to filter zooplankton from the water and suspended sediment.

Despite these adaptations to life in a turbid river, and the species' endurance in North America for 140 million years, fisheries biologists estimate that by the 1980s paddlefish populations in the Mississippi drainage basin had fallen to 10 to 20 percent of historical levels. Commercial fishing for paddlefish caviar was a major culprit in the decline, as was habitat loss. Siltation under the ponded water of reservoirs destroyed spawning areas. Dams reduced upstream and downstream movement of the migratory fish. Channelization mostly eliminated backwater areas important to both young and adult paddlefish and removed the zones of low velocity where paddlefish feed on zooplankton drifting downstream. The passage of a large commercial vessel creates a two- to three-minute recession in water level along a river shoreline that can strand young paddlefish. Turbulence from the vessel's passage kills paddlefish larvae. The beautifully adapted electroreceptors that

7.4 Paddlefish in an aquarium at the National Mississippi River Museum and Aquarium. This superbly adapted native fish is now uncommon in the river but can be seen along with other native species at this museum in Dubuque, Iowa.

detect swarms of tiny zooplankton are swamped by the signals returned from the large metal gates of locks and dams, which the paddlefish try to avoid.

The continued presence but precarious status of paddlefish in the Mississippi is an example of what drives the hopes and anxieties of fisheries biologists. On the one hand, the Upper Mississippi remains quite sinuous despite more than a century of river engineering, and the pools behind the locks and dams provide reasonably complex shorelines beyond the navigation channel. On the other hand, fisheries biologists recently predicted that unless management action is taken, within fifty years the Upper Mississippi River will consist of a main channel bordered by dry land with few shallow marshes and will support limited diversity and abundance of fish. There is much at stake. The habitat diversity associated with the main channel and numerous backwaters supports an entire fishy world. A total of 306 species of fish are known to inhabit various parts of the Mississippi River basin. The river supports an unusually high number of species relative to other large rivers in the temperate zones because the river network is physically complex, with a wide range of aquatic habitats such as backwaters and floodplain lakes. The north-south orientation of the river basin also provided a corridor for escape and subsequent recolonization during glacial advances and retreats.

As on most big rivers, on the Mississippi fish species diversity increases downstream. The headwaters from Lake Itasca to St. Anthony Falls are home to sixty-seven species. The falls were historically an important barrier to upstream fish migration, and the terrain of the upper river was liberated from the glacial ice relatively recently. Headwaters fish must contend with shifting sediments on the streambed, boggy corridors and areas of dense aquatic vegetation, and periodic very low flows. Fishing and other disturbances have been minimal, and fish communities appear to be largely the same as they were in the late 1800s.

From St. Anthony Falls downstream to the Mississippi's junction with the Missouri, 132 fish species are present, and 114 species live in the river segment between the mouth of the Missouri and the mouth of the Ohio. The distribution of these fish also reflects natural barriers such as rapids, falls, and lakes. About half of the species present downstream from St. Anthony Falls simply cannot continue upstream past the falls. Those that can go upstream are mostly lake species that were able to cross drainage divides when glacial lakes connected the headwaters with the St. Croix River and the Red River of the North. Completion of the locks at St. Anthony Falls in 1963 opened another stretch of river to intrepid fishy pioneers.

Channelization and navigation development and pollution have also affected fish communities in the Upper Mississippi. Dam number 19, which is especially high because it is operated for hydropower, prevents upstream migration of many fish species. The ponded water upstream of locks and dams provides habitat for some species, but this is gradually being lost through sediment accumulation. The pool behind lock and dam number 19 has lost 55 percent of its original volume through sedimentation in the years since it was closed in 1913.

Several of the lock and dam complexes create anoxic zones where oxygen dissolved in the water drops to levels that are dangerously low for fish and other aquatic organisms. When lock and dam number 1 was completed in 1917, it collected most of the raw sewage of Minneapolis and St. Paul. Lock and dam number 2, completed in 1930, collected the rest. A Bureau of Fisheries study during August 1927 found that seventy kilometers of the river downstream from St. Paul lacked sufficient oxygen to sustain any fish life. Although sewage contamination has now been reduced by better treatment facilities, the drop in number of fish species from 132 upstream to 114 between the Missouri and Ohio junctions may reflect the extensive channelization and levees that restrict habitat extent and diversity. Habitats in the upper and middle reaches of the Mississippi now include both naturally occurring features such as sloughs and tributary mouths and engineered features such as tailwaters, navigation pools, and areas of slower flow near rock-wing dikes.

The entry of Missouri River waters into the Mississippi further affects fish species distributions. Species such as walleye that are less tolerant of turbidity are not common below the mouth of the sediment-laden Missouri. On the other hand, the Missouri contributes several species such as pallid sturgeon (*Scaphirhynchus albus*), western silvery minnow (*Hybognathus argyritis*), plains minnow (*Hybognathus placitus*), and various species of chub (*Macrhyobopsis* spp.) that are restricted to the middle reaches of the Mississippi.

Yet another significant influence on fish species distribution along the Mississippi is the introduction of exotics. Common carp (*Cyprinus carpio*), grass carp (*Ctenopharyingiodon idella*), rainbow smelt (*Osmerus mordax*), goldfish (*Carassius auratus*), striped bass (*Marone saxatilis*), and other exotic species have been introduced to the middle reaches of the river. Fish stocking began in 1872 along the upper river. Carp introduced themselves during the 1880s. Carp are highly efficient at feeding on the plankton also eaten by species such as paddlefish. Common carp became a major component of the commercial fishery in the river, reaching a mean annual harvest value of $270,000 between 1953 and 1977. During the succeeding decade the carp began to decline, partly because of PCB contamination in areas such as Lake Pepin, and partly because of poor spawning success during high or variable water years in a river that now lacks the sheltered backwater habitats critical to larval and juvenile fish of many species. The dams for the 2.7 meter navigation channel created large amounts of new backwater habitat by flooding the center sections of the navigation pools. The shallow edges of these new backwaters are readily accessible locally from the deep main channel, however, and do not provide the shelter and limited accessibility important to young fish that need protection from predators. Lock and dam 19 at present appears to be limiting the upstream migration of some carp species, and proposals to build a fish passage around it must address the movement of the carp. The potential for migration by exotic carp, like the downstream movement of contaminated sediment in navigation pools, illustrates some of the environmental costs of rehabilitating the Mississippi River system.

The number of fish species present in the Mississippi rises to 150 in the river segment from the mouth of the Ohio downstream to the coast. None of the fish in the main river are endemic, although the tributaries include endemic species. Some of these species are in trouble. The U.S. Fish and Wildlife Service lists three species in the Mississippi drainage as endangered, but other species are under consideration for listing. NatureServe, a network of natural heritage programs, considers twelve Mississippi basin fish species imperiled. That no fish has yet become extinct provides further incentive to restore some of the natural habitats and functions along the Mississippi system while all the components of the river ecosystem remain.

Throughout the Mississippi River basin, from the smallest tributaries to the big river itself, the number of individuals and species making up fish communities has changed as a result of land use and river engineering. Fish must cope with excess nutrient inputs from fertilizer as well as toxic wastes from agricultural and urban lands. Habitat loss from channelization, levees, and dams restricts fish abundance and diversity. Many fish now rely on floodplain habitats that include depressions left by excavating sediment to build levees, water treatment lagoons, and canals, as well as backwaters present at tributary junctions and oxbow lakes formed as meander loops are cut off from the main channel during floods. Increased sedimentation from land use changes and channel engineering further limits habitat availability. Alteration of the seasonal flow of water by dams and tile drains interferes with fish movement, reproduction, and growth. Rising water temperatures from loss of riverside shading stress many fish species. The effects of contamination, habitat loss, and flow regulation are no recent revelation. Careful observers commented on changes in the Upper Mississippi watershed and the resulting changes in species composition of fish as early as the 1890s.

Very costly management actions are now being undertaken to mitigate some of these effects, although it remains to be seen whether these will prove sufficient given the magnitude and scope of the problems. The first step in rehabilitating any river is to monitor it and document what has changed from historical or reference conditions. As of 1999, scientists ranked five of six monitoring criteria as moderately impacted along the upper impounded reach of the Upper Mississippi and highly impacted along the lower impounded reach.

Partly in response to this type of evaluation, several federal and state agencies have come together as stakeholders in the Upper Mississippi basin to coordinate continued monitoring and restoration efforts, including those described earlier for backwaters at locks and dams. As of 1995, site-specific habitat rehabilitation and enhancement projects were being constructed at fifty-four locations throughout the Upper Mississippi. Habitat in the main channel is restored by changing dredge disposal and modifying wing dams. Floodplain wetlands are restored by acquiring lands currently in agriculture and replanting them with native species. In each case, successful rehabilitation depends on understanding the habitats needed by plants and animals, the processes that create and maintain those habitats, and the effects that proposed manipulations of the river have on these habitats and processes. Limited funding, however, has been one of the primary constraints on efforts to restore the Mississippi River system.

Federally funded restoration programs are crucial to rehabilitating the river and to addressing the large-scale limitations to river health posed by navigation structures and maintenance or by point and nonpoint sources of contamination along the river. But citizen groups and nongovernmental organizations are also taking matters into their own hands, removing hundreds of tons of trash from the river and planting thousands of hectares of native bottomland trees along the shorelines and islands of the Mississippi. These undertakings foster a sense of stewardship and responsibility in people living along the river's course that may also generate greater political support for large-scale river restoration.

A River of Grain

Much of the historical and contemporary river engineering on the Mississippi was undertaken to maintain navigability for barges, which move along the Upper Mississippi at regular intervals. An individual grain barge is 4 meters high and 10.6 meters wide. Up to fifteen barges can be moved with one tow on the upper river, although "tow" is a misnomer given that most of the barges are actually pushed from behind by a single boat. Downstream from Cairo, Illinois, each tow is allowed to transport forty-five barges. Each barge can hold fifteen hundred tons, or the equivalent of sixty loaded semitrailers. A fifteen-barge tow thus carries the equivalent of nine hundred truckloads. Some of the barges carry rock aggregate, benzene, coal, and salt as part of the more than ninety million tons of cargo barged annually between St. Paul and St. Louis, but grain fills most of them. Of all the grain exported from the United States, 60 percent moves by barge down the river to terminals

7.5 *Above,* towboat and barges on the upper Mississippi. This is a relatively small collection of barges being pushed by a single towboat. *Below,* grain elevator along the upper Mississippi River. Grain is trucked from surrounding fields for storage in the large silos, then loaded directly onto river barges.

between Baton Rouge and New Orleans, where it is loaded for transport to Asia and other regions.

The Mississippi has been called a river of grain; fifty-five million tons are exported each year from the Mississippi drainage, in addition to what is consumed within the United States. The big river also reflects the flow of grain in more subtle ways. Water evaporated from oceans, forests, and lakes falls as precipitation over the drainage basin, creating the surface and subsurface moisture that nourishes the vast fields of corn, soybeans, and wheat. Nitrogen and phosphorus fertilizers and organochlorine pesticides manufactured across the United States are carried to the Upper Mississippi basin to fertilize the fields of the nation's breadbasket. Rainwater and snowmelt running off the farm fields carry along fertilizers and pesticides attached to the particles of sand, silt, and clay eroding from the fields. Eventually these particles of water, sediment, and attached contaminants make their way through surface and subsurface paths into tributary rivers and then the Mississippi itself.

These by-products of the national grain production industry join the downstream flow of the grain, but they do not move as efficiently or predictably as the grain barged along a channel that has been largely engineered for just this purpose. Some of the sediment accumulates in each navigation pool upstream from a lock and dam complex. Some of the sediment spreads across the floodplain wetlands and is stored there. Some of it moves all the way down the river to settle in the Mississippi's delta, or in the zone where river and ocean waters mix in the Gulf of Mexico. Attached contaminants move with the sediment all along this complex path with its shorter- and longer-term storage sites.

Other contaminants travel dissolved in the river water. Nitrogen intended to produce fields flush with cornstalks now feeds algae in the Gulf of Mexico, creating algal blooms that cover a swath as big as New Jersey. Concentrations of nitrate in the river and some of its tributaries have increased by two to five times since the early 1900s as farmers have applied more and more nitrogen fertilizers to their fields to increase yields. This overapplication means that much of the nitrogen simply runs off the field into surface water and groundwater. The flux of nitrate to the Gulf has tripled in the past thirty years, with most of the increase occurring between 1970 and 1983. Much of this nitrogen comes from the agricultural lands of the Upper Mississippi.

Pesticides such as atrazine, applied to corn in these agricultural lands, also move in solution into surface water and groundwater used for drinking. Pesticides are ubiquitous in the Upper Mississippi. An estimated two-thirds of all pesticides used nationally for agriculture are applied within the Mississippi River basin. The hazards to human and environmental health vary with the land use patterns along the river and with the resulting chemical brew that is present in different portions of the river.

A 1995 study of pesticides throughout the Mississippi River basin by U.S. Geological Survey scientists found atrazine and metolachlor in more than 95 percent of the samples, although maximum contaminant levels and health advisories recommended by the Environmental Protection Agency were exceeded in only a small percentage. In most conditions atrazine degrades rapidly to less toxic compounds (although it can persist for more than a year in dry, sandy soils and cool climates), but it is linked to breast and ovarian cancer in humans. Conservative estimates of mass transport indicate that the Mississippi discharges 176 tons of atrazine, 78 tons of cyanazine, 62 tons of metolachlor, and 20 tons of alachlor into the Gulf of Mexico each year. Scientists from the U.S. Geological Survey concluded that pollution by agrochemical runoff and groundwater may be the most significant recent factor responsible for deterioration of water quality in the big river and its tributaries.

Not only are these poisons present in the waters draining directly from agricultural lands, U.S. Geological Survey samples throughout the Mississippi drainage basin found *detectable levels of multiple pesticides in every rain and air sample collected from urban and agricultural sites.* Samples from a background site near Lake Superior in Michigan, removed from dense urban and agricultural areas, detected compounds including atrazine. These results indicate atrazine's propensity for long-range atmospheric transport.

One bargeload of grain represents 120 hectares of land devoted to agricultural production. It also represents an unmeasured flush of topsoil, fertilizers, and poisonous pesticides all along the river and into the biologically rich Gulf of Mexico.

Mussels

Some of the sediment and nutrients moving down the Mississippi historically were filtered from the water by extensive beds of freshwater mussels, but this cleansing function has been seriously compromised by loss of mussels from overharvest. Native Americans harvested mussels living in the streams of the Mississippi River basin to obtain food and freshwater pearls. As the encroaching Europeans pushed the Native Americans west of the Mississippi, the many species of mussels in the river system had a brief reprieve, and by the late nineteenth century they were abundant along many of the stream channels. The reprieve ended in 1891 when German immigrant John Boepple developed a way to use the mussel shells in manufacturing buttons and opened a small factory in Muscatine, Iowa.

Within a decade, mussel harvest and button manufacture formed an economic mainstay of towns all along the river. Working from north of Prairie du Chien, Wisconsin, to south of Canton, Missouri, the industrious mussel gatherers harvested twenty-four thousand tons of mussel shells in 1899 alone. This level of harvest could not be sustained, particularly because mussel gatherers customarily collected every mussel they could find.

The invention of the "crowfoot" in the late 1890s accelerated the mussel harvest. The crowfoot consisted of numerous four-pronged hooks attached to an iron bar. Mussels rest on the streambed with their shells

7.6 The crowfoot used to harvest mussels, here in an exhibit at the National Mississippi River Museum and Aquarium in Dubuque, Iowa. Widespread use of this harvesting technique severely disrupted mussel communities and led to declines in their numbers.

partially open to filter water through their soft tissues. When the crowfoot was dragged along the streambed, the touch caused the mussels to snap their shells shut on the hooks, making it easy to lift them into a boat. Fishermen using a crowfoot harvested from deeper waters that might otherwise have served as a refuge to maintain breeding populations of mussels. Widespread use of the device led to repeated dragging and disruption of mussel beds, causing gravid females to abort and injuring very small mussels so seriously that they died even if returned to the water. By the 1930s, a combination of falling prices, increasing rarity of mussels, and habitat degradation from repeated dragging and from sedimentation and water pollution finally destroyed the pearl button industry.

The Upper Mississippi River basin today contains sixty-two species of mussels, which represent 20 percent of the three hundred species found within the United States and Canada. The U.S. Fish and Wildlife Service lists five species as endangered, and NatureServe considers sixteen species imperiled, or 26 percent of the Upper Mississippi mussel species. As in the case of fish, habitat loss and water pollution are among the problems contemporary mussel populations face. The introduced and prolific zebra mussel, which literally smothers native mussels by crowding on top of them, may pose the greatest threat. Zebra mussels rode the surging waters of the 1993 flood down the Illinois River into the Mississippi and are now spreading largely unchecked.

The 1993 Flood

A steep flight of concrete stairs rises from the river in St. Louis, Missouri, with an immense silvery arch standing beyond the stairs. This is Gateway Arch, symbol of a national gateway to the West. Here the Illinois and the Missouri rivers join the Mississippi. Much of the westward expansion of the United States followed the Missouri and its tributaries upstream to the Rocky Mountains. Above the steps rising from river level in St. Louis lie the high-water marks from the 1993 flood, which rose to less than a meter below the top of the flood wall. All along the upper river rise levees and flood walls built in response to the floods of 1965 and 1973, but it is the more recent 1993 flood that truly forms the upper basin's high-water mark.

The 1993 flood was the greatest ever recorded at forty-two of the stream gauging stations in the Upper Mississippi basin. The volumes of water pouring out of the Mississippi during August 1993 were so great that the water could easily be traced as it moved southeast around Florida and then up the east coast to North Carolina in September. The flood inundated more than 2.7 million hectares of agricultural and urban lands, and estimates of economic damage range between twelve and sixteen billion dollars. After the flood, the Federal Emergency Management Agency "hazard mitigated" over twelve thousand properties through acquisition, relocation, elevating, or flood-proofing with levees or other structures. The government also purchased two thousand hectares of floodplain for the Big Muddy National Fish and Wildlife Refuge along the Missouri River. As economic costs increase with each major flood despite the levee system and other engineered structures, we will be able to reduce flood damage only by eventually moving as many people and structures as possible out of the historical floodplain.

As on the Danube, the Ob, the Nile, and the Ganges, floods spread out across the Mississippi's extensive bottomlands before people began to channelize the river and confine it within levees. The wetlands and riverine forests of the bottomlands slowed the downstream passage of floods, decreasing the highest discharge at any point along the river by attenuating the flood peak and allowing sediment suspended in the floodwaters to settle out in areas of lower velocity. Invertebrates, fish, and plants living along the Mississippi adapted to the flood pulses and came to rely on seasonal access to the habitat and nutrients provided by floods. Nutrients and organic matter released from newly flooded soils, for example, stimulate microbial activity and the production of creatures that serve as fish food just at the time that fish larvae need to feed on such minute organisms. The small fish can also escape predatory larger fish in the shallow waters of the floodplain. As the Mississippi was progressively confined between levees during the nineteenth and twentieth centuries, the floodplains were converted to agriculture or other land uses, and the floods were contained within the leveed main channel. Levees are built primarily to limit inundation during floods, but their effect has become very controversial in many cases.

Those behind the levees are safe as long as the levees remain intact and continuous. But by conveying water more rapidly downstream, levees can enhance the flood peak dumped onto downstream areas unprotected by levees. And if the levees fail, the site of failure becomes like a firehose that effectively directs rapidly moving water and sediment onto the former floodplain. The intact portions of the levee then keep water from moving back into the channel as the flood recedes.

Levees can fail by being overtopped, by becoming saturated and collapsing through liquefaction, by slumping, or by being undermined. Levee failures during the 1993 flood created sites of rapid flow and intense turbulence that scoured the floodplain and channeled sediment into areas far from the main river channel. Up to 28 percent of the total Mississippi discharge flowed through one large levee break complex at Miller City, Illinois, carrying with it nearly 8.5 million cubic meters of sand.

The floodwaters spreading across the Mississippi bottomlands during 1993 benefited some organisms. Populations of native fish grew after the 1993 flood, reflecting increased production of juvenile fish on newly accessible floodplain habitats. Thirty-six species used the floodplain for spawning and nursery areas along the lower Illinois River. As the floodwaters gradually receded, some of the new fish went to feed predatory fish and birds, which also increased after the flood.

The ecological downside of the 1993 flood was twofold: the spread of pest species and of contaminants. Mosquitoes and introduced zebra mussels thrived during the inundation. Zebra mussels were accidentally introduced to the Great Lakes in the ballast water of ships coming from Europe. The prolific, highly mobile larvae of these mussels followed the canal system of Chicago from Lake Michigan to the Illinois River, and the 1993 floodwaters transported huge numbers downstream into the Mississippi.

Agricultural, urban, and industrial contaminants were also dispersed on an unprecedented scale during the 1993 flood. The Mississippi River basin contains the largest and most intensive agricultural region in the United States. More than 80 percent of the corn and soybeans grown in the country, and much of the cotton, rice, sorghum, and wheat, comes from the basin. To increase yields from these crops, more than 110,000 tons of pesticides and about 6.3 million tons of nitrogen fertilizer were applied to the basin in the early 1990s. Runoff pouring from farm fields through the Upper Mississippi basin in 1993 carried a witch's brew of poisons. Despite the much greater volume of water flowing down the rivers, concentrations of nitrate and herbicides were similar to the highest concentrations measured during the much lower flows of spring and summer 1991 and 1992. In other words, the dilution often seen during floods failed to take place in 1993. The 1993 flood flushed an estimated half a million kilograms of atrazine and 911,800 tons of nitrate nitrogen into the Gulf of Mexico.

Water discharging from the Mississippi River plays a critical role in the ecosystem of the Gulf. Discharge from the Mississippi and Atchafalaya rivers forms the Louisiana Coastal Current, and biological productivity in the Gulf is highest along this river plume because of the nutrients carried into the ocean with the river water. Nitrogen is the nutrient that most commonly limits the productivity of estuarine and coastal waters. But excess nutrients can definitely be too much of a good thing.

Increasing amounts of nitrogen have been flushed downstream to the Gulf as the application of nitrogen fertilizers has become more widespread in the Mississippi drainage basin. Here the nitrogen stimulates algal blooms along the Louisiana coast. As in the Lake Victoria algal blooms indirectly caused by Nile perch, when the masses of algae in the Gulf of Mexico caused by excess nitrogen from the Mississippi River die and sink, the decaying organic matter uses up oxygen in the bottom layer of the water, creating anoxic conditions in which other organisms cannot survive. Nitrogen loading was recognized as a growing problem before 1993, when so-called dead zones of eight thousand to eleven thousand square kilometers formed in the Gulf each year. During 1993 the dead zone grew to nearly seventeen thousand square kilometers, and this greater extent persisted in 1994. The dead zone has continued to fluctuate between thirteen thousand and twenty thousand square kilometers, with no sign of decreasing.

The only effective way to limit the flush of nitrogen into the Gulf, and thus the existence of the dead zone, is to reduce nitrogen loss at the source; the farm fields of the Mississippi River basin. Suggestions include reinventing the agricultural system of the Corn Belt. The current system features low economic returns and high nutrient and sediment losses, but it could be altered to create a more ecologically based landscape emphasizing nutrient sinks and a legume base for supplementing fertilizer nitrogen (something more like the corn-beans-squash crop rotation of the Native Americans that relied on nitrogen-producing legumes to fertilize the soil). Natural and created wetlands and riverside belts of vegetation also buffer rivers by absorbing excess nitrogen coming from farm fields and urban wastewater treatment plants. Allowing Mississippi River water to spread across the floodplain of the river's delta would result in some nitrogen uptake and storage by deltaic wetland plants. The combination of changed farm practices and riverside, wetland, and delta buffer zones has the potential to substantially reduce nitrogen inputs to the Mississippi River. This would improve water quality all along the river and allow the ecologically and commercially important Gulf fisheries to recuperate.

The Illinois and the Missouri

The Illinois and Missouri rivers, two of the Mississippi's primary tributaries, enter the main river within a few kilometers of each other. Each river basin reflects, on a smaller scale, many of the environmental changes present on the Mississippi. The Illinois River joins the Mississippi from the east near St. Louis. The river drains 80,300 square kilometers, most of them in the state of the same name. The river basin parallels the Upper Mississippi River basin as a whole in that historically extensive floodplain forests and wetlands that supported abundant fish, waterfowl, and other organisms were systematically converted to farmlands during the nineteenth and twentieth centuries. Levee construction, land drainage, agricultural and urban pollutants, and channelization, along with increasing sediment loads from upland farming, altered the riverine corridor and caused substantial declines in the abundance and diversity of the river's plant and animal communities. Despite these massive losses and accumulating problems, a 1992 National Research Council report identified the Illinois River as one of the country's few large floodplain river systems that still retain enough natural characteristics to allow for restoration. The Nature Conservancy, the Illinois Department of Natural Resources, and other private and government groups are acquiring forests and wetlands along the river, restoring a functioning ecosystem hectare by hectare.

The Missouri River enters from the west a few kilometers downstream from the junction of the Mississippi and Illinois rivers. Of all the Mississippi's tributaries, the Missouri carries the greatest load of sediment because it crosses sedimentary rocks that weather and release abundant sediment in the semiarid climate that characterizes much of the Missouri's drainage, as well as erodible glacial deposits on the Great Plains. The loads of sediment carried in suspension by the Mississippi have decreased by half since the Mississippi valley was first settled by Europeans. This decrease has occurred mainly since 1950, because the numerous large reservoirs built in the Missouri River drainage now effectively trap about 80 percent of the sediment coming downstream. The Missouri, however, still contributes more than half the Mississippi's total sediment load. This load is not matched by the Missouri's water flow. Because of the great dry swath created by the rain shadow of the Rockies, the 3,860 kilometer Missouri drains 43 percent of the Mississippi's total area but contributes only 12 percent of the Mississippi's total discharge.

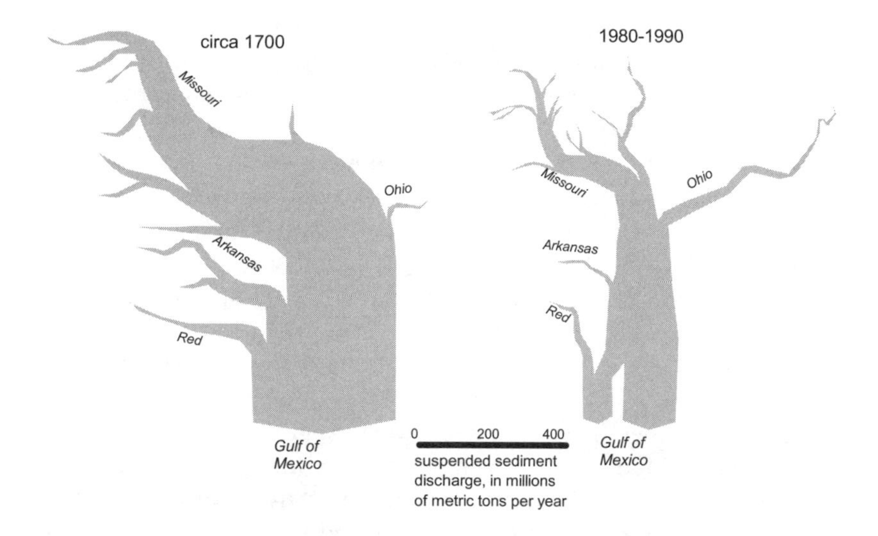

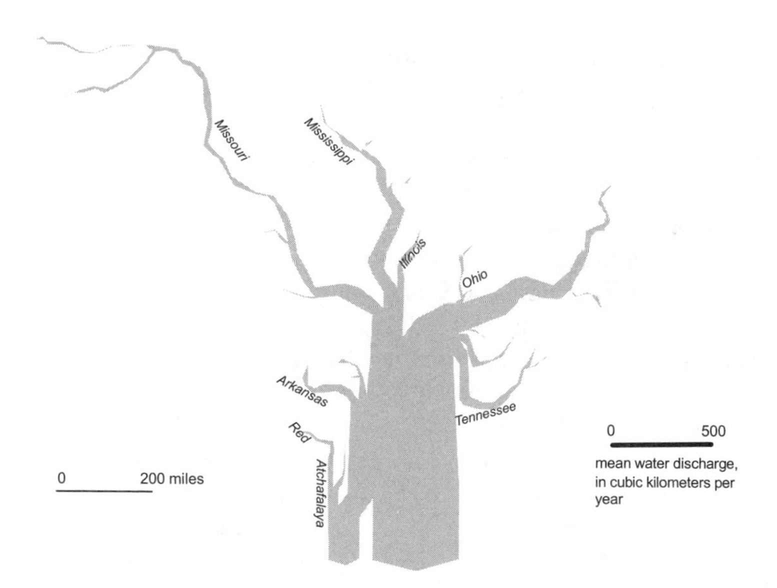

7.7 Schematic illustration of historical and contemporary contributions to suspended-sediment discharge (*above*) and water discharge (*below*) of the Mississippi River. (After Meade 1996, figures 5 and 6.)

As on the Mississippi, channel alterations accompanied nineteenth-century steamboat traffic on the Missouri. A history of frequent flooding also led to piecemeal construction of urban levees and encouraged the Corps of Engineers to build the first dam on the mainstem Missouri in 1938 at Fort Peck, Montana, although the dam is principally used for navigation and hydropower. Today six major dams, from Fort Peck downstream to Gavins Point, on the Nebraska–South Dakota border, regulate flows on the main channel. The course of the Missouri between these dams has been transformed into a series of lakes separated by short river segments, whereas the river downstream from Gavins Point has been altered more by straightening, bank stabilization, and levee construction—all the changes necessary to create a 2.7 meter deep navigation channel 1,225 kilometers long. Approximately 10 percent of the original floodplain is now inundated during the average annual flood because levees confine the lower Missouri River to a width of 180 to 330 meters. The river engineering favored the growth of farming in the river's former floodplain, as well as a network of barge traffic. Nearly 95 percent of the Missouri River basin's landmass is devoted to agriculture, and farmers, along with barge operators, are now vocal opponents of scientists and environmentalists who seek to change river management along the Missouri.

Environmentalists question whether the barge traffic, which peaked in 1977 and has been declining ever since, is worth the loss of much of the Missouri River ecosystem. Fisheries scientists attribute an 80 percent decline in the commercial fish harvest during the past century to river engineering. Sixteen fish species are listed as imperiled. Natural riparian vegetation has been nearly eliminated.

Localized restoration projects that attempt to halt or reverse this trend of species loss include controlled flooding on small parcels of intensively managed public lands along the lower Missouri River. Although these artificially flooded wetlands provide bird habitat, they remain disconnected from the main river by levees, at least in part to restrict wetland access by carp. Native, floodplain-dependent fish are also excluded from the wetlands. As on the mainstem Mississippi and the Danube, these restoration projects do not seek to re-create a fully functioning river ecosystem but rather aim to provide conditions that will benefit specific species or groups of species. Management of these isolated floodplain parcels reflects the "string of beads" restoration concept, in which not all of the river's floodplain needs to be reopened in order to revitalize the ecosystem. Instead, river health can be improved by acquiring key floodplain habitats such as flood-prone areas near tributary confluences or remnant backwaters that form "beads" along the river corridor. Acquiring and managing individual beads is much more economically and politically feasible than attempting to restore the entire river corridor, and it can build public support with demonstrated benefits.

The Middle Mississippi

Along the middle segment of its course, the Mississippi flows 313 kilometers from the mouth of the Missouri to the mouth of the Ohio, the source of nearly half the Mississippi's water. For the first 170 kilometers the Mississippi flows in a trench five to seven kilometers wide eroded more than a hundred meters into Paleozoic bedrock. Limestone bluffs form valley walls well back from the main channel. Secondary channels, islands, sandbars, and a few abandoned channels continue to create complexity along the river's course, but the engineering present in the Upper Mississippi seems almost gentle compared with the increasingly restrictive system of levees the river encounters downstream. The Middle Mississippi is extensively diked to maintain the 2.7 meter navigation channel, and flood control levees narrow the extent of high water. By 1968, the surface area of the Middle Mississippi was 39 percent less than the area measured in 1888, before much of the channel engineering.

The Mississippi does not meander broadly until it passes through the Thebes Gap area of Missouri and Illinois, a gorge cut through Shawneetown Ridge by higher river flows about nine thousand years ago. Downstream from the gap the floodplain widens abruptly to nearly eighty kilometers, and the river's course grows more sinuous. From Thebes Gap downstream the river flows on the Mississippi Embayment, a deep trough of downward-folded bedrock filled with sediments deposited over the past few million years. As the Mississippi has meandered back and forth across its floodplain during thousands of years, it has created a complex, subtle topography of undulating low ridges and swales that represent former natural levee ridges and abandoned channels filled with ponded water or sediments. Crescent-shaped oxbow lakes lie among expansive, largely flat backswamps. On aerial photographs the lakes, ridges, and swales give the landscape a texture or grain, as though the surface is alive and growing.

Cairo, Illinois, sits at the junction of the Ohio River and the Mississippi, 288 river kilometers south of St. Louis. The name Ohio comes from Iroquois and means beautiful river. The French called the Ohio *la belle rivière*. The Ohio River and its tributaries drain 567,000 square kilometers, only one-sixth of the total area drained by the Mississippi, but contribute nearly half of the water discharged to the Gulf of Mexico. The Ohio contributed very little of the Mississippi's sediment load before European settlement, but the load increased by five to ten times as Europeans brought deforestation and row-crop farming to the drainage basin. As sediment clouded the water and accumulated along the channels, construction of a progressively deeper navigation channel mirrored the history of river engineering on the Mississippi. A 2.7 meter channel completed in 1929 now extends the full 1,635 kilometers from Pittsburgh to the Ohio's mouth and along 2,170 kilometers of tributaries. As on the Illinois River, the history of land use, river engineering, and loss of habitat and species along the Ohio mirrors that along the Mississippi.

The Lower Mississippi

Downstream from Cairo, Illinois, the river becomes officially the Lower Mississippi for the final 1,590 kilometers of its journey to the Gulf of Mexico. Almost everything beyond the main channel is hidden by levees in this flat landscape, and the riverscape is dominated by the constant barge traffic. In its lower reaches the Mississippi more than ever looks like an industrial river.

The natural floodplain beyond the levees includes nearly ninety-two thousand square kilometers and varies in width from thirty to two hundred kilometers. The Lower Mississippi was historically a very sinuous river, with a complex topography of abandoned channel remnants forming wetlands across the broad floodplain. Natural floodways formed by channels running roughly parallel to the Mississippi for some distance carried excess water during high floods. Flow into these channels is now mostly blocked by an extensive network of levees that has restricted the active floodplain to about ten thousand square kilometers. Today the Lower Mississippi is fairly straight, particularly along its lowermost portions from Baton Rouge, Louisiana, to the Gulf.

The levee system of the Lower Mississippi grew in fits and starts, much like the locks and dams on the upper river. French settlers at New Orleans began to construct levees to protect the city from floodwaters in 1717. In 1850 the Corps of Engineers began a survey of the Lower Mississippi that considered levees, natural and man-made outlets from the river, and cutoffs between river meanders as means of flood control. Another massive flood in 1858 hastened the Corps's study, which resulted in an 1861 report recommending a "levee-only" policy of flood control.

Predictably, the floods continued to come. Severe floods in 1903 and 1907 were followed by floods in 1912 and 1913 that destroyed many of the nonfederal levees and caused many deaths. This led to further pressure for government intervention in flood control. In 1917 Congress passed a Flood Control Act that called for further levee building. In his book *Rising Tide*, John Barry provides a fascinating account of the struggle between civilian and Army engineers to develop a flood-control policy for the Lower Mississippi. The Army's emphasis on levees won the day. Belief in the levee-only approach was reinforced when a near-historic flood in 1922 was contained by federally designed levees. In a classic example of hubris, the commanding general of the Corps of Engineers announced in 1927 that the Lower Mississippi levee system was ready to withstand the worst of floods. Two months later, the greatest flood yet on the Lower Mississippi overwhelmed the levees and spread over a hundred kilometer swath from Memphis southward to the Gulf. That was the end of the levee-only policy.

Like the 1993 flood on the upper river, the 1927 flood was preceded by months of rain. The coup de grâce was heavy rain over several hundred thousand square kilometers stretching from Illinois to Missouri and Texas. The 1993 flood on the upper river crested at 28,100 cubic meters per second in St. Louis. The 1927 flood crested at an estimated 84,270 cubic meters per second in New Orleans.

After 1927, flood control in the lower valley became completely a federal responsibility. Emphasis shifted from levees-only to recognition that other structures such as spillways and dams were also needed. The Corps built seventy-six reservoirs in the Upper Mississippi valley and forty-nine on the Missouri River and its tributaries. The Bureau of Reclamation added another twenty-two flood-control reservoirs on the Missouri. A nonstructural approach emphasizing reduced use of floodplain lands grew steadily more popular starting in the 1970s, but the floods continued to come and the levees continued to be built. At present, the channels of the Lower Mississippi drainage basin flow between 3,532 kilometers of levees.

This enormous riverine straitjacket created multiple changes in the Lower Mississippi. Major natural floodways were eliminated. The land area of the regularly flooded bottomlands was reduced by more than 90 percent. The remaining bottomlands are highly fragmented and have lost much of their ecological function. Floodplain lakes are isolated. Lateral channel migration, which historically created backwaters in the form of abandoned channels, largely ceased, causing deterioration of the diverse habitat necessary to support a wide variety of riverside vegetation, fish, birds, and other organisms.

The highest densities of larval fish along the Lower Mississippi are found in backwaters, which also serve as nurseries for juvenile fish. Rivers like the Ohio have large backwater areas where tributaries enter, but the Lower Mississippi has few tributaries. As existing backwaters along the lower river gradually fill with sediment, habitat for many river and floodplain organisms becomes progressively more limited.

Water surface elevation of the channel rose as flows along the lower river became progressively more confined between levees. Engineers severed fifteen meander bends between 1933 and 1942 to create a straighter channel with swifter flow to reduce water surface elevation. These artificial cutoffs shortened the river by 228 kilometers and did indeed create swifter flow--so swift that erosion was accelerated and the channel had to be stabilized with 1,368 kilometers of rock and other hard materials. River energy that would previously have gone into natural meander cutoffs and bend enlargement now goes into attempts to build midchannel bars, so that the Corps is kept busy dredging the main channel and removing the bars. A naturally functioning river is self-maintaining. Once you start engineering a river, you inevitably take on a constant maintenance project.

With continuing help from engineers, the channelized river flushes sediments and dissolved contaminants fairly efficiently into the Gulf of Mexico. Like the Twin Cities and St. Louis, Baton Rouge and New Orleans contribute high levels of urban contaminants to the Mississippi because of increased population densities. Industrial contaminants such as hexachlorobenzene enter the river from the industrial corridor along the lowermost four hundred kilometers of the Mississippi. This reach of the river is sometimes called cancer alley because the death rate from cancer and other diseases here is above the national average. As with contaminants in humans living along portions of the Ob and the Ganges, these statistics unfortunately are not difficult to understand given the context of toxic waste production. At least 136 major industries line the banks of the river along the 250 kilometers between Baton Rouge and New Orleans, mostly petrochemical plants that manufacture pesticides and plastics or refine oil. In 1987 these industries discharged eighty-nine million kilograms of toxic chemicals directly into the Mississippi.

The Mississippi passes New Orleans flowing at a higher level than the city. The steamboats so important to history along the rest of the river were first introduced here in 1812, and the region's economy continues to rely on commercial navigation. South of New Orleans the river has such uniform width and controlled bends that it looks artificial. The flat, wide river continues on through the Bayou, named from the Choctaw word *bayuk*, "creek," to its delta.

"Delta" can have different meanings in relation to the Mississippi. The word is used to describe the culture of a broad region that extends along the river from Vicksburg to Memphis. It is also used for the huge depositional zone the river occupied over the past few millennia and the relatively small zone where active deposition occurs today. During the past seven thousand years the active delta of the Mississippi has swung back and forth across a swath of coast nearly 420 kilometers wide. The river historically deposited so much sediment at its meeting point with the ocean that each lobe of the delta built up above the surrounding coastal wetlands. At some point during this process, the Mississippi would rapidly shift to an adjacent lower elevation and begin depositing another delta lobe. The contemporary delta, which projects into the Gulf like the long, skinny leg and foot of a bird, has now been active for about a thousand years.

Left to its own devices, the Mississippi would probably have shifted course sometime between 1965 and 1975 and begun flowing down the Atchafalaya River to create a new delta sublobe. The Atchafalaya, which drains forty-seven hundred square kilometers of land adjacent to the Mississippi River, became a distributary channel of the Lower Mississippi about four hundred years ago when a looping meander of the Mississippi intercepted both the Red and Atchafalaya river courses. Wood and sediment formed a plug at the entrance to the Atchafalaya that prevented it from capturing the flow of the Mississippi, despite the Atchafalaya's steeper downstream slope. In 1831 Henry Shreve created a cutoff on the Mississippi, intending to increase river velocity enough to eliminate siltation at the mouth of the Red River, which entered the Mississippi immediately upstream from the Atchafalaya. The cutoff meander became known as the Old River. Cutting off the big bend reduced water and logs entering the Atchafalaya, allowing private citizens and the State of Louisiana to begin dismantling the logjams over the next three decades. But clearing the logjams caused the Old River meander loop to become a new connection between the Atchafalaya and the Lower Mississippi.

The Atchafalaya has a three-to-one advantage in slope over the Lower Mississippi. As more and swifter water flowed down the new pathway, the Atchafalaya was well on its way to becoming the Mississippi until the Corps stepped in. In *Designing the Bayous,* Martin Reuss describes in detail the long history of river engineering as the Corps attempted to prevent the Atchafalaya from capturing the Mississippi's flow. The Corps interfered with the process of river capture because the infrastructure of coastal and lower river navigation existing along the Mississippi by that time could not be replicated along the new channel without enormous expense. Internationally respected engineers including Hans Einstein, Albert Einstein's son, worked with the Corps for years to design the low dam that now controls flow into the Atchafalaya. The Old River Control Structure is a concrete sill within the main channel that extends more than a kilometer onto the adjacent overbank areas. The structure was completed in October 1963, but problems with its integrity and function began in 1964 and continue to the present. The Corps is fighting a battle that the river will eventually win. Meanwhile, navigation interests, farmers, and environmentalists seeking to preserve the Atchafalaya's wetlands argue over river engineering and flow regulation in this region where wetlands are becoming scarcer and scarcer.

The Atchafalaya wetlands adjoin a much larger, discontinuous wetlands complex that spreads along the margins of the Gulf of Mexico. The Mississippi River delta alone has seven hundred thousand hectares of coastal marsh. This is the largest continuous wetland in the nation and constitutes part of one of the world's largest estuarine areas. The region is also exceptionally rich ecologically. The marshes and bayous shelter eleven threatened and endangered species and provide winter habitat for fifteen million waterbirds.

The problem with this region for which superlatives of size and richness so readily come to mind is that it is also disappearing at a rate easily described with superlatives. Louisiana has been losing coastal wetlands faster and faster throughout the twentieth century. Estimates of average loss rise from 17 square kilometers a year in 1913 to 117 square kilometers a year during the 1960s, before declining to 69 at present. That last figure might not sound so bad by comparison, but it translates to 0.8 hectare of wetlands lost every hour, part of a cumulative loss estimated at 3,900 to 5,300 square kilometers since the 1930s.

It is not difficult to find potential explanations for this hemorrhaging of wetlands. First, the barrier islands that provide the first line of defense against the erosive power of hurricanes and tropical storms are themselves eroding. Oil and gas pipelines and access canals affect the islands' stability, and pollution of many types reduces the ability of coastal vegetation to stabilize them. Jetties and seawalls intercept sediment moving along the coast by longshore transport (just offshore parallel to the coast) that would normally replenish sediment removed from the barrier islands by wave erosion.

Second, sea level at the Mississippi delta is rising by 1.2 to 4.3 centimeters each year. About 20 percent of this rise is caused by melting of the world's remaining glaciers. The remaining 80 percent is actually relative as the coastal sediments subside through combined normal compaction of sediments and human removal of water, oil, and natural gas. The input of sediment to the delta also declined substantially during the twentieth century as upstream dams stored sediment along the Missouri and the Mississippi. Engineered bypasses of the Lower Mississippi River and closed distributary channels on the delta further direct the sediment that does reach the coast away from the delta and into the deep water beyond the continental shelf.

Third, most of the land loss is not from the periphery and shoreline erosion but from internal breakup of marsh vegetation as a result of physiological stresses. Levees built throughout the wetlands during the 1930s ended the spring floods that replenished local soils and water tables. The nutria (*Myocastor coypus*), a large rodent from South America, escaped from captivity and began devouring the roots of marsh plants that had not evolved to withstand such herbivory. The Corps built fourteen major ship channels to inland ports by the 1960s. Wind and water erosion along the channels magnify saltwater intrusion that kills the freshwater wetland plants. More than thirteen thousand kilometers of canals thread back and forth through the great wetlands complex, exacerbating erosion and saltwater intrusion. Natural gas and oil pipelines, and associated maintenance paths, also disrupt the wetlands and contribute to their degradation.

Despite the many recent changes in the flow of water, sediment, and nutrients, the Louisiana wetlands continue to support a commercial fishery worth three hundred million dollars annually. More than two million people live in and around the coastal wetlands, and they are very much aware of land loss. Coastal areas are eroding so rapidly that changes are readily perceived over a decade or two. Engineering responses to date include releasing freshwater to mimic spring floods. Engineers also divert freshwater into the delta to decrease salinity. They rebuild marshes with dredge material and salt-tolerant plants and attempt to increase the sediment released to the delta. Stabilizing the shoreline and restoring barrier islands can also help reduce coastal erosion.

Despite spending half a billion dollars in the past decade, none of these experiments has effectively counteracted the regional trend of land loss. At present the Corps and other groups have a fourteen billion dollar unfunded plan to save the rest of the coastal wetlands. The situation will not change unless the level of funding and the integration among individual components of the plan are comprehensive enough to address a problem that has been building for decades and that extends from the headwaters dams to the offshore drilling rigs. As on the Ob, the Danube, the Nile, and the Ganges, changes associated with land use and river engineering throughout the Mississippi's huge drainage basin combine to cause major problems in the river's delta and coastal zone.

Restoring the River

It is tempting to conclude this chapter by writing, "Thus the Mississippi ends, not with a bang but a whimper." This giant among rivers that once built a huge delta far into the Gulf of Mexico now recedes daily, its flow tapped and regulated, its sediment trapped all along the river's path and contaminated with chemicals that will have unpredictable effects for decades to come. This is a grim picture for the great national river, but it can be a snapshot along the way, not the final portrait.

The National Research Council identifies the Upper Mississippi as an ecological rarity, for its size, that still preserves enough floodplain and flood pulse to allow restoration. In 2004 the Nature Conservancy developed a list of forty-seven priority conservation areas within the Upper Mississippi. These segments of river can be revitalized by protecting undeveloped lands and restoring natural processes of flooding and channel movement. But time is running out as population and land use inexorably increase. This is why many of the scientists who study the Mississippi River system and the citizens who live along it feel a sense of urgency in protecting and restoring the river.

The Mississippi is not exceptionally polluted compared with major rivers in other industrialized countries. Sources of pollution have been and can continue to be reduced. Polluted sediments can be remediated. Towns and cities throughout the drainage substantially improved sewage treatment, for example, after the Clean Water acts of the late 1960s and early 1970s. Oil slicks are now rarely seen on the river, despite in-stream refueling of barges, because the Coast Guard monitors in-stream polluters and enforces regulations. Phosphate pollution and the attendant sudsy patches floating on the rivers have decreased following changes in detergent chemistry in the early 1990s.

Dams and other engineering infrastructure can be removed or operated in ways less disruptive of natural flow cycles. Although most dams operated for navigation along the Upper Mississippi are unlikely to be removed, the partnerships established between the Corps of Engineers, the Nature Conservancy, and other government agencies and environmental organizations point the way toward balancing commercial and environmental needs along the Mississippi.

Damaging floods can also provide opportunities to restore natural landscapes in the floodplain. Approximately eighty-one hundred hectares along the Lower Missouri are now in public ownership as a result of buyouts since the 1993 flood. Waterfowl use the river's side channels, and cottonwood and willow forests are regenerating in the resulting Big Muddy National Wildlife Refuge.

The restoration of floodplain wetlands and the reconnection of the main river and secondary channels is not an all-or-nothing proposition. Acquiring and managing critical "beads" along the rivers can create the strings of restored habitat that will serve as lifelines to many plant and animal species and to the health of the river ecosystem as a whole. This concept is particularly amenable to a gradual approach and localized actions that can help give residents of the river corridor a sense of stewardship and that can be effectively pursued by nongovernmental organizations working in concert with government agencies.

The Mississippi River and the surrounding landscape are still present, still vital, and still capable of being restored. The Mississippi, as described in the title of a 2000 publication, is a working river and a river that works (Upper Mississippi River Conservation Committee 2000). Like the countries along the Danube, the United States has the technical knowledge and the monetary resources to restore the river corridor. If we choose, we can once more connect our greatest national river, from its headwaters to the ocean, with the adjacent floodplains and uplands that keep the river corridor vital.

Interlude

Having fallen as April rain on the central headwaters of the Mississippi River and flowed down the great river to the Gulf of Mexico, the Mississippi water droplet resumes its global journey the next summer. Fishing vessels are active in the Gulf, relentlessly pursuing the fish and shellfish that survive despite the pollution from offshore oil rigs and the excess nutrients and other contaminants coming down the Mississippi.

Evaporated from the warm, turbid waters off the mouth of the Mississippi, the water droplet is carried southwest toward the equator. The droplet is swept high into the atmosphere along the northern margin of the Intertropical Convergence Zone, jostled by violent thunderstorms and updrafts as it rises. Falling from the turbulent confusion of the convergence zone in a brief but intense thunderstorm, the Mississippi droplet briefly rejoins the ocean before being evaporated once again. Carried aloft, the vaporized water droplet crosses the equator and becomes incorporated into the Southern Hemisphere Hadley cell before it finally descends once more, far to the east of the longitude where it was initially evaporated from the ocean's surface.

As the water droplet descends, its path crosses that of the migratory birds moving north through eastern Australia. Scarlet honeyeaters (*Myzomela sanguinolenta*) are moving to their wintering grounds in the coastal forests of central northeastern Australia. Golden bronze cuckoos (*Chalcites lucidus*) are moving between New Guinea and Norfolk Island. In the waters offshore, some three thousand humpback whales (*Megaptera novaeangliae*) are traveling north to their breeding grounds near the Great Barrier Reef after feeding far off in the waters of Antarctica. Surrounded by these purposeful movements in the atmosphere and the ocean, the Mississippi water droplet falls as July winter rain on the southern headwaters of Australia's Murray-Darling River basin.

EIGHT

The Murray-Darling: Stumbling in the Waltz

Australia's Great Dividing Range extends more than three thousand kilometers from the Cape York Peninsula at the northern end of the continent to Port Philip Bay in the south. The individual peaks and high plateaus that compose the range form the continent's dominant high-elevation region, although even the highest peak, Mount Kosciusko, rises to only 2,228 meters. East of the range the land falls away steeply to a narrow coastal fringe. Westward the drop is more gradual, but just as final; portions of the continental interior lie below sea level.

Air masses sweeping inland from the Pacific drop moderate amounts of rainfall as they rise over the Great Dividing Range. Annual precipitation averages 100 centimeters in the eastern highlands and locally exceeds 250 centimeters, but by the time they have cleared the range and descend westward, the air masses bring only dry winds; precipitation drops to less than 25 centimeters in the semiarid interior. The continental interior changes quickly along an east-west transect from forested peaks to semiarid steppe and then desert whose reddish soils give rise to the metaphor of the red heart of a continent.

The interior is the classic Australian landscape of long, open views across plains of golden dried grass. Spread widely among the grasses stand white-trunked gum trees whose stiff, dusty green leaves hang unmoving amid shimmering heat. Little rain falls, and its coming is unpredictable. Life is dominated by the need to adapt to long

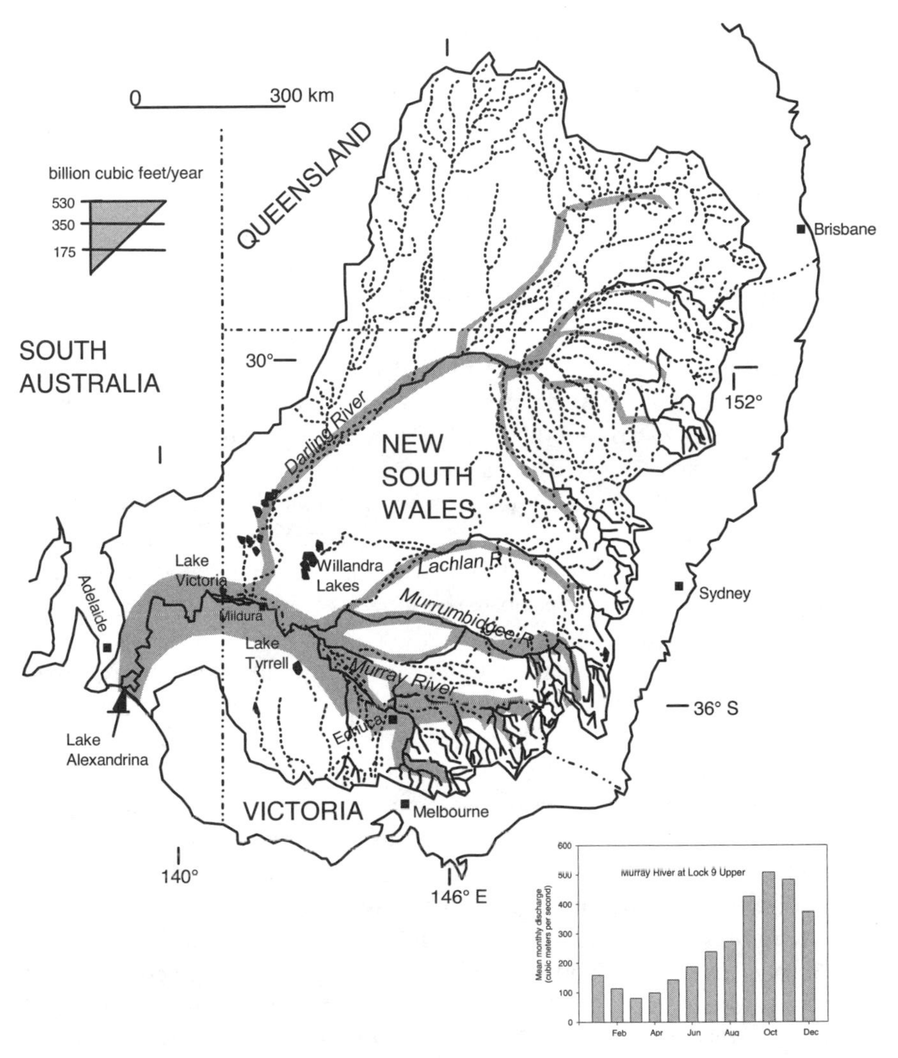

8.1 Map of the Murray-Darling drainage, with dashed lines indicating channels that periodically dry up. Shaded bands indicate average yearly river discharges in subcatchments of the drainage. (After Chen 1995, figure 1, and Jensen 1998, figure 13.1.) Inset at lower right shows mean monthly discharge at lock 9 on the lower Murray River.

dry spells. When people first entered the Australian mainland sometime before forty thousand years ago, they either settled in the wetter portions of the continent or developed adaptations such as nomadism. The second great wave of human immigration to Australia, which began at the end of the eighteenth century, reversed this long tradition of adaptation and survival. People of European descent instead began to adapt the land to themselves. The rivers of the Murray-Darling basin were crucial to their visions.

The Murray, Darling, and Murrumbidgee form the three primary rivers of the Murray-Darling basin. They rise among the western summits of the Great Dividing Range at elevations of twelve hundred to fifteen hundred meters and flow down into the flat desert basin, combining for the last three hundred kilometers and more before flowing into the Pacific through Lake Alexandrina and Encounter Bay. Like Africa's Nile, the Australian rivers receive only minimal tributary contributions once they enter the lower portions of the drainage basin, and 80 percent of the basin contributes no surface runoff most of the time. Much of the Murray-Darling flow originates as winter rains in the southeastern highlands. The Murray headwaters, in particular, cover less than 5 percent of the basin's total area but produce more than half of its discharge. Stream flow generated by the winter rains is then steadily diminished by evaporation and infiltration, uptake by wild plants, and diversion for human uses along the length of the river courses.

Despite its large area, the combined Murray-Darling drainage is small in flow by world standards. The more than one million square kilometers of the basin on average produce only about 300 cubic meters per second of flow. Like other dryland rivers, the variability from year to year is enormous. A century of historical records include flows of 20 to 1,560 cubic meters per second.

The relatively small flows of the Murray-Darling basin nonetheless support more than half of Australia's gross primary production and a substantial portion of contemporary Australian society. A quarter of the nation's cattle and dairy farms rely on Murray-Darling water, as do 50 percent of sheep and cropland. The rivers support three-quarters of Australia's irrigated land, which produces 90 percent of irrigated field crops. Sixteen cities and 50 percent of the water needs for the state of South Australia come from the Murray-Darling basin, as well as the water for three million of Australia's approximately twenty million people. This consumption severely taxes the limited water supplies of the river system. Total annual diversions of surface water for irrigation in the basin over the past two decades have been about 80 percent of the natural outflow from the Murray River during the twentieth century.

Changes in the flow of the Murray-Darling basin now begin right at the headwaters, where water from the eastward-draining Snowy River is diverted into tributaries of the Murray. The Snowy River contributes only 5 percent of the Murray's flow in wet years but up to a third in dry years. Built between 1958 and 1968, the Snowy Mountains Hydroelectric Scheme first impounds headwaters of the Snowy River in two artificial lakes and then diverts the impounded water through tunnels up to twenty-three kilometers long. Because these tunnels have large entrances and low gradients, pioneering fish species have migrated along with the diverted waters.

Continuing downstream from the highest elevations, the headwater tributaries of the Murray-Darling basin flow swiftly over cobble beds. Only limited portions of the tributaries are classic mountain streams cut into deep, narrow valleys, because the eastern highlands of Australia include extensive upland plateaus that are fairly flat. The basement rocks of eastern Australia were repeatedly folded into a broad asymmetrical arch during the long Paleozoic era of geologic time from 540 to 245 million years ago. Upwarping of the rocks continued when the Australian tectonic plate split from Antarctica and moved northward starting approximately 65 million years ago. The history of deformation created a saucer-shaped topography in which the folded rocks now form low mountain ranges that flank a basin covering three hundred thousand square kilometers west of the mountains. Marine sedimentary rocks in the basin record repeated invasions of the adjacent ocean across the southwestern portion of the lowlands, but the eastern portion of the basin is filled by terrestrial sediment. Some of this has been carried down by the rivers of the Murray-Darling system, which have been in existence throughout much of the past 65 million years, in striking contrast to the comparative geologic youth of rivers such as the Nile and the Mississippi.

Final retreat of the ocean from the Murray-Darling basin was accompanied by uplift in the southwestern basin that effectively dammed the ancient Murray River. This formed a freshwater lake covering sixty-eight thousand square kilometers, which geologists call Lake Bungunnia. Erosive processes established a permanent outlet for Lake Bungunnia after about eight hundred thousand years ago. The immense lake drained fairly rapidly and fragmented into several smaller ones, of which Lake Tyrrell in Victoria is the largest modern remnant. Regional climate became more arid about five hundred thousand years ago, with fields of sand dunes forming in the southwestern part of the basin. There is thus some geological underpinning to the common European perception of Australia as a timeless, ancient land. The eastern highlands, the western sedimentary basins, and the Murray-Darling drainage are all geologically older than comparable major mountain and river systems on other continents.

8.2 *Above*, Wimmera River, Little Desert National Park, in the southern headwaters of the Murray River. Widely spaced eucalyptuses, or gum trees, line the riverbanks, with an open, grassy understory beneath the trees. *Below*, west of Kosciusko National Park, which includes the headwaters of the Murray and the Snowy rivers, the land drops away to lower-elevation plains with sparse woodlands.

European perceptions of Australian antiquity arose partly from the flat, open, apparently changeless appearance of the landscape and partly from the simple technologies of the Aboriginal peoples the Europeans encountered. Aborigines likely entered Australia from the northwest and were present by at least forty thousand years ago. As they traveled from wetter regions into the desert of central Australia, they adapted to their arid new land.

Australia is the world's driest continent. Surface rainfall is exceeded by potential evaporation on over three-quarters of the continent's surface at present, although climate has fluctuated substantially since the Aborigines first arrived. Despite the challenges of a dry landscape, some of the earliest widely accepted evidence of human occupation in Australia comes from the Willandra Lakes region in the semiarid belt that fringes the desert core of the central part of the Murray-Darling basin. The Willandra Lakes are among a series of interconnected lake basins along the courses of various tributary rivers just upstream from the junction of the Murray and the Darling. These lakes have been dry for more than fifteen thousand years, but they once had a surface area of a thousand square kilometers of freshwater. When the climate changed and the lakes dried out fifteen thousand years ago, people living around them moved to the large rivers along which the first European explorers found them. The great majority of the more than ten thousand recorded Aboriginal sites in the Murray-Darling basin are along rivers.

Middens along rivers in the Murray-Darling basin indicate that Aborigines ate a variety of plants and animals. They did not simply use what they found; they also manipulated the river environment to meet their needs. Systematic burning of grasses and sedges around bodies of water at the onset of rain stimulated fresh growth. These fires, along with deliberate burns to flush game and accidental bush fires, kept the ground open and killed the seedlings of fire-sensitive species such as river red gum (*Eucalyptus camaldulensis*). Periodic small burns kept fuel loads lower and created tight mosaics of variously aged vegetation. When Europeans and Aborigines met in the late eighteenth century, Aboriginal culture had not changed dramatically from prehistoric times. Communities of one hundred to two hundred people were present along the Murray-Darling rivers. An estimated twelve hundred people lived in the Barmah Forest, which is today the largest river red gum forest remaining in the Murray-Darling basin. The Europeans nonetheless perceived Australia as unoccupied and unused and therefore free for the taking.

The portion of the Great Dividing Range known as the Blue Mountains limited initial exploration inland from the first European coastal settlement at Botany Bay (near present-day Sydney) in 1788. Surveyor John Oxley led an expedition into a portion of the Murray's headwaters in 1817. Following a river almost blocked in places by fallen timber, Oxley ranged widely. His party came close to dying when vast impassable marshes halted their downstream progress and diverted them south onto the arid plains.

Piecemeal exploration of the Murray-Darling continued through the 1820s with discovery of the Murrumbidgee, Murray, and Darling rivers, but Charles Sturt was the first European to systematically explore the Murray-Darling basin, during 1829–30 and 1844–46. His travels formed the final stages of a long, slow progression inland from the southeastern coast by the European settlers. Sturt believed he would find a huge inland sea west of the Great Dividing Range. Instead he found vast deserts through which the sometimes dry river meandered in a tenuous lifeline. Sturt was surprised to find that the river periodically entered extensive reed-covered swamps and disappeared so effectively that he was unable to find the downstream outlet.

The undependable nature of rivers in the Murray-Darling basin, which could flood wildly and then go dry, came as a great surprise to the first European explorers and settlers, as did the extensive marshes along portions of the river courses. They also could not conceive of adapting to the flood or drought nature of flow on inland Australian rivers and quickly set out to alter the natural flow regime to permit permanent settlement based on irrigated agriculture. Just over a decade after Sturt finished his explorations of the Murray-Darling, settlers were attempting to regulate and enhance water supplies in the basin. In 1859 enterprising farmers cut a channel between the Lachlan and Murrumbidgee rivers.

Into the Flatlands

The headwaters portion of the Murray-Darling basin is quite short. The forests of the upper basin quickly give way downstream to the more open mallee scrub. Mallees are a type of eucalyptus that branch into several trunks near the ground. Often growing widely spaced between grasses and shrubs, their trunks sometimes have a reddish orange hue that matches the sandy soil. Lines of larger eucalypti such as the tall, wide-spreading white ghost gums and river red gums mark the courses of rivers across flat plains that shimmer like a mirage during the heat of midday. Towns are small and widely spaced. Sheep and cattle pastures dominate the riverine plains, along with vineyards, orchards, and fields of grain or cotton.

The Murray flows 2,560 kilometers in total and the Darling 2,740 kilometers, but within about 300 kilometers of their origin they and their tributaries become gently sloping channels formed in the silts and sands of Australia's flat interior, because most of the catchment is below 180 meters in elevation. River gradients become less than five centimeters per kilometer. The rivers slow down and spread out as they enter the interior. Some of the flow filters into the sandy bed, reducing the ability of the remaining flow to transport sediment. Plugs of sediment form across the channels, forcing the flow sideways to a new channel. These "floodouts" form on the lowland portion of tributary channels throughout the Murray-Darling drainage basin.

Over the past few thousand years, individual channel segments have migrated back and forth across basins up to twenty-five kilometers wide, leaving behind broad floodplains and anabranching channels that carry flow only during floods. The migrating rivers also leave behind billabongs—abandoned channels forming ponds that provide the only water on the floodplain during extended droughts. In some portions of the interior the rivers form single-thread channels that meander sinuously across the floodplains. In other areas they break up into multiple channels that branch and rejoin unpredictably or spread into shallow lake basins or extensive swamps. Long-abandoned and buried channels provide geologists with a record of past river flow and climate.

Hill slopes supply differing amounts of runoff and sediment to rivers as climate fluctuates. Where rivers are narrowly confined within a valley, each new river configuration overprints and at least partly obliterates older channels. But where the rivers can move back and forth through time, as on the seventy-seven thousand square kilometers of the Riverine Plain in the eastern Murray basin, abandoned channels can be buried in sediment and preserved. Four generations of older channels that together span most of the past hundred thousand years are present on the Riverine Plain. Meandering channels that carried fine sediment suspended in the water column along with larger sediment along the streambed alternated with straighter channels that mainly transported coarser sediment on the streambed. These numerous smaller channels seem to have been part of a large alluvial fan on which rivers divided into smaller, distributary channels. All four generations of ancient channels carried much larger discharges than modern

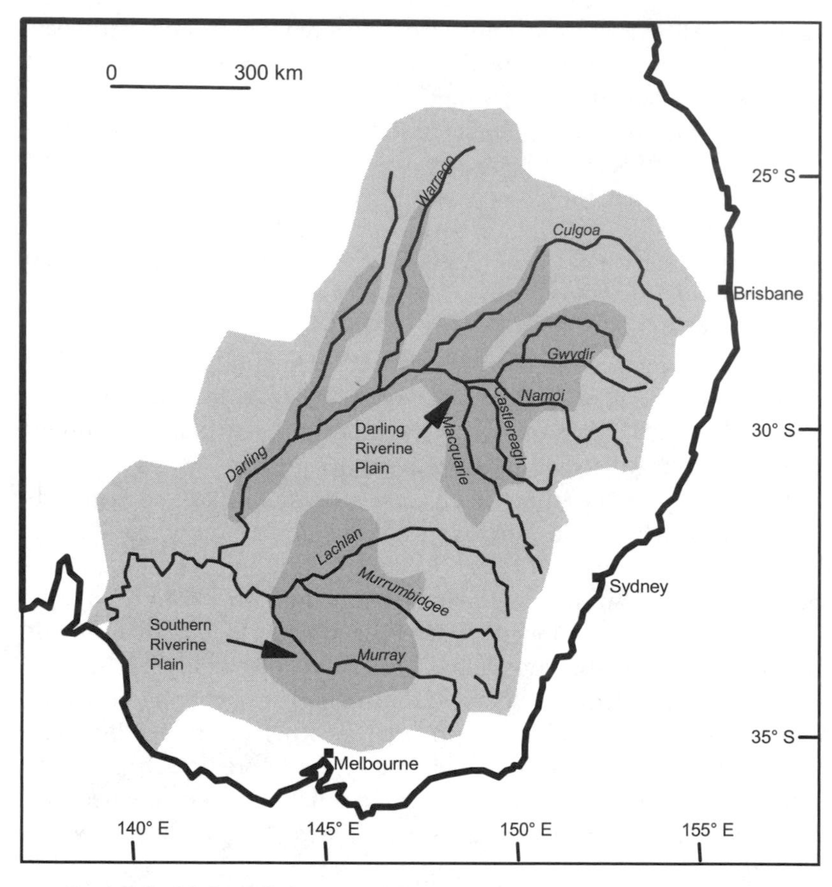

8.3 Riverine alluvial plains (darker gray) of the Murray and Darling catchments (lighter gray). The large extent of these plains that flood irregularly supports diverse environments. (After Young et al. 2002, figure 1.)

rivers do. The transition to smaller modern rivers began about twelve thousand years ago. Scientists have interpreted these changes in river configuration in the context of past climate, as indicated by the size of the river channels and by pollen records of past vegetation and sedimentary records of dune activity and past lake levels. Conditions were wetter between approximately fifty thousand and twenty thousand years ago, then drier until about fourteen thousand years ago, with active dunes in much of the central and lower Murray basin.

At present, variability of stream flow increases westward across the basin as the rivers proceed farther from the relatively reliable moisture of the eastern highlands. Australia generally has more variable annual river flows than any other continent. Variability in stream flow also increases with drainage area, reaching its greatest values in the Murray-Darling basin. Climatologists are puzzled about this, because other areas of the world with equally variable rainfall do not have quite so much variability in river flow. The unpredictable nature of flow in the Murray-Darling rivers seems to result from the combination of highly variable rainfall, exceptionally high evaporation potential in the hot, dry interior, and unusually flat terrain that produces slow flow with much time for evaporation and infiltration as the water spreads into multiple shallow channels. The middle and lower Murray-Darling basin has no segments of steep, narrow channel equivalent to the cataracts of the Nile, for example.

As if to emphasize the timelessness of the landscape, water takes a long time to get through the Murray River system. Water takes six weeks to travel from Albury to the Murray-Darling junction and another two weeks to reach the sea. Floods may take still longer, because the floodwaters spread widely into secondary channels and low-lying ponds. The Murray and Murrumbidgee rivers naturally peak in spring to early summer, whereas the Darling is more likely to flood in summer. Flood characteristics are often unpredictable. Peak flows vary from short, sharp peaks of a few weeks to high flows that extend over many months. Flood recession can be rapid or attenuated as water lingers on floodplains and in billabongs. Everything depends on the pattern of rainfall in a particular season, as well as the combined effects of rainfall and river flows in previous years.

The tendency of floods to spread across broad wetland areas nurtures a rich riverine ecosystem that depends as much on intermittent drying as on flooding. Periodic drying is particularly important in areas with clay soils. Nutrients become bound to the saturated clay during periods of extended inundation. Drying and rewetting release the nutrients, which are then rapidly taken up by plants and soil microbes that fuel the floodplain food web through new successional stages and breeding cycles.

In the branching, diverse channels of the Murray-Darling system, secondary channels that remain dry between floods accumulate leaf litter and other organic matter. When the rising waters enter these secondary channels, the leaf litter releases dissolved organic carbon as microbes, bacteria, and aquatic invertebrates consume it. Freshly dropped litter provides higher-quality food for riverine microorganisms than do dried, partially degraded leaves. Frequent smaller floods that wet the secondary channels thus provide higher-quality food, albeit in smaller quantities, than do infrequent larger floods.

Even a relatively brief flood can trigger a cascade of changes in the floodplain. Release of dissolved organic carbon initiates growth of bacteria and microalgae. Within a few days to a couple of weeks, microinvertebrates such as copepods that lie dormant within dry sediment emerge to feed on the bacteria and algae. The microinvertebrates in turn fuel secondary consumers such as macroinvertebrates (primarily aquatic insects), fish, and waterbirds. The dry floodplain and secondary channels effectively act as a seed bank from which abundant life surges at the cue of a flood. If the flood lasts long enough, organisms from insects to fish utilize the higher waters to disperse across the floodplain and move upstream and downstream to new breeding and nursery sites where their offspring can wait until the next flood rejuvenates them.

Part of the richness of these cycles of lateral and longitudinal connectivity depends on the presence of native plants on the floodplain and along the river channels. Riverside vegetation is dominated by river red gum along the riverbanks and in low-lying areas and by black box (*Eucalyptus largiflorens*) in less frequently flooded higher areas and along the margins of the floodplain. Although the red gums form dense forests in locales such as Echuca, they more commonly grow as open woodlands. Trees fifteen to twenty-five meters tall create an open canopy under which herbaceous species and woody shrubs such as lignum (*Muehlenbeckia florulenta*) do well.

Older trees shed bark and branches that provide habitat for numerous small animals. One study found that densities of yellow-footed antechinus (*Antechinus flavipes*), the only native small mammal inhabiting the floor of red gum forests, increase when wood loads on the forest floor exceed forty to fifty thousand kilograms per hectare. Antechinuses are small, carnivorous marsupials that eat invertebrates and occasionally small lizards. Although they are only ten centimeters long, their high metabolic rates and activity levels keep them eating, and they profoundly affect the numbers of invertebrates in floodplain forests. Their survival partly depends on the regeneration of red gum forests, which in turn largely depend on flooding, unpredictable as the floods may be. The threatened brown treecreeper (*Climacteris picumnus*) is similarly reliant on floodplain dynamics. The density of this native bird increases substantially as wood loading on floodplains increases because it depends on dead timber for foraging and nesting sites. Scientists have studied antechinuses and brown treecreepers. These species can be taken as representative of dozens, if not hundreds, of other species that also depend for their survival on fallen trees, branches, and bark in floodplain forests.

River red gums also provide an important source of nutrients to the river and floodplain. Decaying plant litter and fallen logs provide carbon, nitrogen, and other nutrients. Red gum wood typically supports a diverse bacterial community. The leaves of most Australian eucalypti provide only low-quality sources of nutrients. The leaves have low nitrogen content, a thick, waxy cuticle, and high levels of lignin and tannins, all of which makes them unpalatable to the microbes and invertebrates that feed on leaves and hasten their decay. Red gum leaves, however, provide the most available source of carbon to organisms inhabiting the river and floodplain.

The mechanisms by which red gum leaves leach carbon into the river ecosystem support the idea that flow variability, as well as channel and floodplain complexity, is critical for ecosystem health. Plant litter that accumulates along the upper banks of the main channel and across the floodplain during times of low flow becomes available to aquatic organisms both as habitat and as a food source during floods. Consistently low water levels created by flow regulation inhibit river organisms from using the nutrients stored in plant litter, whereas consistently high water levels prevent the plant litter from accumulating and becoming available for microbial decomposition.

Variable flows, muddy water, and constantly shifting bed and banks prevent much rooted vegetation from growing in the main river channels of the Murray-Darling basin, but more stable low-lying areas can have large stands of common reed (*Phragmites australis*) and cumbungi or bulrush (*Typha* spp.). Low light levels and shifting sands also limit the abundance and diversity of aquatic invertebrates in the river channels. Under these conditions, woody snags are particularly important for creating shelter from the current and providing a base for algae, fungi, and invertebrates such as freshwater shrimp that graze on the algae. The snags also provide a refuge, spawning site, and feeding ground for native fish and larger invertebrates. Freshwater mussels are common throughout the Murray basin. Murray crayfish (*Euastacus armatus*) live within the river channels, and yabbies (*Cherax destructor*) inhabit the still-water floodplain environments.

Compared with many of the other great rivers of the world, the rivers of the Murray-Darling basin flow through a landscape of disconnectivity. Rivers alternately flood and go completely dry. Channels branch abruptly or end in a floodout. Billabongs remain cut off from the main channel for much of the year. The health of the Murray-Darling ecosystem nonetheless depends on maintaining the limited connectivity between the main channel and adjacent floodplains, secondary channels, and wetlands. Artificially increased connectivity can be ecologically detrimental to rivers, as when it helps invasive exotic species move from adjacent river basins into the Murray-Darling, but ecologists generally place a high value on naturally occurring connectivity within river basins. The current troubled status of many native fish species in the Murray-Darling system illustrates the ecological effects of altering connectivity.

Tough Fish, Tougher Times

The changeable, often harsh climatic conditions and long geological isolation of the Murray-Darling basin limit the number of fish species, most of which are endemic. Only about twenty-four native freshwater fish and another fifteen to twenty-five marine and estuarine species are present in the basin, a low level of diversity by world standards. Fisheries biologists interpret the broad salinity tolerances among most of the freshwater species as evidence that they are recent evolutionary derivatives of marine families.

Given the challenges of life in the channels of the Murray-Darling, native fish have evolved different life strategies. Golden perch (*Macquaria ambigua*) and silver perch (*Bidyanus bidyanus*) spawn only during floods that occur when the water is warm enough. These highly mobile species are able to take advantage of localized flooding and then disperse widely in search of the next warm flood. Adult golden perch move as much as a thousand kilometers upstream and nine hundred kilometers downstream. Their offspring are equally mobile, developing as the buoyant eggs float downstream. The fish live twenty-six years or longer and grow opportunistically when conditions are right, both characteristics well suited to the uncertain floods of this river basin.

Other fish species do not depend on a flood to spawn. Small, short-lived crimson-spotted rainbowfish (*Melanotaenia splendida fluviatilis*) and western carp gudgeon (*Hypseleotris klunzingeri*) spawn seasonally, although they may spawn repeatedly when they are lucky enough to have flooding. Larger Murray cod (*Maccullochella peelii*) and freshwater catfish (*Tandanus tandanus*) also spawn seasonally in response to temperature and day length. Although they spawn independent of floods, the young fish depend on food from flooded wetlands and may not survive without a flood. These larger fish lay eggs on snags or in shallow gravel nests. A few species such as galaxiids and lampreys make a long journey downstream and rely on estuarine or marine environments for spawning and growth of the young.

Most of the native fish species in the Murray-Darling have declined in abundance and range during the past century, and several are now threatened or endangered. Historical records indicate the former abundance of these fish, which provided a reliable food source. Aboriginal men speared fish in the main channel throughout the year. The plentiful fish became concentrated when the waning stage of a flood left shallow water across the floodplain. Women and children then joined the fishing, scooping up the vulnerable fish in specially made baskets. So important were the large Murray cod to Aborigines living along the river that at least one tribe featured the fish in their creation stories.

Europeans also quickly realized the importance of the cod, writing glowing descriptions of the fish. Surveyor George Evans named the first large river that he encountered west of the Great Dividing Range the Fish River because of the Murray cod. Historian Eric Rolls quotes Evans as writing that "if you want a fish it is caught immediately; they seem to bite at any time." Evans also wrote of the cod, "Nothing astonished me more than the amazing large Fish that are caught; one is now brought me that weighs at least 15 lb." (1981, 38). Oxley and Sturt ate Murray cod. Other explorers wrote of catching large numbers of cod weighing more than fifty kilograms each. The perceived endless abundance of fish led to abuse. Historian Eric Rolls describes the devastation of fisheries accompanying construction of a railroad along a tributary of the Darling: "At the beginning of that drought [1918] the rail came through to Coonabarabran. The Castlereagh River was a series of waterholes and the railway gang dynamited every hole as they progressed. . . . There have been no cod caught in that part of the river since" (1981, 344).

Murray cod, the top fish predator in the Murray-Darling system, quickly became the fish of choice for both commercial and recreational fishermen. The cod retained this status into the twenty-first century. High regard, however, did not guarantee protection. Commercial fisheries were established by the 1840s, but by the late nineteenth century precipitous declines in fish catches hampered the industry. Only four native fish species now support commercial fisheries. When the South Australian government banned commercial fishing for cod in the Murray River in 2002, the decision touched off a storm of protest and was overturned the next year. The brief ban reflected the precarious state of a species that many Australians consider a symbol of restoring the health of the nation's premier river system. The difficulty in such restoration arises from the multiple causes of decline in cod and in the river as well. These include habitat degradation, pollution, reduced flooding, barriers to migration, introduced exotic fish species, and overfishing. Detailed scientific studies of Murray cod and other riverine fish were not undertaken until river alteration was well under way in the basin, and scientists continue to debate the relative importance of the various types of changes in the rivers.

One of the most important forms of habitat degradation to affect Murray cod was removing snags along the rivers to improve navigation and flood conveyance. The cod strongly prefer snags, particularly when the wood forms logjams, which they use as shelter from fast water. The big fish, which can reach sixty centimeters at five years and continue growing for decades, do not like big water. Instead, they remain close to the riverbanks and shallow areas, moving out into secondary channels during times of flood and swimming upstream to spawn. Individual cod are territorial, spending much of their time around a specific snag or small pool, where they eat crustaceans and other fish. As happened on the Mississippi, in 1858 Europeans began to remove snags from the rivers of the Murray-Darling system to aid navigation. Snagging continued into the 1980s to reduce overbank roughness and loss of irrigation waters.

Snagging substantially reduced critical habitat for native fish such as the Murray cod, but the most important aspect of habitat degradation along the Murray-Darling system, as on most rivers, was that multiple changes occurred nearly simultaneously. During the 1830s, settlers moving into the basin started clearing native vegetation and importing herds of domestic grazing animals. The era of paddle steamer travel along the river during the 1850s to 1890s was accompanied by snagging and by clearing river frontage for landing sites and fuel wood. Channelization also used structures designed to concentrate flow in a narrow section of the channel so the swifter current would scour and remove naturally occurring sandbars.

The first steamer was launched in 1853. By 1870 hundreds of the ships moved up and down more than five thousand kilometers of river channels during eight to nine months in years of normal flow. Simultaneous development of river transport, communication, and farming created a mutually supportive environment for further development. River navigation improved market accessibility for agriculture and encouraged settlement along river networks, which in turn supported the steamers and associated trading posts. Accessible stands of river red gums were steadily cut to fuel the steamers, some of which burned a ton of wood an hour, and to supply timber mills at Echuca. When trains gradually replaced river navigation, red gums were in demand for railroad ties. One source estimates that twenty billion trees have been removed since European settlement of the Murray-Darling basin. More than 30 percent of the floodplain along the Murray River has been cleared of native vegetation, a lower percentage than on other rivers in the basin.

From the 1850s onward, mining resulted in further clearing of hill slope and floodplain vegetation for fuel and building supplies, as well as dredging and sluicing of sediments in metal-bearing tributaries such as the Kiewa, Ovens, and Mitta Mitta rivers. Settlers who stayed to farm after the ores played out needed timber and pasture. The first European settlers encountered open woodlands with a few large trees and occasional dense belts of smaller trees and shrubs. The grasses began to disappear first, as the cloven hooves of introduced cattle and sheep compacted the soil and the Aborigines ceased regular burning. The trees followed as droughts combined with overgrazing. Bare, powdery soil that no longer supported stock was taken over by government forestry commissions, which so successfully encouraged reforestation by quick-growing pines that what had once been frequent small grass fires became devastating forest fires. Each of these fires was followed by dense regrowth that eventually supported another massive fire.

All these changes altered the amount of water and sediment coming into the river channels. In *A Million Wild Acres,* Eric Rolls notes that sand-filled creeks were common in inland Australia long before cattle and sheep altered soil characteristics. The first explorers described chains of ponds along the watercourses during the dry season. Animals and Aborigines dug for water beneath the channel sands. But early European settlers noticed that the channels dried more quickly each year. By the early 1900s waterholes that had persisted throughout the year along many channels were drying out. Within decades the forest grew over some of these sandy creek beds.

Locking Up the Murray-Darling

As the naturally variable flows became even more uncertain, settlers began to engineer the river flows. Water for irrigation was first diverted at Mildura during the 1860s. Major diversions of water for irrigated agriculture and municipal drinking supplies began during the 1890s. These larger diversions resulted in part from a severe drought across northeastern Victoria during 1880–82. When members of the national government started to investigate irrigation systems in 1883, they found that little had been written on the topic. The government responded by forming a commission chaired by Alfred Deakin to study irrigation systems in California.

Deakin's observations of irrigation practices in five American states led to irrigation and water-supply acts that established district irrigation trusts in much of the Murray valley within Victoria, starting in 1886. One of the lessons Deakin drew from his American trip was that government support of private enterprise was required if the drylands of Australia's interior were to be permanently settled and agricultural productivity was to improve. Deakin ultimately rewrote Australian water law in ways that mimicked the prior appropriations system of the American West, which protects the water rights of users irrespective of proximity to a river channel.

Extensive, dependable irrigation systems require dependable supplies of water, not the boom-or-bust flows characteristic of much of the Murray-Darling basin. The first water-control structure across the main channel of the Murray was built in the 1920s. Fourteen locks and weirs went up along the Murray between 1922 and 1937. (Weirs store water that spills into or out of locks as boats use them.) Meanders were cut off during the 1930s. The first large dam created Hume Reservoir (three billion cubic meters) in 1936. Other major weirs and tributary dams appeared during the 1940s. Irrigation demands increased especially rapidly with the introduction of irrigated cotton farming during the 1960s. Cotton is a water-intensive crop; the spread of irrigated cotton fields has been possible only because of parallel expansions in dam building and water storage.

At present, water flowing down the Murray-Darling basin is interrupted by four major reservoirs, fourteen lock and weir structures, more than five hundred weirs on river main stems and numerous smaller weirs on tributaries, and five coastal barrages. Together these structures earn the Murray-Darling a "strongly impacted" rating on a 2005 evaluation of flow regulation of the world's large rivers. The Kiewa River remains the only Murray tributary that does not have flow-control structures. The net effect of these changes was to transform a river characterized by highly variable flows into a series of permanently stable pools with much less frequent floods and less water moving downstream. More than 140 years of building flow-regulation structures on the Murrumbidgee River, for example, has reduced median outflows at the junction with the Murray to a quarter of natural levels and increased the percentage of dry years from 5 percent to 57 percent. Whatever water does move is much more highly controlled by structures. Flood peaks are reduced and delayed. Floodwaters recede abruptly. Overbank flooding is less frequent and less extensive. Previously temporary wetlands throughout the Murray-Darling basin that were inundated once every two to three years have become either permanent wetlands or largely dry. Connectivity of the main river channels with their floodplains and secondary channels has been substantially reduced, as has longitudinal connectivity. At least ten of the Murray-Darling's fish species migrate at some stage of their life cycle, but migration is now hindered by more than 150 barriers to fish movement within the basin, and only 22 make some provision for fish passage. The least regulated rivers still maintain a diverse native fish fauna, whereas the more highly regulated channels are largely populated by alien species such as carp. As Walker and Thoms summarized the situation in 1993, "It is not melodramatic, therefore, to consider the lower Murray as a river in crisis" (177).

Billabongs

The history and contemporary status of billabongs, which once formed the oases of the Australian inland, illustrate some of the environmental effects of extensive flow regulation and historical changes in land use. Billabongs were the most abundant standing freshwater on the Murray-Darling floodplains before European settlement. Along some reaches of river, the billabongs provided more than 60 percent of the available water surface. Billabongs create refugia for aquatic species during periods of increased aridity. During normal times the billabongs provide critical habitat for animals that need sheltered, still waters or abundant food supplies in nursery areas for development of young. Billabongs contain snag habitat along the shoreline where small fish can hide from predators among fallen logs and feed on the aquatic insects that prefer to live on submerged wood rather than sediment. Dense beds of aquatic plants such as reeds and water lilies provide habitat in which small fish can shelter from predators, and they furnish food for a variety of aquatic organisms. The plants also reduce the velocity of water flowing through the billabong, trap sediment in litter and root mats, and affect both nutrient cycling and the degree of vertical mixing and levels of dissolved oxygen in the water.

Because the physical and chemical environment of billabongs is especially heterogeneous, the biota they support is equally heterogeneous. Temporary billabongs harbor far more species and individuals of invertebrates than either slow-flowing secondary river channels or the unvegetated margins of the main channel. Tens of thousands of species, from bacteria and algae through insects and vertebrates, use the billabongs at some point in their lives. For example, scientists have recorded more than a hundred bird species around the billabongs on the upper Murray River floodplain at Wodonga.

Kangaroos and wallabies grazed floodplains and billabong margins before European settlement. When hoofed cattle and sheep replaced these native soft-footed mammals, floodplain soils were compacted, reducing infiltration and increasing runoff. Increased runoff, along with other land use changes, promoted gully erosion. Water and sediment flowing through the swiftly eroding gullies carried excess fertilizers and pesticides that accumulated in billabongs. Early European settlers did not recognize the ecological importance of billabongs, and many were filled or drained to increase the supply of arable land. Others became garbage or waste dumps or sheep dips. Although there are no accurate estimates of original or remaining extent of billabongs across the drainage, losses of wetlands in general range from 30 percent in Victoria to 50 percent in more degraded portions of the Murray-Darling basin. Scientists now propose that to maintain a high level of ecological diversity on floodplains, management practices must maintain the historical diversity of water flows into billabongs so that some are permanently flooded whereas others flood only seasonally or irregularly.

Impoverished River Ecosystems

As might be expected, the many changes in river processes eventually led to substantial changes in plant and animal communities. Scientists documented widespread ecological disruption throughout the Murray-Darling basin by the 1980s. Observed changes include poor recruitment in native fish and reduced habitat for waterbirds. The lower, regulated Murray valley demonstrates less vigorous and less frequent regeneration in key floodplain plant species than occurs in upstream reaches. Floodplain trees have died back along areas with reduced overbank flows. The diversity of aquatic food webs and vegetation is limited in the extremely stable environment of weir pools.

The status of gastropods and platypus further illustrates the ecological losses of the Murray-Darling basin. The basin fauna includes numerous small gastropods, among them snails, clams, and mussels. More than 430 species are known, of which about 99 percent are endemic and nearly half remain undescribed. Many of these species are confined to small areas such as springs. The native gastropods feed on periphyton, the "scum" that coats sand and gravel on riverbeds and riverbanks. Periphyton used to be composed primarily of microbes and algae that produced good snail food. Filamentous green algae, which provide an inferior snail food source, have become much more common as a result of stabilized water levels. Consequently, gastropods abundant in the Murray-Darling basin during the first half of the twentieth century have declined markedly since the rapid expansions of water storage and diversions at midcentury. Four taxa are officially recognized as endangered at present, although several others qualify. Two species once widespread in the basin (*Notopala sublineata* and *Notopala suprafasciata*) are on the verge of extinction. Only freshwater limpets remain common, although even these are distributed in patches. Some mollusks, however, find refuge in irrigation pipelines fed from Murray-Darling channels and become numerous enough to block pumps and irrigation equipment.

The distribution of the platypus (*Ornithorhynchus anatinus*), one of Australia's wonderfully unique animals, has shrunk in the lower reaches of the Murray and the Murrumbidgee. The platypus finds the bottom-dwelling river insects it feeds on by sensing their electric fields. This ability is hampered by saline water, and salinity in the rivers of the Murray-Darling is increasing as a result of change in groundwater levels. Several introduced fish species may also compete with the platypus for food. Persistent organochlorine residues in the fatty tissues of platypuses indicate other challenges to the animals' survival.

Introduced species have further stressed the Murray-Darling ecosystem. Domestic cattle and sheep grazing the floodplains have largely eliminated successful recruitment of any floodplain plants they find edible, destroying habitat for native birds, small mammals, and reptiles. One-third of the more than seven hundred plant species on the Murray River floodplain are exotics. As Australian ecologist Tim Low noted in *Feral Future*, the Murray-Darling basin is particularly prone to invasive pests because of its vast size and degraded condition. Several introduced aquatics, including the pervasive *Eichhornia* and *Elodea*, are now weeds that thrive in the sheltered waters of billabongs.

Introduced carp (*Cyprinus carpio*), European rabbits (*Oryctolagus cuniculus*), and willows (*Salix* spp.) have substantially altered floodplain and river environments. Carp native to Asia were introduced to Australia as early as the 1860s. With an annual breeding cycle and few predators, carp have spread throughout the Murray-Darling river network. The carp's suction method of bottom feeding stirs up sediments and uproots aquatic plants, increasing turbidity and the mobility of bed sediments. Carp also compete with native fish for habitat and food. Equally prolific rabbits have also spread throughout the drainage basin. Their food requirements create a major threat to any regenerating native plant species, particularly eucalyptus seedlings. The rabbits can also ringbark and kill mature trees. Their burrowing degrades highly erodible soils. Although it may be difficult for people from other continents to envision rabbits as a plague, during parts of the twentieth century rabbit populations in Australia exploded to produce tens of millions of animals.

Starting in the 1830s willows were planted to distinguish the main channel from secondary channels for navigation and to stabilize the banks of irrigation canals across the floodplain. The willows have spread rapidly on their own, particularly upstream from weirs with stable water levels. Hundred kilometer portions of riverbank along the lower Murray are now entirely covered in willows, which exclude native vegetation and provide poor habitat for insects and birds as well as limited wood in the channel for native fish. The foliage and roots of willows growing densely along streambanks reduce water velocity, increasing sedimentation on the banks, and the willows shade the river margins. This shade decreases primary productivity associated with microscopic photosynthesizers and thus reduces summer food supplies for other river organisms. Willows also support a less diverse fauna of aquatic invertebrates than do native plants such as river red gums. Decaying willow leaves have lower density and diversity of bacteria and diatoms than decaying red gum leaves. Freshwater shrimp common in the Murray-Darling basin prefer the microorganisms that colonize the red gum leaves to those that colonize the willow leaves. Clearly not all riverside plants are equal in ecological function and importance in a specific environment.

Rabbits, livestock, other human-induced removal of plants, and exotic plants have caused more than two-thirds of the Murray-Darling's vegetation to become severely to moderately degraded since European settlement. Changes in floodplain vegetation are especially important for native animal species because, as in other river environments around the world, floodplains in the Murray-Darling basin are disproportionately rich in animal species. Amphibians, reptiles, birds, and mammals from the small echidna (*Tachyglossus aculeatus*) to the big western gray kangaroos (*Macropus fuliginosus*) and red kangaroos (*Macropus rufus*) depend on the floodplain forests and wetlands.

Salinization and Algal Blooms

Given these extensive and substantial changes in their greatest national river system, Australians are now trying to develop methods to restore at least some measure of river function and revive the robustness of native river ecosystems. Salinization first caused broad public awareness of the ills of the Murray-Darling and remains one of the most visible and highly discussed unforeseen consequences of twentieth-century land use patterns in the basin. Natural processes in the Murray-Darling basin create high salt content, although land use practices have been exacerbating river salinity for decades. Only 4 percent of the annual precipitation delivered to the basin makes it out as river discharge. The rest evaporates from soils or water bodies or transpires from crops or wild plants. Evaporating water leaves behind any dissolved chemicals, but excess salt does not result simply from large amounts of evaporation. The additional salt flux out of the Murray basin appears to derive from saline groundwater that has been accumulating for millions of years. This salinity is exacerbated by mobilizing of salt stored in the soil as irrigation water percolates into the soil and then back into the river. This latter addition is significant; about half of the chloride the Murray acquires as it flows through the middle third of its basin during summer months comes from return flows of irrigation drainage.

Another factor in salinization of the surface water is the clearing of native vegetation and its replacement with shallow-rooted plants for grazing and crops. Native eucalyptus trees can send their roots down more than eighteen meters. Eucalyptus species generally permit less than 1 percent of annual precipitation to pass the root zone. These plants return 99 percent of precipitation to the atmosphere through transpiration and thus severely limit groundwater recharge. Much of the Murray-Darling basin now experiencing salinity problems was historically covered with mallee woodland. Clearing for dryland cropping has removed about 70 percent of the original mallee cover. The imported shallow-rooted plants are not nearly as thorough in returning water to the atmosphere. Groundwater recharge in some areas of the basin has consequently increased by several orders of magnitude, causing groundwater levels to rise. Rising groundwater mobilizes highly saline moisture in the soil above the former groundwater level, leading to greater discharge of saline groundwater into surface waters.

The native fish of the Murray-Darling basin fortunately are extremely tolerant of high salinity. Ecologists have found that the responses of other organisms are more complex. Increased salinity is associated with lower diversity but higher abundance of several forms of invertebrates that form the base of aquatic food webs. Humans, and many crops, are not especially tolerant of salinity.

Any actions taken to reduce salinity in the Murray-Darling surface waters may take a long time to have an effect. Lags of thirty to two hundred years occurred between initial clearing of native vegetation and an increase in groundwater recharge. Hydrologists predict that similar time lags will occur between revegetation with native plants and a reduction in recharge. Australian scientists now predict that the volume of salt mobilized into the rivers of the Murray-Darling system will double in the next hundred years. About 60 percent of this salt will be stored again on floodplains, in wetlands, and in irrigated areas, at the worst producing vast salt-encrusted flats unable to support any but the most salt-tolerant vegetation.

Despite the growing public awareness of salinity problems, water quality issues in the Murray-Darling drainage came to a head only when people were scared by the largest bloom of blue-green algae ever recorded. The algal bloom affected a thousand kilometers of the Darling River in 1991 and threatened several rural water supplies. High concentrations of blue-green algae are dangerous because they can produce neurotoxins and hepatotoxins as well as causing symptoms in humans ranging from skin and eye irritations through hay fever to acute gastroenteritis. Losses of sixteen hundred sheep and cattle were attributed to the 1991 blooms.

Sturt noted in 1830 that water in the Darling River tasted of vegetable decay and had a slight greenish tinge. Algal blooms have been recorded in the Murray-Darling basin since algal monitoring began in 1947 and were becoming increasingly common before the record-setting 1991 bloom. Although the species of algae involved in blooms are naturally occurring members of the freshwater community in Murray-Darling rivers, changing river conditions allow them to become dominant and create the population explosions known as blooms.

Blooms tend to occur during summer, when low, turbid flow creates a warm surface layer in which the algae can grow rapidly. Increased phosphorus in the water, largely from runoff of agricultural fertilizers, enhances the blooms, as do the pools of nearly stagnant water created by control structures in the rivers. Manure from sheep and cattle can also contribute phosphorus. Discharge from billabongs used as stabilization ponds in sewage treatment, or as recipients of organic pollution from abattoirs or paper mills, contributes more phosphorus. Even waterbirds, increasingly concentrated in declining numbers of wetlands, can add nutrient pollution. Water quality analyses during the 1991 Darling River algal bloom, however, indicated that the tributary Namoi River, one of the largest cotton-growing areas in Australia, contributed 66 percent of the total phosphorus load to the Darling River system.

Irrigated cotton fields also increase contaminant loads of adjacent rivers, because cotton growers use up to seventeen chemicals, including insecticides, herbicides, and fertilizers. DDT and its metabolites are still detectable in river sediments, even though these chemicals have not been used since 1981. Pesticide monitoring in the central and northwestern Darling basin indicates that less than 30 percent of water samples taken at sites downstream from cotton farms meet guideline concentrations. The maximum concentrations recorded at these sites exceed the acute toxicity thresholds for some native Australian fish species, and fish kills have occurred near cotton-growing areas along major tributaries of the upper Darling River.

Strategies proposed to minimize algal blooms in the Murray-Darling basin include setting a minimum discharge ensuring that the river maintains sufficient flow to reduce pools of warm surface water rich in phosphorus. Pulsing the discharge to alter velocity would also reduce warm pools. Altering the depth of water withdrawal from weir pools would permit drawing the water supply from the lower levels that have less blue-green algae. Another method of reducing algae concentrations in drinking water involves extracting groundwater from wells adjacent to the river, essentially using the riverbank to filter the water, as is done along the Danube. But this might not work along portions of the Murray-Darling where soil salinization causes water filtered through shallow wells to also become saline.

Algal blooms increased awareness of a broad range of other contaminants in Australian drinking water. Australia had no mandatory drinking water standards as late as 2001, and many water systems did not meet basic water quality criteria. Many communities still do not have regular monitoring or testing as required by government-authorized drinking water guidelines. Many writers quote the description of the Murray River as Australia's longest sewer.

In his 1849 book, Charles Sturt wrote, "No doubt but that future generations will see that fine sheet of water [the Murray] confined to a comparatively narrow bed, and pursuing its course through a rich and extensive plain." The area covered by irrigated agriculture in Australia as a whole increased by 26 percent between 1983–84 and 1996–97, but the use of water for irrigation increased by 76 percent as water-intensive crops such as cotton and rice became more widespread. Half of Australia's rice crop grows in the Murrumbidgee basin. Australia currently has the highest per capita storage of water in the world at forty-six hundred cubic meters per person, as well as the economic base and lifestyles that go with such an enormous capacity to store water. Now that Sturt's vision has come to pass, contemporary Australians struggle to undo the negative effects that have come from creating an agriculturally rich and extensive plain.

Clawing Back the Water

A primary challenge to any comprehensive form of water management in the Murray-Darling basin is integration across political boundaries. The Murray-Darling crosses parts of the four Australian states of Queensland, New South Wales, Victoria, and South Australia as well as the Australian Capital Territory. Water is divided among the states by parliamentary agreement, but as late as 2001 there was no legislative provision for environmental flow needs. Environmental conditions in the Murray-Darling basin have continued to deteriorate even though water diversion from rivers in the basin was capped in 1995. A study in 2001 suggested that a 40 percent reduction in the existing cap would be needed to restore critical flow volumes and floods of ecologically relevant size and frequency. It is clear that some of the natural seasonality and year-to-year variability in flows along the rivers must be restored if ecosystem function is to be maintained. This is easier said than done, however, in a basin in which all water supplies are already heavily allocated.

Proposed restoration measures include increasing overbank flows and manipulating water levels to benefit waterbird breeding. Bottomland habitat can be restored by controlling livestock grazing in riverine areas and protecting the natural regeneration of riverside species, as well as by artificial planting of these species. Creating additional temporary wet-

8.4 The Murray River near Echuca, just downstream from the Barmah Forest.

land habitat would further help the Murray-Darling ecosystem. Changes in the timing of water releases from reservoirs could maximize environmental and water quality benefits, although this new timing may not coincide with irrigation or municipal needs. Each of these potential solutions requires changing water management schemes that agricultural and municipal water consumers have come to depend on.

Various partial solutions to environmental deterioration in the Murray-Darling basin will become publicly acceptable only if people perceive their benefits as equaling or exceeding the benefits from existing water consumption. In other words, Australians must weigh the relative benefits of longer-term sustainability against the short-term economic status quo. As part of this balancing act, one pilot study attempted to assign an economic value to the 29,500 hectares of the Barmah Forest in Victoria, the largest remaining river red gum wetland forest on the Murray River floodplain.

High flows historically inundated the Barmah floodplain after early spring rains and snowmelt in the highlands. Now the high flows are smaller and less frequent. What high flows remain come during summer and early autumn in response to downstream irrigation demands: the river and floodplain are essentially used as a giant canal to get water downstream to farmers. To improve the efficiency of this downstream conveyance, many of the secondary, ephemeral channels that create a dense network among the floodplain wetlands have been blocked. These smaller channels used to distribute high flows across the floodplain and then drain the forest as flows declined, leaving behind a mosaic of low-lying wetlands that would retain water for longer or shorter periods. The changes in flow through the forest are now changing the distribution and composition of plant and animal communities. Plant communities have also changed in response to grazing and timber harvest. Summer grazing of sheep on the grass plains present throughout the Barmah Forest began in 1841. Cattle replaced sheep by 1885, and both cattle and horses continue to be grazed in the forest. Rabbits were present by the 1880s. River red gum has been extensively cut within the forest since 1870. Development upstream, as well as the changes in flow, has also increased the input of sediment to the Barmah Forest, particularly over the past forty years. Heavy metals move downstream attached to silt and clay, and metal loads to the forest have increased by up to 300 percent since the 1950s.

The Barmah Forest remains a biologically rich environment despite these many alterations. The forest is used by at least 206 species of birds, 25 native mammal species, 27 reptile species, and 10 species of amphibians. There are 54 waterbird species that breed in colonies within the forest. Ornithologists consistently estimated breeding populations in the millions before the commissioning of Hume Reservoir, which significantly reduced flows into the forest floodplain. Flow regulation and other changes in land use have caused an 80 percent reduction in the frequency of successful breeding in most waterbirds compared with the natural precedent. Scientists identified the interval between breeding episodes during extended periods of drought as the most critical factor likely to affect the long-term survival of these species. As a result, the first environmental water allocation agreed to within the Murray-Darling basin was established for the Barmah Forest in 1993.

The first so-called environmental flow of one hundred million cubic meters occurred in 1998, but it produced negligible ecological benefit. Modeling studies now suggest that accumulating this environmental flow provision for several years to allow a really large flood less frequently would produce better ecological results than releasing one hundred million cubic meters each year. This model was validated during 2000–2001, when a large natural flood, combined with the largest environmental flow yet released in Australia (341 million cubic meters released in three pulses of water) enabled at least fifteen thousand pairs of twenty or more waterbird species to breed successfully in the forest. One endangered egret species bred for the first time since 1975.

The pilot economic study concluded that the value of the Barmah Forest, with all its conservation, recreation, and landscape attributes, lay between $A76.5 million and $A97.5 million (1998 values). Although it is extremely difficult to accurately place a monetary value on clean water and all the benefits that accrue from it, this study indicates that a more comprehensive valuation of rivers as ecosystems, rather than simply as water supply systems, might change current priorities of water routing in the Murray-Darling.

Shifting public attitudes toward recognizing the importance of healthy river ecosystems is difficult at least in part because of Australia's long-standing folk perception that hard work and determination can overcome climatic challenges. This attitude emphasizes using river engineering to overcome drought and limited river flows rather than reducing water consumption or changing land use patterns to reflect dry conditions. The history of land use in Australia, however, is more complex than implied by a story of resolute agriculturalists taming a recalcitrant landscape. Australia has the highest year-to-year variability in precipitation of any continent. When people of European descent set out to overcome this dryness and unpredictability, they essentially applied brute force rather than attempting to understand realities of climate and soils and to work within them. One history of irrigation in the Murrumbidgee valley succinctly noted, "Because of the great magnitude of the scheme, and the relative inflexibility of the plans, it was only through the bitter experience of hundreds of farmers and the expenditure of very large grants of state money that some measure of stability was achieved" (Langford-Smith and Rutherford 1966, 24). Stability for the irrigators has clearly come at the expense of the rivers as ecosystems.

As scientists, water users, and government agencies grow increasingly aware of and concerned about salinization, blue-green algal blooms, and deterioration of the river ecosystem, people are gradually getting behind measures such as the Living Murray initiative. This initiative aims to restore ecosystem function in the river basin by setting guaranteed minimum flow levels, known as "environmental flows." Three targets for yearly average flows are currently being evaluated, and restoration projects are being started throughout the basin. Restoration projects already under way make it clear that it is far cheaper to prevent river degradation than it is to restore degraded rivers. The Snowy Mountains Hydroelectric Scheme cost $A800 million to complete, for example, and the Snowy River Restoration Project will cost at least $A643 million.

At the heart of any initiative to restore the rivers of the Murray-Darling basin lie the questions, How much water does a river need? and How can this water be clawed back from other users? (to quote the authors of a 2003 paper on flow restoration and protection in Australian rivers; Arthington and Pusey 2003, 377). The answer to how much water a river needs depends on the specific characteristics of the river and on how completely ecological function will be restored. Some studies in Australia suggest that 80 to 90 percent of natural flows may be needed to maintain a low risk of environmental degradation, yet restoration of slightly over half of natural flows may also maintain many ecosystem processes and communities at lower levels. The caveat to restoring lower percentages of natural flows is that the long-term sustainability of such minimal maintenance remains unknown.

Answers to how water can be "clawed back" from existing consumptive users include water conservation. Cities consume 8 to 10 percent of Australia's stored water, with up to 80 percent of this water used on lawns. Agriculture uses another 79 percent of the stored water, and three-quarters of this goes to the one-quarter of farms that grow water-sucking crops such as cotton. Imposing water pricing that reflects the true value of water by including all the environmental benefits from water in rivers would render much of this wasteful consumption of water economically unpalatable to current users. This issue of the true value of water is not unique to the Murray-Darling and is a problem that many countries now grapple with.

The End of the Murray

The Murray gradually loses its identity as a distinct river in its approach to Encounter Bay. First the river water diffuses into Lake Alexandrina and the adjacent, smaller Lake Albert, then it spreads into the saline Coorong Lagoon that stretches eighty kilometers along the coast. The coastline here has gradually subsided during the past two million years. As the coast has sunk, global sea level has alternately fallen and risen in association with the great continental-scale glacial advances and retreats at higher latitudes. Superimposed on the complex interplay of sinking coastline and fluctuating sea level are the dunes formed as winds and waves rework the sand supplied by longshore transport. The shape and location of the Murray's mouth are constantly changing in this dynamic environment. Shifts of fourteen meters have been recorded in a twelve-hour period, and historical records indicate that the river mouth has migrated more than a kilometer since the 1830s.

In the early days of European settlement, the 750 square kilometers of the Murray estuary were a mixture of brackish water and freshwater that shifted with changing river flows and tides. Aboriginal people lived along the lower Murray and the lakes, where extensive reed swamps supported abundant fish, birds, and mussels. Sandy heath and mallee scrub alternated with patches of pine forest and stands of gum trees on higher, drier ground.

The region began to change when smallpox arrived from New South Wales about 1814, decimating Aboriginal populations. The founding of the colony of South Australia in 1836 accelerated the influx of European settlers. Within a few decades the newcomers had built locks on the river and drained the lands surrounding Lakes Alexandrina and Albert. Construction of coastal barrages prevented incursion of marine water and changed the lakes to freshwater bodies with permanently elevated water levels. The barrages, along with upstream water use, have now reduced the freshwater discharge from the Murray River to the sea by 75 percent. The mosaic of swamps and forest gave way to a pastoral landscape of dairy farms and small towns linked by roads and railroads.

Much of the Murray-Darling basin is the classic Australian outback: dry, wide open, sparsely populated. This is the landscape of the country's traditions and dreams that A. B. "Banjo" Paterson immortalized in his poem about the jolly swagman camped by the billabong. The swagman, like the American cowboy, was an itinerant laborer. A swagman carried his bedroll, or swag, across his shoulders—waltzing his matilda—as he walked from job to job among the dispersed outback stations. The Australian slang phrase waltzing your matilda implies grace, freedom, and rhythm of movement. All of these the rivers of the Murray-Darling system have largely lost. Like the Nile, the Murray-Darling basin is sucked nearly dry by diversions and water consumption of people living in this dry landscape. And like the Danube, the Mississippi, and the Ganges, the wide-spreading flows that once nourished extensive bottomland ecosystems are now largely stored in reservoirs and parceled out along channelized rivers. But like the countries of the Danube and the Mississippi, Australia also has the economic ability and the public awareness to restore the Murray-Darling. As contemporary Australians reevaluate the distribution of flowing waters in their arid landscapes, perhaps they can return some of the rhythm and grace to their great national river.

Interlude

The Murray water droplet reaches the Pacific Ocean off south-central Australia late in September. Local currents carry the droplet slightly north and then west along the margins of the Great Australian Bight, the huge gentle curve in the southern margin of the continent.

The bight is a region of great beauty. In places along the coast, cliffs sixty meters tall rise abruptly from the blue water. As the cliffs retreat unevenly, they leave behind dramatic sea stacks, arches, and caves. The surf breaks on sandy beaches along the base of the cliffs and on bedrock platforms that lie offshore. Surfers and swimmers share the bight with numerous sharks and several species of whales.

Rounding Cape Leeuwin at the southwestern edge of Australia, the water droplet moves swiftly north as part of the Capes Current. The southward-flowing Leeuwin Current weakens during this time of year, when winds tend to blow northward. The Murray water droplet moves north for nearly a year. In July it is picked up by the South Equatorial Current and carried westward south of the equator for another seven months or so. By mid-February the droplet reaches a portion of the Indian Ocean, from which it is evaporated and carried farther westward and north in the turbulence of the Intertropical Convergence Zone. The convergence zone is gradually migrating northward at this time of year, and the Murray water droplet moves along with it before falling as late winter rain over the northern portion of Africa's Congo River basin.

NINE

The Congo: River That Swallows All Rivers

This river begins in obscurity. First there is the question of what to call it. Most people in the world recognize the river by the name Congo. To Europeans this name conjures images from Joseph Conrad's *Heart of Darkness*. These are images of impenetrable jungle, enervating heat and disease, and cannibalism. To Africans the name conjures equally dark images of slavery and colonial exploitation. Some people prefer to call the river Zaire, a Portuguese corruption of the African word *nzadi* or *nzere*, meaning the river that swallows all rivers.

Across an expanse of more than 3.8 million square kilometers of central Africa, an area larger than India, the Congo swallows dozens of tributary rivers that anywhere else in the world would count as major drainages in their own right. At the southeastern end of the Congo's drainage basin, among the highlands bordering the great East African Rift Zone, smaller rivers come together to form the Congo headwaters. Geographers still disagree about where the great river really begins. Some place the headwaters south of Lake Tanganyika in Zambia. This river flowing from the highlands between Zambia and Tanzania begins as the Chambezi but is reborn each time it enters and leaves a lake. The Chambezi flows southwestward for a few hundred kilometers before it enters Lake Bangweolo, reemerging from the marshes at the lake's southern shore as the Luapula River. Turning northward, the Luapula flows more than five hundred kilometers before it enters Lake

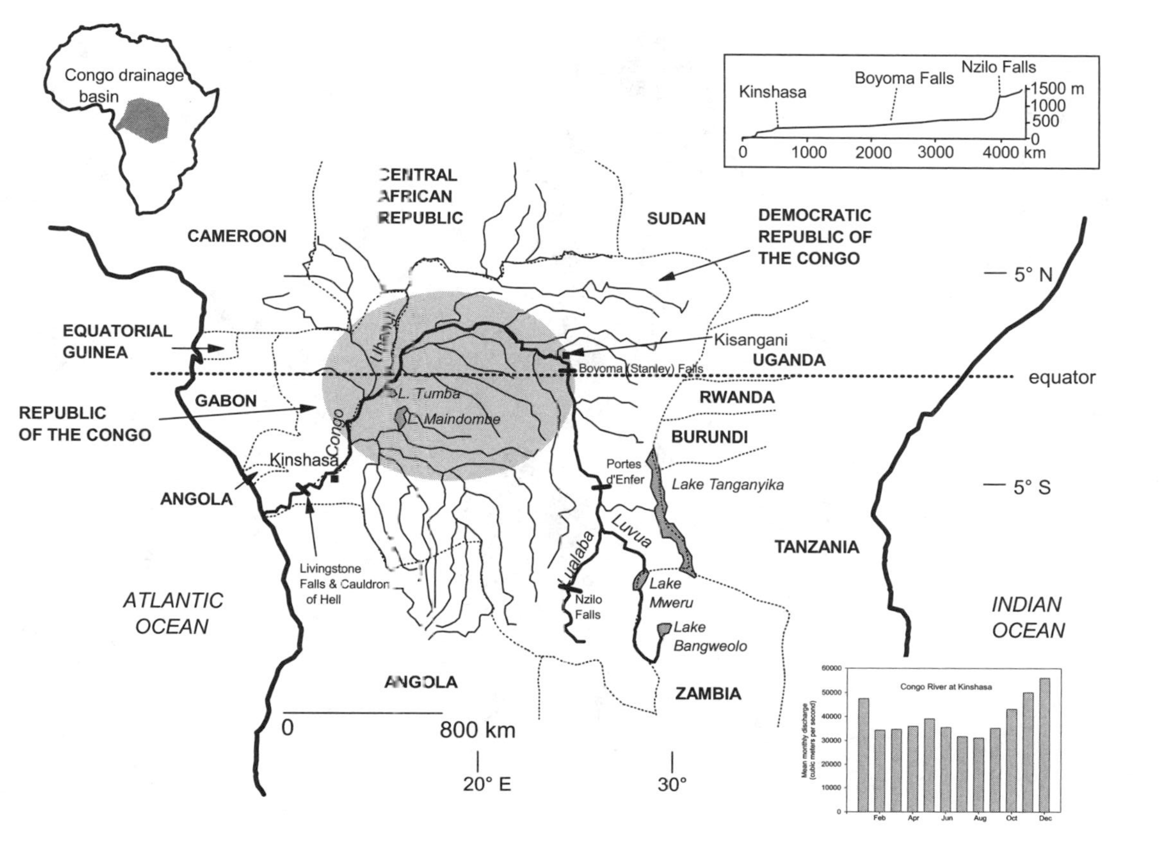

9.1 Location map of the Congo drainage basin, indicating places mentioned in the text. Country borders are indicated by dashed lines, main channel by a heavier line. The central gray oval indicates the approximate extent of the cuvette, the central lowlands and rain forest of the Congo. The inset at upper right shows the longitudinal profile of the Lualaba-Congo system; the inset at lower right shows mean monthly flows at Kinshasa.

Mweru, from which it emerges as the Luvua River. The Luvua continues northwestward for another couple of hundred kilometers before it joins the Lualaba River.

Other geographers assign the Congo's source to the Lualaba itself. The Lualaba begins only about four hundred kilometers upstream from the confluence with the Luvua, but at this confluence the Lualaba has twice the flow of the Luvua. The combined Lualaba and Luvua flow northward toward the equator, acquiring water from rivers draining the western flanks of the Rift Zone.

Two of the world's greatest rivers have their sources here in the highlands bordering the Rift Zone. Where the African and Arabian tectonic plates pull apart from one another, mountains and deep, steep-sided valleys complexly deform Earth's surface. Only a few tens of kilometers

9.2 View of the Oso River, a right-bank tributary to the Congo-Lualaba River that drains the eastern slope regions of the western Central African Rift. (Photograph courtesy of Jürgen Runge.)

of separation determine whether rain falling here drains south or west to form the Congo or into the Rift Zone to form the headwaters of the Nile. The surprising proximity of the Congo and Nile headwaters helps to explain why many of the first Europeans exploring this region remained confused about which river's headwaters they were on.

The Lualaba flows from the highlands down the hundred kilometers of rapids that form the Boyoma (Stanley) Falls and into the *cuvette*, "basin," the huge central valley that is lower than the highlands rimming it on all sides. The famous rain forest of the Congo, which covers nearly 2.8 million square kilometers, lies within the cuvette. The transition from the savanna highlands on the margins of the river's drainage into the lowland rain forest also marks the boundary between the upper Congo and the middle Congo. Now the Lualaba is known as the Congo. The Congo turns northwest, west, and then southwest within the cuvette, forming a broad arc on the map as it crosses the equator twice and collects the tributary Ubangi River from the north and a series of lesser rivers from the south and east.

From the cuvette the Congo flows west-southwest through limestone gorges cut into the Crystal Mountains. As it leaves the mountains the river plunges down the Livingstone Falls, which comprise at least thirty-two cataracts extending along 350 kilometers. The name Europeans bestowed on the final downstream cataract, the Cauldron of Hell, suggests how much this cataract impeded their navigation of the Congo. The lower Congo flows another 160 kilometers from the Cauldron of Hell before entering the Atlantic Ocean.

On a map, the Congo and its tributaries trace a series of irregularly curved channels like sickles harvesting the water and sediment of central Africa. The Congo takes a circuitous and seemingly inefficient path in its journey from the headwater savannas down through the vast rain forest and out to the Atlantic. But the path of least resistance between two points may not be a straight line, particularly where a river crosses many types of rocks and topography. The cuvette reflects underlying forces in Earth's mantle, and the odd route of the Congo River reflects stream capture during relatively recent uplift along the western flank of the Congo basin.

Continents like North and South America are shaped by their proximity to the boundaries of tectonic plates. Uplift and volcanism create huge mountain ranges where the Pacific Plate plunges into Earth's interior beneath the western edge of the Americas. Africa, in contrast, has seen little active uplift related to plate margins since the assembly of Gondwana, the southern supercontinent, some 600 million years ago.

Since that time, the African Plate has mostly been bounded by rifts where plates pull apart from one another. The tensions within Gondwana resulted in the opening of the South Atlantic approximately 150 million years ago and the separation of Africa and South America about 100 million years ago.

Geologists hypothesize that the sinking of dense, subducting oceanic plates at plate convergent boundaries is the dominant mechanism driving plate movement. If this is true, then nothing drives Africa. When a plate moves, it tends to smooth out convective motion in the underlying mantle. If a plate is stationary, underlying convective cells produce crustal swells over hot areas (high South Africa) and depressions over cold areas (low Congo basin). The Congo basin may have formed a low-lying region long enough to support a very large lake, of which Maindombe and Tumba lakes in the central basin are the modern remnants. Geologists hypothesize that a short stream draining to the coast eroded back through the higher ground at the western rim of the basin, capturing the drainage of the lake and creating a through-flowing river from the eastern highlands to the Atlantic. This hypothesis is supported by evidence of a dramatic increase in the amount of terrestrial sediment being deposited just offshore approximately twenty million years ago.

Sediments deposited over eons in a broad basin create a series of rock units that underlie the Congo River catchment and outcrop in roughly concentric rings forming a sort of bull's-eye centered on the young rocks of the cuvette. Underlying the sedimentary rocks and appearing along the edges of the Congo catchment are crystalline rocks more than 570 million years old that represent the oldest core of the African continent. At the three points where the Congo crosses particularly hard rock, the river's downward path is interrupted by steep rapids and cataracts. The upstream set of cataracts begins at the Portes d'Enfer, or Gates of Hell, and from here the river descends steeply for 125 kilometers. Resistant outcrops of slate and granite form the Boyoma Falls farther downstream, and the hard limestone of the Crystal Mountains forms the Livingstone Falls, the lowest set of cataracts.

The cataracts were one of the primary barriers to initial European exploration and colonization of the Congo drainage basin. The Portuguese were the first Europeans to explore the area. Diogo Cão discovered the river's mouth in 1482, and three years later he sailed a caravel upstream to the Cauldron of Hell. Africans living along the lower river called the river the Zaire, but the Portuguese named it after the Kongo tribe living in the estuary. The Cauldron of Hell formed the upstream limit of European travel for nearly four centuries.

The masses of water moving through the Cauldron of Hell helped limit upstream travel. The sheer volume of water pouring out of the Congo basin, an average 1.25 million cubic meters per year, is second only to the flow from the Amazon. Although the 3.8 million square kilometer Congo drainage is only 472,000 square kilometers larger than that of the Nile River, with which it shares a drainage divide, the Congo's discharge is fourteen times greater thanks to the abundant rains over the Congo's equatorial catchment. This moisture nourishes the rain forest that has also been a major barrier to human exploration and occupation of the central Congo drainage.

The immense Congo rain forest stretches east to west across eight degrees of latitude centered on the equator. Some 3,500 kilometers of the Congo's length lie within the 2.8 million square kilometers of this forest. This is Joseph Conrad's Heart of Darkness, of which he wrote at the end of the nineteenth century, "Going up that river was like traveling back to the earliest beginnings of the world, when vegetation rioted on the earth and the big trees were kings. An empty stream, a great silence, an impenetrable forest. The air was warm, thick, heavy, sluggish."

To those deeply familiar with the forest and the river, the Congo is a landscape of light and life. Giant mahogany trees emerge from the uneven forest canopy, providing a platform for epiphytes, vines, mosses, and all the other plants competing for light. The constant hum and shrill of insects hints at the abundant life thriving among the network of plants, where, as in the Amazon, a single large tree hosts its own ecosystem of invertebrates, amphibians, and reptiles. Moving down through the tangle of trunks and vines, deeper sounds replace the calls of insects and birds. Widely traveling bands of forest elephants follow their own network of trails, announcing their progress by low-frequency rumbles. Fruit trees seeded from their dung line the elephant trails. Chimps drum on the flaring buttress roots of the big trees and call to one another as they move between the ground and branches overhead. Small deer move more quietly through the forest. Roving bands of ants add to the complicated pathways spreading across, over, under, and through the forest. When the rivers spill out of their banks and across the surrounding lowlands, fish swim above the forest floor. The stream is not empty, the forest is not impenetrable, and the Congo pulses with sound.

Early Human History

Portuguese and other European explorers were not the first to be intimidated and turned aside by the challenges of travel within the central Congo drainage basin. The mighty river and its extensive floodplain forests at least partially excluded humans for millennia. When Bantu-speaking people began their great migration south and southeast from today's frontier between Cameroon and Nigeria during the first century AD, they left a savanna highlands environment in which they grew millet and sorghum grains. For many decades archaeologists assumed that because the hot, wet lowlands of the Congo rain forest could not support these crops, the Bantus kept moving, seeking the drier grassland environments in which they could practice their traditional slash-and-burn agriculture. The Bantus became the dominant people of the southern third of Africa as they spread out across an ever-broader region. Near the mouth of the Zambezi River, they encountered the banana, Asian yam, and taro, food plants of southeastern Asia that grew well in hot, moist environments. Acquiring these plants allowed some of the Bantu peoples to move into the Congo lowlands during the tenth century. By the thirteenth they had effectively absorbed or displaced the indigenous Pygmy people as they settled throughout the Congo rain forest.

Recent archaeological work is modifying this interpretation of human history in the Congo lowlands. A period of drier climate approximately three thousand years ago appears to have created savanna conditions with more open woodlands and grassy areas in parts of the Congo basin. This may have helped Bantu peoples move into and through the Congo lowlands before their primary period of settlement during the tenth century. The Congo River and its tributaries also provided migration corridors for people traveling by boat.

Sometime during the early decades of the fourteenth century, an outcast young Bantu chieftain decided to carve out a territory for himself. He and his followers built a kingdom of more than 555,000 square kilometers and four to five million people through combined military conquest and religious alliance. This was the celebrated Kingdom of Kongo, a militaristic kingdom of subsistence farmers. The Kongo people knew how to forge iron and copper, and they formed an extensive trading network. The Congo drainage basin, which Europeans perceived as trackless wilderness as late as the nineteenth century, had in fact been crisscrossed by thousands of kilometers of river and land trails for many centuries.

The Portuguese began to explore the lands of the Kingdom of Kongo only a century and a half after it was established. African peoples in the interior and along the coast, as well as European merchants, discovered that abundant profits could be made in supplying slave labor for European plantations in the Indies and the Americas. As profits grew, slavers increasingly treated their captives as commodities rather than as people. This steadily more vicious slave trade and the power struggles it engendered tore the Kingdom of Kongo apart by 1665. The export of humans from the Congo region continued into the mid-nineteenth century, and Arab slavers remained active into the early twentieth century.

As the slave trade gradually declined in importance and the ivory and rubber markets increased, various European countries sought to map Africa so they could determine resources and markets to be exploited through colonies, and also as part of a wider program of exploration aimed at enhancing knowledge of geography and natural history. David Livingstone reached the Congo's Chambezi headwaters in 1867, then vanished from contact with the outside world until Henry Morton Stanley found him alive and well in 1871. The details of the Nile and Congo headwaters remained unknown until a brutally determined Stanley explored the length of the Congo during 1876–77 and cleared up once and for all the confusion about the relations among the Nile, Niger, and Congo. King Leopold II of Belgium commissioned Stanley to build a road around the notoriously impassable lower rapids of the Congo, opening up the interior for more thorough exploitation and colonization. In 1885 the European powers divided the Congo basin among the French, the Portuguese, and Leopold's personal African colony, the Congo Free State, which became the Belgian Congo in 1908.

Savanna Headwaters

Headwater streams throughout the Congo's broad drainage rise in highland savannas more than fifteen hundred meters above sea level. At the northern end of the Congo basin these savannas form a narrow transition zone between extensive rain forest to the south and the grasslands of the Sahel and Sudan to the north.

A mosaic of soils and vegetation patterns the northern savannas. Gallery forests grow along the watercourses. Tallgrass savannas and wooded grasslands grow on the drier areas between rivers. This mosaic reflects a complex history of climatic fluctuations and human land use. Fossil evidence of plant distribution suggests that significant changes in regional climate have occurred more than twenty times within the past ten million years. Plant distributions responded to each climate change. Drier climate savanna communities invaded the Congo basin during dry periods, and rain forests grew more extensively during wetter periods. Because individual plants do not pick up their roots and migrate to an area of more favorable climate, the climate shifts that allowed seeds dispersed by wind, water, and animals to germinate in more favorable locations resulted in different associations of plant species through time. The dissected topography of the plateaus forming the northern savannas also introduced variability into the environmental mosaic, as did an increase in the frequency of naturally generated wildfires approximately twenty-five thousand years ago when forest environments grew drier and more open as a result of interspersed grasslands.

Humans further changed the fire regime during the past three thousand years by introducing slash-and-burn agriculture. Traditional low-intensity agricultural burning maintained soil fertility and secondary forest growth, but recent increases in population have shortened the average fallow period from twenty years to five years. Five years is proving too short for forests to regenerate and soils to regain fertility.

Among the gallery forests along the banks of headwater streams grow tall stands of papyrus and sedges that shelter some of the fish inhabiting these clearwater channels. Scientists have identified approximately 690 species of fish within the Congo River basin. Much less research has been done in the Congo, however, than in other large rivers. Fish biologists estimate that the basin may contain a thousand species of fish. This is the richest fish fauna in Africa, and 80 percent of these species are found only in the Congo drainage. The rapids spreading widely across the drainage basin act as a barrier for some fish species, resulting in geographically distinct fish communities from the headwaters to the mouth of the river. Fish diversity is highest in the lower parts of the river basin that have the most diverse habitat. What is perhaps most remarkable in comparison to other great rivers of the world is that none of the fish species in the Congo are at present threatened with extinction.

Fish populations in some of the headwater lakes of the Congo basin have undergone the type of recent dramatic declines that characterize fish populations in headwater lakes of the Nile drainage. In *Harvest for Hope*, Jane Goodall describes witnessing the effect of changed fishing methods in Lake Tanganyika. Fishermen traditionally lured sardine-sized dagaa to small boats at night using a lamp, then scooped them up in a net with a long wooden handle. Goodall saw the abundant nightly catches of dagaa spread out on the lakeshore to dry each morning during the early 1960s. The dried fish were shipped to Kigoma, the nearest town, and throughout the region. Fish catches immediately increased when the Food and Agriculture Organization of the United Nations introduced seine nets. The catches became progressively smaller with time, however, partly because of the unsustainable level of harvest and partly because the small mesh of the seine nets was catching such young fish that none escaped to breed. By the 1990s the nets often returned almost empty, and in 2000 the Tanzanian government banned their use.

Grazing animals are widespread across the highland savannas. Herds of buffalo, antelope, zebras, giraffes, and gazelles attract leopards and other large predators. Many of these animal populations are at risk from increasing human population as well as from the poverty and ongoing civil wars that drive poaching and deforestation even in protected areas. Few animals now remain in the highland Copperbelt, where malachite ore has been mined since the fourteenth century. Mining increased in extent and intensity during the twentieth century when Zaire's dictator Joseph Mobutu oversaw a copper boom that lasted into the 1980s. Strip mining during this period left a blasted landscape devoid of plants and animals, from which chemicals leaching into the rivers spread contamination downstream for tens of kilometers.

Mining in the Congo

The Congo drainage is blessed and cursed with economically valuable mineral deposits. Producing mines are scattered along the eastern and southern margins of the basin, primarily in the highlands. Many of these are part of the central African Copperbelt that lies along the eastern margin of the Congo drainage. The economic mineral deposits here were initially developed during the 1920s. The mineral deposits were emplaced hundreds of millions of years ago as a complex mixture of sediments eroded from the adjacent highlands and volcanic and intrusive rocks was squeezed and baked during a new episode of plate movement and mountain building. Movement through the metamorphic zone of superheated fluids rich in dissolved minerals left behind ores of copper and cobalt that now form half the world's cobalt reserves, as well as lesser amounts of lead, zinc, and uranium.

The minerals are a blessing in that they underpin the contemporary economies of the countries where they occur. The mining industry accounts for about 25 percent of the gross domestic product and 75 percent of total export revenues in the Democratic Republic of the Congo, for example. This country, which encompasses most of the Congo River drainage, is approximately one-fourth the size of the United States and has a population of about fifty-two million. The GDP was $710 per capita in 2000, making the country one of the world's poorest, though it is richly endowed with coal, cobalt, columbium-tantalum, copper, diamonds, germanium, gold, manganese, petroleum, tin, uranium, and zinc. International mining companies would be only too happy to exploit these resources, but repeated and prolonged military conflict in the region has made continuing operations difficult.

In most regions of the world, the curse of mining is the toxic pollution that remains for centuries to poison soils and streams as the contaminated materials are gradually disseminated across wider areas by hill slope and river erosion. In the Congo drainage the "collateral damage" from mining is exacerbated by direct destruction of human and animal lives.

Coltan mining is among the forms of Congo mining now creating an international outcry. Coltan is a source of the element tantalum, which forms an essential coating for components of many electronic devices, including cell phones and computers. Coltan mining exemplifies nonindustrial mining. Industrial-scale mining such as copper-silver mining in the southeastern portion of the Congo basin requires massive infrastructure, an extensive transport network, and water and electricity to process the minerals. The Shaba Province of the Democratic Republic of the Congo, the site of most of the country's mining, historically used about half of the nation's generated electrical power. Some of this power was delivered by a transmission line running more than sixteen hundred kilometers from the Inga Dam on the Congo River south of Kinshasa to the city of Kolwezi in Shaba. Industrial-scale mining also leaves behind enormous open-pit mines and thousands of tons of contaminated tailings and other wastes.

Nonindustrial mining might seem more benign in that individual miners or small groups move in with picks and shovels, but when large numbers of miners enter an area they need food and supplies. This creates a widening zone of impacts as the miners and those who come to supply them create a network of roads that helps them penetrate the forest for food and timber. Roads give rise to settlements that give rise to further roads in a process that inexorably reduces outsiders' perception of the forest as a threatening and mysterious place to be avoided. Refugees from Rwanda's civil war; impoverished and desperate people from densely populated eastern Africa; and opportunists following jobs in mining, hunting, and logging all move into the Congo's forest along the newly developed network of roads and trails.

As the demand and the price for coltan rise, illegal miners move into areas such as Kahuzi-Biega National Park in the eastern portion of the Congo drainage or the Okapi Wildlife Reserve in northeastern Congo. The national park is a United Nations World Heritage Site that contains rich deposits of the tarry black mineral that individual miners can prospect on their own using only shovels. When civil war erupted in 1998, the price of coltan rose quickly on global markets, and thousands of miners entered the park. They cut trees for charcoal and built huts from the flagpole-sized bamboo that the park's mountain gorillas love to eat. The miners slaughtered gorillas and other wildlife for their meat and hides as the bushmeat trade increased. Researchers from around the world who work in Africa identify the bushmeat trade as being among the greatest threats to the survival of tropical wildlife. The number of miners rises and falls with the price of coltan, but the damage they cause remains.

Many of the miners are not simply individuals trying to get rich. Outside troops from Rwanda and local rebel bands use the revenue from coltan, gold, diamonds, and illegal timber harvest to finance their operations. A United Nations panel listed coltan and these other commodities as the most prominent reason for continuing the various civil wars in the region as of 2001. The civil wars have extensive effects on the landscapes where they occur. Refugees from Rwanda and Burundi, for example, caused significant deforestation, soil erosion, and wildlife poaching in the eastern Congo beginning in 1994. Indigenous peoples in the Congo region are particularly vulnerable as these wars of the outside world come to their homelands. As of 2008, military operations and the resulting social and environmental destruction continue to increase in the Congo drainage.

Despite the increasing international attention to the destructive side effects of mining in the Congo drainage, no studies have yet been published documenting the effects of mining wastes on stream ecosystems. This presumably reflects the danger of undertaking scientific studies in the region at present. Some sources mention waste from mining operations in passing. The Bakwanga Mining Company, for example, dredges for industrial diamonds and deposits six to seven million tons of fine sediment into streams tributary to the Congo. No one has quantified the effects of these tailings on the receiving streams. In this respect the Congo is unique among the rivers mentioned in this book. Although it can be logistically challenging to work in the Amazon or politically difficult to work in the Ob, the dangers of working in the Congo have severely limited scientific research in the drainage.

The Cuvette

The principal tributaries of the Congo flow steadily downward from the highland savannas and headwater lakes, occasionally cutting through a gorge in a series of rapids, but mostly proceeding as clear, gently sloping streams fed by rainfall. It nearly always rains somewhere in the Congo drainage basin, because the basin extends both north and south of the equator. Close to the equator 200 centimeters of rain falls each year, spread over all twelve months. The abundance of rainfall diminishes progressively to the north and south, dropping to 150 centimeters in parts of the basin. Distinct wet and dry seasons also occur farther from the equator as the Intertropical Convergence Zone follows its massive annual migration from latitude 5° north during December-February to 20° north during June-August. The northern portions of the drainage

9.3 Looking south along the Ubangi River, Central African Republic, during the rainy season, September 2003. (Photograph courtesy of Jürgen Runge.)

receive their heaviest rainfall during the Northern Hemisphere summer and autumn, whereas the southern portions are wettest during the Northern Hemisphere winter and spring.

The Congo flows some 850 kilometers as it descends from the highland savannas to the rain forests of its great central basin. Vegetation grows more abundant along this descent: grasses become thicker, and trees form more closely spaced pockets of woodland. The stream becomes wider, and people living along the channel use dugout canoes shaped from single tree trunks. Fishermen trail nets from the dugouts, paddling with long-handled blades as they stand in the boats. Then the river abruptly funnels into a gorge a kilometer deep and 100 meters wide as it enters the Gates of Hell. For nearly 120 kilometers the river remains impassable to boats and then, growing quieter, it enters the rain forest.

Beyond the series of rapids that form the Gates of Hell at the upstream end of the cuvette, the Congo flows through lowland rain forest for approximately thirty-five hundred kilometers, dropping only three hundred meters in elevation en route. About half of this drop occurs along the hundred kilometers of rapids known as the Boyoma Falls. The churning brown water of the falls forms a dramatic interruption to the middle Congo's otherwise placid flow. Henry Stanley wrote that the river at the farthest downstream cataract does not merely fall but is precipitated downward "into a boiling and tumultuous gulf, wherein are lines of brown waves 6 feet high leaping with terrific bounds, and hurling themselves against each other in dreadful fury."

Stanley also noted the bravery of the native fishermen who ventured into the swift flow near the cataracts. Fishermen from the Wagenya tribe still paddle their diminutive dugouts to the very edge of the turbulent water. There they erect complicated wooden scaffolding. Men and boys scramble about on the intertwined wood as nimbly as spiders, reaching down with pole baskets to catch fish swimming past the falls. Fish living in this portion of the river are specially adapted to fast, turbulent water. Some species have small, slender bodies that allow them to live in the narrow spaces between rocks on the streambed. Other species have suckers or hooklike spines so they can fasten on to rocks and vegetation.

Away from the falls, the Congo meanders in broad bends or breaks up into multiple smaller channels intertwined around large, forested islands. Stanley wrote of the river's gradually increasing in breadth from three thousand to four thousand meters. Along this portion of the river lies the city of Kisangani, the usual upstream terminus of barge traffic.

9.4 Larger ship on the Congo River at Kisangani.

The seventeen hundred kilometers of the Congo between the Boyoma Falls and the Livingstone Falls forms the river's longest navigable stretch. By including tributaries such as the Ubangi, the network of navigable waterways stretches to more than fourteen thousand kilometers. More than a million people at present live within the Congo basin. These people speak more than four hundred different dialects, yet the close linguistic relationships among dialects suggest that centuries of trade along the river have created a language zone that is more cohesive than any other in Bantu Africa. The languages of this zone reflect the importance of water. The central Congo basin is flat and saturated. People spend their lives in boats, fishing and collecting swamp plants, and they choose ground that is not too wet and not too dry to grow their crops. They have highly developed aquatic vocabularies, just as people living in high latitudes have a wide array of words for snow and ice.

Henry Stanley also wrote of the intense rainstorms along the central portion of the Congo River. Storms are brief but powerful, accompanied by winds that can take down even the giant trees of the rain forest. Much of the rain falling on the central basin remains on the surface. The claylike soils that cover large portions of the cuvette limit the ability of water to soak into the ground. The region is so flat that water ponds in even slight depressions rather than draining away.

9.4 (*continued*) Traditional canoes (pirogues) carved out of tropical trees along the Mpoko River near the confluence with the Ubangi River, Central African Republic. (Both photographs courtesy of Jürgen Runge; photographs taken August 1992.)

Water flowing down from the surrounding highlands joins water falling onto the central basin to inundate the vast floodplains adjoining the river channels. Approximately a quarter of the central basin is permanent swampland, and a substantial portion of the rest becomes swamp during high-water periods. Smaller marshes, bogs, and stream channels generously interlace the areas considered to be dry land. These *bai* wetlands support mineral-rich grasses and sedges that attract birds and animals.

The region's generally soggy condition helps to buffer rises and falls in water level along the Congo River during times of flood because the floodwaters move more slowly downstream as they spread across the swamps, lakes, and floodplains. The seasonal distribution of rainfall that keeps water feeding into the main channel from either the northern or the southern portion of the basin throughout the year also buffers the flood levels. The Congo River consequently does not have the large fluctuations between high and low flow that are common on many rivers. Water level in the river may rise three to four meters during the floods of November and December. This is an impressive change on a very large river, but water level in the Amazon River rises more than thirteen meters during the flood season. Water in the main channel and larger tributaries of the Congo remains turbid and milky brown with suspended silt and clay, but in some of the forested rivers the water more closely resembles brown tea because it is rich in organic acids leached from the forest floor.

The turbidity of the main channel and the larger tributaries makes it difficult for fish to see predators, prey, or anything else. In response, many species of Congo fish have evolved strategies other than visual orientation, as have aquatic species in the turbid Amazon and Mississippi. Fish of the mormyrid family spend the day under cover and feed at night, generating their own electric field and using electrolocation just as bats use sonar to echolocate. More than a hundred of the Congo basin's nearly seven hundred named species of fish are mormyrids. Taking the electric strategy to extremes, the electric catfish (*Malapterurus electricus*) can emit a discharge that stuns humans. The catfish is only a sluggish swimmer, but it can eat well by emitting a powerful volley of high-frequency discharges close to a school of prey fish and paralyzing a number of the prey simultaneously.

Distinctive communities of fish occupy the various habitats of main river, marginal waters along shores, flooded forest, swamps and permanent pools, and tributaries that are present in the central basin of the Congo, although most fish use different habitats as they grow larger and as water levels change. Shallow marginal waters along banks and islands, and the areas over these features during high water, contain far more fish than do open waters. Young fish of numerous species like the shallow waters, where dense vegetation supports food such as invertebrates and provides cover from predators.

Most fish species migrate upstream and then away from the main channel into the flooded forest as the water level starts to rise during the annual flood. One major fish group, the silvery "whitefish," migrate long distances between their feeding and breeding sites. Some whitefish use the floodplain for feeding during the flood season; others pass their whole life in the main channel. Most fish species in the Congo basin breed at the start of the September-October floods.

More than eighty thousand square kilometers of forest are permanently or seasonally flooded in the central river basin alone. Forested rivers often have negligible primary productivity from algae, bacteria, and other microorganisms that grow on the streambed or in the water column and form the base of the food web in other rivers. In the absence of these minute creatures, inputs of leaves, twigs, fruits, insects, and other terrestrial nutrient sources are critical. Foods are most abundant and varied during the flood season, which forms the main feeding and breeding period for many fish. Fish eggs hatch within several hours to two days of being laid, and young grow quickly in the flooded zone.

As water level falls after the flood season, adult fish and larger young fish move back to the river channels, where they are restricted to feeding on bottom debris. Among those moving back to the main channel are the "grayfish," which, unlike the whitefish species, do not migrate long distances upstream and downstream for spawning. Some species, as well as young fish, may stay behind in swamps and permanent pools that become isolated as water level declines. Oxygen levels in these isolated water bodies can decrease to the point where the fish are killed, but species with special adaptations to low oxygen levels seek out the waters of seasonal swamps. These "blackfish," which include various species of catfish and lungfish, tend to be dark and have accessory respiratory organs such as lungs.

The Congo and the Amazon are unique in having such extensive flooded forests, but fish in the two rivers are surprisingly different. The Congo has a much higher percentage of endemic species that are unique to this river basin. Most ecologists attribute this high rate of endemism to the more stable hydrological regime of the Congo, where water level varies much less over the year than in the Amazon, and to the more numerous rapids and cataracts that form migration barriers to many fish.

Human alterations of the forest and fish communities are also more evident in the Congo, where population density averages fifteen people per square kilometer, as compared with four people per square kilometer in the Amazon. Higher population density creates wetland loss and deforestation. Estimates by nation range from 40 percent in Cameroon, through 50 percent in the Democratic Republic of the Congo, to as high as 70 percent in Liberia. These wetlands store and recycle nutrients and provide habitat for many species of invertebrates, fish, and birds. The Congo has an estimated deforestation rate of 7 percent (5 percent in the Amazon), and nearly half of the original forest in the Congo has been lost (13 percent in the Amazon). This translates to more than a million square kilometers that have been deforested, an area roughly equivalent to the combined size of the U.S. states of Florida, Alabama, Georgia, and the Carolinas or to a country like Venezuela or Tanzania.

Deforestation indirectly threatens fish by degrading water quality and changing seasonal flow regimes, and it directly harms fish communities by removing a major source of nutrients. It also destroys subtly interrelated processes among fish and terrestrial animals. In Gabon, elephants tramping through the floodplain forest leave large footprints that form temporary pools where little killifish lay their eggs. The eggs remain dormant during drier periods, and the new fish hatch once water levels rise again. Forest elephants also periodically visit the papyrus swamps that grow thickly along forest river valleys. As the elephants wallow in the swamps, they create open pool habitat that is quickly colonized by swamp fish, and local fish productivity increases. Fish density declines again as the pools subsequently fill with dense swamp vegetation, so that the swamp fish are dependent on these periodic disturbances.

The most insidious danger of deforestation is the potential for massive losses of nutrients from the river basin ecosystem. The abundant warmth and moisture of tropical environments promotes rapid chemical breakdown of rocks and sediment, producing thick, deeply weathered soils. Nutrients are rapidly mobilized under these conditions, so that water in tropical soils and streams should have high nutrient concentrations, with nutrients being flushed downstream. Nutrients recycled by fungi, bacteria, and invertebrate animals, however, are taken up by trees and other plants. The plants very effectively trap nutrients and store them within the ecosystem so they can be used by insects, fish, and other organisms eating the plants. When individual trees fall naturally, the stored nutrients are quickly recycled by decomposing microbes and insects. Henry Stanley was one of the first naturalists to note the absence of dead wood or snags in the Congo River, despite the very large and abundant trees all along its banks. But massive rates of tree removal overwhelm the natural recycling process, and the nutrient storage mechanism provided by trees and other plants is lost. The ecosystem becomes less able to support diverse organisms.

As European and Asian loggers bulldoze roads into previously remote and inaccessible portions of the Congo basin, they open these regions to hunting, trading, and commerce. Hunters with modern weapons supply a human population that is at present doubling every twenty-three years. The effects of hunting on many wildlife populations are now apparent. Long-lived animals that are slow to reproduce, such as the great apes, are particularly harmed. Because the difficulty of living in or traveling through the rain forest protected much of the region's wildlife well into the twentieth century, however, diversity and abundance of many animals remain high despite recent logging and mining. Manatees and tortoises, rare elsewhere, are still present in the Congo basin. The waterways support a rich bird fauna of cranes, storks, pelicans, ducks, herons, and parrots. Scientists have identified more than 250 species of birds within the rain forest, including species that migrate seasonally to Europe and elsewhere.

The Lower Congo

Flowing from the cuvette, the Congo enters the Malebo Pool, a thirty kilometer reach of quiet water up to twenty-five kilometers across. Downstream from the Malebo Pool lie the cities of Kinshasa on the left bank and Brazzaville on the right bank. The cities bracket a reach where the Congo distends, like a snake eating an egg, around the huge girth of M'Bamou Island. No bridge connects these cities on opposite banks of the river. If they wish to visit Brazzaville, the million and a half inhabitants of Kinshasa must go by boat.

Boats play a critical role in Kinshasa's history. More than a million tons of cargo and ten million passengers a year now pass through this port, which is either the beginning or the ending point for all the Congo's boat traffic. Kinshasa forms the start or the end of a boat journey because just downstream lie what Stanley named the Livingstone Falls. These falls that formed the single most important obstacle to exploration and exploitation of the Congo drainage basin are still an impassable barrier to boat traffic. Here the Congo roars down thirty-two sets of rapids spaced along approximately 370 kilometers of a narrow gorge cut through the Crystal Mountains.

When the Congo emerges from the Crystal Mountains onto the coastal plain, it flows for another hundred or more fully navigable kilometers into the Atlantic. The seaport of Matadi stands at the base of the rapids, just downstream from the Cauldron of Hell. Huge oceangoing ships sail up the Congo estuary to this point. Villages of the Bakongo people, descendants of the Kingdom of Kongo, line the river from Matadi to the coastal city of Banana.

The Congo does not immediately lose its vigor as it merges with the Atlantic Ocean. River discharge peaks in December, but a plume of river water extends up to 250 kilometers into the Atlantic through February and March. The Congo discharges an annual average of nearly 42,100 cubic meters per second as it enters the ocean, a volume of flow second only to that of the Amazon River. This flow can swell to more than 70,200 cubic meters per second during floods. Sediment carried along with this water settles from suspension as the river flow loses its velocity and mixes with ocean water. This sediment piles up along the continental shelf, the shallow zone immediately offshore, until the pile becomes unstable and collapses in a turbidity current. Repeated turbidity currents over thousands of years have cut a submarine canyon into the ocean floor. This canyon now extends sixteen hundred kilometers out to sea and can reach depths of twelve hundred meters, so that whenever a turbidity current occurs, the sediment is effectively funneled far out onto the seafloor.

The enormous canyon off the Congo's mouth was first discovered in the 1920s, but the magnitude of the effects of turbidity currents was fully understood only once deep-sea imaging with sonar and other techniques became available after World War II. Before that, the dynamics of the ocean floor off the Congo's mouth provided a mysterious hazard for deep-sea cables such as those used for telegraph lines. Cables crossing the Congo's offshore canyon broke approximately once every two years, whenever the river discharged a particularly heavy sediment load and triggered turbidity currents.

Like the great fans off the mouths of the Nile, Mississippi, and Ganges, deep-sea sediments off the mouth of the Congo record fluctuations in the river's discharge of water and sediment through time. Microscopic creatures living near the surface of the ocean off the mouth of major rivers are sensitive to changes in temperature and salinity that reflect changes in the river's flow. As these creatures die, their remains sift down through the water and accumulate along with pollen grains from the river's drainage basin and sediment carried out to sea by the river. The abundance of certain species of microorganisms and plant pollen in cores of these sediment mixtures records a major increase in flow from the Congo River approximately thirteen thousand years ago. Similar increases recorded in the ocean sediments off the mouths of the Niger River and the Nile reflect an increase in monsoon rains over these basins as the Intertropical Convergence Zone migrated to its present range of positions from the more southerly position it occupied during the last ice age.

A Dark Future?

For now the mighty flow of the Congo River remains largely uninterrupted. The more than fourteen thousand kilometers of channel within the Congo drainage contain an estimated one-sixth of the world's hydroelectric potential. Twenty-five dams exist on the upper river and its tributaries, mostly to generate power used by mining companies, but the drainage basin as a whole has a moderately impacted "yellow" rating on the compilation of global impacts to rivers at the start of the twenty-first century. In contrast, African rivers such as the Nile, with a long history of human exploration and use, rate an imperiled red. The Congo's water is also relatively clean, with only localized pollution sources from mining, cities, and agricultural erosion. Schemes are afoot, however, that would seriously alter the Congo's integrity.

A group of five African power companies has formed NamPower to promote a six billion dollar scheme to build the world's largest hydroelectric power plant 225 kilometers downstream from Kinshasa. The proposed Grand Inga plant would be able to produce thirty-nine thousand megawatts of electricity. Project proponents claim it will take only a small dam to generate three times more energy than any other hydropower plant in the world, because electricity will be generated using "run of river" technology in which turbines are anchored in the riverbed. Environmentalists familiar with a long history of river ecosystem disruption are skeptical. They note that even a small dam will likely block some of the Congo's abundant sediment load that helps to nourish coastal areas with nitrogen, phosphorus, and other nutrients, just as the High Aswan Dam disrupted sediment and coastal ecosystems on the Nile and the Djerdap Dam on the Danube altered Black Sea ecosystems. Questions also remain about the effects the proposed Congo dam and its turbines would have on fish populations.

Other proposals for the Congo River involve water export. Water shortages will reach crisis levels in several countries of southern Africa in the next few decades. Anticipating this, some Congolese businessmen propose to pump water from the Congo River into two long-distance pipelines across the African continent. One of these so-called Solomon pipelines would run a thousand kilometers south from the Congo's mouth to Namibia. A second would cross several zones that are currently the sites of ongoing civil wars, en route to Port Sudan and the Middle East. In addition to being prohibitively expensive, the proposed projects face political problems and engineering challenges. A competing proposal by the Southern African Development Community focuses on drawing water from one of the Congo's southern tributaries and piping it across the Angolan highlands into the Kunene River, which forms the border between Namibia and Angola, as well as into rivers that feed the environmentally sensitive Okavango delta. Although these schemes currently sound unrealistic and extravagant, history suggests that even these projects may become feasible if water shortages become severe enough.

Distance and difficulty have protected the ecosystem of the Congo River into the twenty-first century: long distances from any large concentration of human population or industry, and difficulties of human exploration and residence associated with climate, disease, and terrain. But distance and difficulty will no longer provide sufficient protection unless they are fortified with conscious and enforced policies to protect at least some portions of this remarkable river system. The Congo is rich in resources that people around the world crave: potable freshwater; river power that can generate hydroelectricity; timber and minerals; and meat and fish. The primary tropical forests of the Congo basin are second in size only to those of the Amazon basin, covering an area nearly seven times the size of California. This region contains the most diverse grouping of plants and animals in Africa, including rare and endangered species such as the eastern lowland gorilla, the mountain gorilla, the white rhinoceros, and chimpanzees. The Congo's tropical rain forest was largely untouched as late as the last decade of the twentieth century, but by 2003 an area estimated at twice the size of Rhode Island was being cleared every year as commercial timber harvest for the Asian and European markets accelerated. Portions of strangely silent forests, locally called "empty forests," have already been denuded of wildlife by the bushmeat trade.

A 2005 article in *National Geographic* magazine quotes American biologist John Hart as noting that population densities east of the Congo drainage exceed 360 people per square kilometer (Salopek 2005). Population densities drop dramatically in the forests that cover much of the Congo. As people move inexorably into Africa's last frontier, the great challenge for protecting the river ecosystem becomes that of identifying and then preserving the most vital islands of habitat, the string of beads that will keep the Congo River vibrant. Scientists and environmentalists who work in the Congo respond to this challenge in part by internationalizing the region through programs funded by organizations outside Africa.

Even as deforestation and mining accelerate, much of the Congo drainage basin remains so dangerous that few scientists conduct research there. Limited portions of the basin, such as existing national parks, are relatively well known, but no comprehensive inventory or analysis exists that can provide a baseline for assessing ongoing change. Partly in response to this lack of knowledge, American Michael Fay set out during 1999–2000 to survey plant and animal species and landscape condition. Fay's thirty-two hundred kilometer route began in the northwestern Congo drainage and ended at the Atlantic coast north of the Congo's mouth. Like Stanley's nineteenth-century journeys through the Congo, Fay's expedition was a grueling experience, but it resulted in the designation of thousands of hectares of new protected areas (Quammen 2006). The challenge now is to enforce protection in a region plagued by civil war, poverty, and rapid population growth.

The "dark heart" of Africa has a long and cruel history of human violence and ruthless greed. But the landscape itself is anything but dark. It is a natural wonderland that sustains a wealth of plant and animal species adapted to the many environments present within the compass of these waters draining a huge portion of the African continent. The darkness lies within ourselves, but it can be overcome if we accept the right of other organisms and their habitats to exist apart from their ability to satisfy immediate human desires. Perhaps then we will recognize the sun reflecting from the emerald green forests and placid brown waters of the Congo as the heart of light and life.

Interlude

Its journey down the Congo River complete, the water droplet once more returns to the Atlantic Ocean off the Congo's mouth. It is springtime in the Northern Hemisphere. The South Equatorial Current picks up the Congo water droplet and carries it west across the Atlantic and north of the equator, where the droplet again joins the Guiana Current, bringing it to the Caribbean by the following winter. Late in December the Congo droplet is evaporated from the surface of the Atlantic and carried aloft in the Hadley cell. Near the uppermost elevations of the Hadley cell, it spins off and goes west again with the jet stream. The major jet stream in the Northern Hemisphere is found most commonly between latitudes 30° and 70° north, but a lesser jet stream also occurs between 20° and 50° north. It is this lesser jet stream that carries the Congo water droplet north of the Tibetan Plateau.

The Congo water droplet again encounters many airborne contaminants as it passes over Europe, and these levels increase as the droplet approaches China. Asia as a whole contributes about 56 percent of the global mercury emissions in the atmosphere. Most of these contributions come from coal combustion in China, India, and South and North Korea. China is also at present the world's largest contributor of chlorofluorocarbons, or CFCs, and other ozone-depleting substances. CFCs receive a great deal of attention because they can deplete the ozone layer in Earth's stratosphere that filters the sun's ultraviolet radiation. Unfiltered radiation can cause cellular damage and mutations. CFCs were developed in the 1930s and were soon used in a variety of industrial, commercial, and household applications. When measurements in the early 1980s indicated that "ozone holes" were developing in the atmosphere as a result of CFC use, most countries dramatically reduced their production of ozone-depleting chemicals. China, however, continues to release these chemicals into the global atmosphere.

The air mass carrying the Congo water droplet meets a cold front over the Tibetan Plateau. The water droplet freezes and falls as snow in the headwaters of China's Chang Jiang River. As it is buried by other snowflakes, the ice crystal melts once more to a water droplet, filters into the ground and joins the shallow groundwater that moves slowly downslope into the headwaters of the great river system. By the following spring, the droplet resurfaces and enters a river, joining the rain-driven floods that occur from March to September in the Chang Jiang's headwaters.

TEN

The Chang Jiang: Bridling a Dragon

The watershed of the Chang Jiang, "long river," known to Westerners as the Yangtze, extends a long, narrow finger up around the northeast margin of the Tibetan Plateau as though searching for something. The finger lies beside the equally elongated headwaters of the Mekong and the Salween rivers, each expanding into a broad drainage as it drops down from the plateau. The upper drainages are crowded to one side by the uplifting mass formed as the Indian and Eurasian tectonic plates collide. As the Indian Plate forces northward into the Eurasian Plate, rocks in the collision zone fold and break and rear upward in the Himalaya and the Tibetan Plateau. One of the fault systems associated with this deformation bends southward around the uplifting mass. The rivers follow this zone of weaker, more erodible rocks as the rocks plunge to lower elevations.

In its upper reaches the Chang Jiang is known as Chinsha Chiang, "river of golden sand." Into the river of golden sand flow melting snow and glacial ice from the surrounding uplands. The uplands are cold and dry, a spare landscape; only about forty centimeters of precipitation falls there during an average year. This semiarid region of steep terrain is highly productive of sediment, and the upper basin provides most of the sediment that gives the Chang Jiang its characteristic café au lait appearance. Nearly all of the sediment the river transports is fine sand, silt, and clay carried in suspension in the water column. When Chinese speak of the river as being polluted they are usually

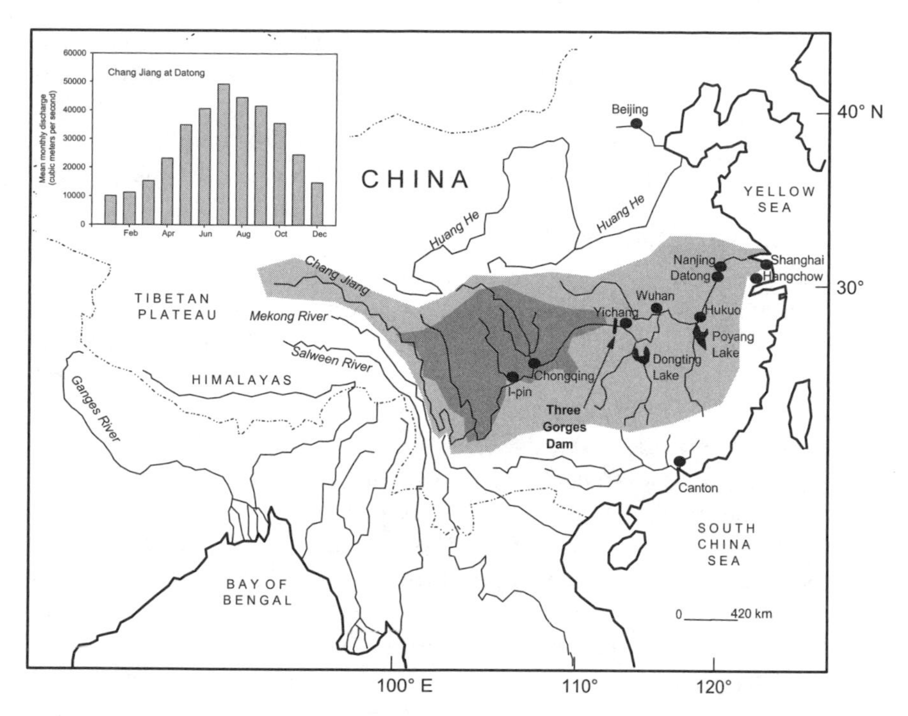

10.1 Location map of the Chang Jiang drainage basin (lighter gray). Darker gray area indicates the approximate location of Sichuan. Inset shows monthly average discharge at Datong on the lower Chang Jiang.

referring to this. An average 480 million tons of sediment are carried by the Chang Jiang into the ocean each year, with yearly values fluctuating from 754 million tons down to 321 million tons. Only four other rivers in the world—the Amazon, Ganges, Brahmaputra, and Huang He (Yellow River)—carry greater average volumes of suspended sediment to the sea each year.

Having begun at elevations of close to six thousand meters, the headwater tributaries drop precipitously through sheer gorges. Much of the Chang Jiang's drop from its headwaters to its mouth in the Yellow Sea occurs in this upper portion, where bedrock walls thinly mantled with trees drop almost vertically hundreds of meters down to a river that is churned into a continuous mass of white. The rugged, snow-whitened peaks of Tibet are visible from river level in stretches such as Tiger's Leap Gorge, where the river has carved such a narrow passage through the bedrock that people imagined a tiger might leap across. The river's steep, rocky course is largely guided by the strength and structure of the rocks it flows over until it reaches the city of Yichang. Downstream from Yichang the river meanders at a much gentler gradient across broad plains with large depressions containing lakes. Flowing hundreds of kilometers across the fertile plains of the middle and lower basins, the Chang Jiang drops very little except for slightly steeper sections at the Three Gorges, known in China as San Xia.

The river's physical downstream gradient is paralleled in reverse by a socioeconomic gradient. At the river's headwaters in Tibet, the gross national product per capita is only 51 percent of the national average, but it rises steadily downstream until it reaches 391 percent of the national average in Shanghai at the river's mouth. Climate, water supply, agricultural fertility, and access to trade have long favored the river's mouth over the headwaters.

Of China's internal waterborne traffic, 80 percent moves up and down the Chang Jiang below the city of I-pin, where the river enters the fertile, populous Red Basin of Sichuan. Sichuan, also commonly spelled Szechwan, means "four rivers," although so many large rivers feed the Chang Jiang that scholars disagree on which four it refers to. The Red Basin of Sichuan lies below the meeting point of the Indian and Pacific monsoons. Monsoon rains swell the river's flow during the March to September wet season, making possible the crops that provide 40 percent of China's agricultural output. Precipitation increases steadily from the 40 centimeters of the arid uplands to an average 140 centimeters in the middle basin between Yichang and Hukuo and 165 centimeters from Hukuo to the sea. In striking contrast to rivers such as the Nile, Ganges, and Murray-Darling that are supplied predominantly by headwaters precipitation, approximately half of the Chang Jiang's flow is derived from the middle and lower portions of its catchment. A growing season of nearly eleven months further enhances agriculture downstream from the Three Gorges. The lower valley of the Chang Jiang has long been China's best rice-growing region, producing nearly three-quarters of the national output and about one-quarter of the world's rice crop.

A third of China's total population lives in the Chang Jiang drainage basin, mostly in the lower portions. With a population of 402 million, the river's basin holds more people than the entire United States. If the Chang Jiang watershed were a country, it would be the world's third most populous. The lure of fertile agricultural lands has been set against the massive destruction of the Chang Jiang's immense floods for millennia, and the concentration of population has led to some of the world's great natural disasters when the river floods. Forty-two million people live on the river's floodplain, and the largest floods, such as those in 1931 and 1935, have each killed over 140,000.

The historical record of floods along the Chang Jiang leaves hydrologists from other countries envious. Records in the United States seldom include floods before AD 1800. The Chinese were keeping systematic records of natural disasters by the middle of the Han dynasty in 100 BC. High- and low-water marks along the Chang Jiang date to before AD 763. These records indicate that the river averages a major flood every ten years and a catastrophic flood every century.

Taming the River Dragon

The Chinese have tried to reduce flood damage for centuries. In Chinese legend, a river is a dragon and a river in flood is a dragon on the rampage. Dragons are dangerous only if you choose to live near them. The Chinese choose to live along the Chang Jiang for the obvious advantages of rich soils and easy transport, but the dragon repeatedly goes on a rampage and kills many of its neighbors. The earliest human response was to build walls to contain the river dragon. More than three thousand kilometers of levees between the headwaters and Yichang contain floodwaters within the main channel of the Chang Jiang. Early versions of some of these levees date to AD 835. The whole northern bank of the middle Chang Jiang was protected with an integrated levee by 1548.

The walls built to contain the river dragon produced unintended consequences. As on the Danube, the Mississippi, and the Murray-Darling, before the fifteenth century the middle reach of the Chang Jiang was a branched river with multiple channels that meandered over a broad plain. Floodwaters historically spread across the plains of the middle and lower river, moving slowly downvalley while depositing the sediment that formed the floodplain. Lakes adjacent to the river became connected to the channel during floods, creating rich habitat as nutrients and fish moved between river and lakes. As levees confined the floodwaters within the main channel, the ameliorating effect of the floodplain was lost. Embankments built downstream from Yichang so confined the river flow that sediment that might otherwise have been spread across the floodplain was instead deposited in the river channel, raising it nearly twelve meters in some places. In addition to confining floodwaters within levees, during the 1960s and 1970s Chinese engineers further hastened the waters downstream by cutting off a series of large meander bends.

Levees are not one-time construction projects even in the best of circumstances. They must be constantly maintained against the river dragon that gouges their sides, creeps over their tops, or insinuates itself through them. And the levees must be raised correspondingly when the entire riverbed rises from sedimentation. Changes in land use across the drainage basin of the Chang Jiang further exacerbated changes on the river caused by levees. Land clearing reduced forest cover by 50 percent in the upper basin, producing more sediment that was now confined within levees. Restricting the flow area increased the depth of flow resulting from a given amount of rainfall, exacerbating flooding. The 1998 flood, the largest since 1954, produced higher water levels in the river's middle reaches than the 1954 flood, even though the flood discharge was less.

Levees also cut off connections between the river and adjacent lakes, reducing lake fertility and eliminating the buffering effect created when the lakes stored floodwaters. By 1873 four gaps had to be created in the southern bank to allow southward discharge into Dongting Lake. But the lakes continued to be reduced in size by land reclamation. Following Mao Zedong's dictum to "plant grain everywhere," massive agricultural reclamation projects at Dongting and Poyang lakes further reduced the lake area and aggravated downstream flooding. These lakes and other low-lying areas along the Chang Jiang, although designated as flood-control basins, are nonetheless used for farming and settlement. Settlement has especially grown since the Chinese economic expansion of the 1980s. One flood-control basin of 890 square kilometers at present has a planned population of 215,000.

The Chinese also regulate flow with dams and reservoirs to control river processes. Numerous dams were constructed in the upper Chang Jiang catchment during 1950 to 1980. Sediment discharge at Datong increased by 10 percent from the 1950s to the 1960s, primarily in response to deforestation and cultivation of hillsides. Sediment discharge then decreased by 34 percent from the 1960s to the 1990s as the dams in the upper reaches were completed and trapped sediment. More than forty-six thousand dams throughout the Chang Jiang basin can now store an estimated 125 billion cubic meters of water.

Three Gorges Dam

Chinese attempts to engineer the Chang Jiang may reach their apogee in the building of the Three Gorges Dam. This dam galvanized opinion around the world and inspired predictions of dire consequences ranging from increased flood hazards, through extinction of species, to changes in regional climate. Completion of the dam in June 2003, like the 1970 completion of the Nile's High Aswan Dam, forms an important point on humanity's progression toward ever-larger dams on the world's great rivers.

The idea for a dam at Three Gorges received support from Chinese leaders beginning with Sun Yat-sen in 1919 and remained alive during succeeding decades, but years of civil chaos and limited economic resources prevented anything beyond feasibility studies for the project. Meanwhile the Chinese continued to "improve" the river within the Three Gorges. More than a hundred hidden bedrock protrusions were blasted out between 1950 and 1970, producing four million cubic meters of rock. Removing Goose Tail Rock, a huge boulder at the mouth of Qutang Gorge, took seven tons of dynamite.

The Chinese completed Gezhouba Dam, forty kilometers downstream from the Three Gorges dam site, in 1988 as a prototype for Three Gorges Dam. Gezhouba is also an enormous dam, 2,550 meters long and 40 meters tall, with ship locks and facilities for generating hydroelectric power. Unfortunately, Gezhouba created a prototype in unforeseen ways. One source of opposition to the bigger dam is the increased risk of a disastrous flood if the dam fails as a result of shoddy construction, excessive sediment accumulation, or overtopping by a landslide-generated flood wave. Critics of Three Gorges Dam cited Gezhouba as an example of substandard construction.

The government argued that Three Gorges Dam, which generated substantial controversy within China despite the national government's heavy lobbying and intolerance of criticism, will improve flood control and navigation and generate clean hydroelectric power. These assertions did not silence opponents of the dam, who emphasized the problems with environment, sedimentation, navigation, and flood control that they predicted the dam would cause. Plans for construction of a dam in Three Gorges nonetheless began to coalesce in the late 1980s, just as international opinion grew steadily more negative toward large dams. The groundbreaking ceremony took place in December 1994 despite growing opposition abroad and within China. The World Bank and other international financial institutions withdrew because of environmental concerns, but private companies in the United States, Europe, and Asia financed the project.

With time, Three Gorges Dam came to represent many things. The great reservoir behind the dam provides water storage and flood control.

Millions of kilowatt-hours of hydroelectric power are generated by the force of water pouring through turbines built into the dam. Navigation improves as the challenging rapids of the Three Gorges drown beneath reservoir waters and locks lift ships over the dam. The world's largest dam provides a symbol of national will and ability that inspires the Chinese.

Everything associated with the Chang Jiang occurs on an impressive scale. The river drains 1,940,000 square kilometers. At 6,380 kilometers, the Chang Jiang is the world's third longest river, behind only the Amazon and the Nile. Where the huge river funnels into a channel only 100 meters wide in Three Gorges, the water level can rise 6 meters in twenty-four hours, and the annual flood can be up to 53 meters higher than low-water level. Where tectonic forces have sheared and fractured the rock, the swirling, abrasive currents of the river have carved long, narrow troughs at least 40 meters below the lowest river level, and in some cases 80 meters below. These troughs cover nearly half of the riverbed within the Three Gorges. The tough rock along their sides is fluted and grooved like the wax of a melting candle. Pools in the massive Three Gorges can be up to 140 meters deep during floods.

The Three Gorges Dam continues the Chang Jiang's tradition of massively impressive scale. Its size alone is difficult to grasp: 2,310 meters long and 185 meters tall, with a difference of 112 meters in the river level upstream and downstream from the dam. The reservoir above the dam extends 670 kilometers upstream to Chongqing. When operated at the projected normal reservoir level of 175 meters, the dam has a total storage capacity of forty billion cubic meters. The ship lock for the dam was carved through a mountain bordering the river.

The riverscape within the three gorges of Qutang, Wu, and Xiling is dominated by the vertical. Canyon walls climb steeply and the river's bends cut off views upstream and downstream. The annual rise and fall of the Chang Jiang at Yichang averaged fifteen meters before Gezhouba Dam was built. Every town along the river has steep pathways or flights of steps extending from low-water level to where the community sits above annual high-water level. High-water marks line the gorge where vegetation and loose sediment have been swept from the steep valley walls below the level of the flood crest. Each tributary has a mound of sand along its banks where the backflooded waters of the Chang Jiang slowed enough to drop their load of suspended sediment. These sand mounds provide reminders of the tremendous volume of sediment carried downstream each year by the Chang Jiang, which will accumulate within the reservoir once the dam is closed.

10.2 *Above*, the channel of the Chang Jiang narrows down into a gorge along a portion of the river known as Three Gorges. This 2002 photograph was taken before completion of the Three Gorges Dam. *Below*, scale model of Three Gorges Dam at the visitor center. Main dam at left, stepped ship lock at right.

The city of Chongqing, which lies at the head of the projected Three Gorges Dam reservoir, is within the length of river where engineers anticipate that problems may result from sedimentation. Thirty million people—more than the entire population of Canada—live in the greater metropolitan area of Chongqing. The city is reputed to have the worst air quality in China. Densely packed skyscrapers rise through thick smog where the city stands at the confluence of the Chang Jiang and Jialing rivers in Qutang Gorge.

One of the environmental justifications for building Three Gorges Dam is that it will reduce emissions by substituting hydroelectric power for forms of power generation, such as burning coal, that produce large amounts of air pollution. China is the single most populous nation on Earth. The Chinese require enormous amounts of electricity as they industrialize and bring their material standard of living up to the highest contemporary levels. When fully operational in 2009, Three Gorges Dam is projected to have twenty-six hydroelectric generating units with a combined capacity of 18.2 million kilowatts, or 84.7 billion kilowatt-hours annually. This is projected to replace other forms of power generation that could be built in the future, thereby avoiding the use of fifty million tons of coal or twenty-five million tons of crude oil annually, as well as one to two million tons of sulfur dioxide and three hundred to four hundred thousand tons of soot

10.3 Approaching Chongqing from downstream on the Chang Jiang in 2002. The view of the city is nearly lost in thick haze.

produced as pollutants from burning coal or oil. To put these numbers in perspective, since 2006 China's inefficient use of energy and huge population have made it the single greatest contributor to global carbon dioxide emissions.

Motorcycles and taxis, steel, iron, and chemicals are manufactured in Chongqing, where industries dump heavy metals and synthetic chemicals into the Chang Jiang. Dilution of these pollutants depends in part on the downstream movement of water and the particles of silt and clay many contaminants attach to, but sedimentation extending upstream from the Three Gorges Dam will reduce such movement.

Chinese engineers are world leaders in dealing with the difficult problems of passing accumulated sediment through a dam, because all the major Chinese rivers have extremely high concentrations of suspended sediment. If sediment is not passed through a dam built on a river with high sediment loads, the reservoir rapidly fills. This filling reduces the storage capacity behind the dam, shortens the dam's productive life, and increases the chances that the dam may fail and cause massive downstream flooding. Engineers have calculated that within eight years Three Gorges Dam will fill to a level at which sediment can be bypassed. Once this level is reached, sediment continuing to enter the reservoir will be passed through the dam and carried downstream. The bypassing of sediment is particularly difficult to engineer. If the system does not work efficiently, pressure on the dam face will build as sediment accumulates.

Calculating sediment accumulation and bypassing within a reservoir is complicated and uncertain. Some engineers use mathematical models based on the Huang He's Sanmenxia Reservoir (constructed in the early 1960s) and the Chang Jiang's Gezhouba Dam to suggest that 30 to 40 percent of incoming sediment will be discharged from the reservoir during the Three Gorges Dam's first thirty to fifty years of operation. Sediment deposition and discharge will equalize after eighty years. Other scientists question these predictions, noting that the unprecedented scale of the Three Gorges project makes it difficult to extrapolate from existing projects, whose state also raises concerns. Sedimentation in the Sanmenxia Reservoir has exacerbated floods in upstream reaches. By 1998, Sanmenxia produced less than a third of the promised hydropower, its turbines were damaged by sediment, and it could not fulfill its flood-control function until the massive Xiaolangdi Dam was built downstream.

The fundamental issue with respect to sedimentation in the reservoir is balancing the dam's power output against sedimentation. The larger the reservoir, the more power the dam can generate. Increasing the reservoir depth, however, also increases sediment accumulation. If the Three Gorges reservoir is operated at a level of 175 meters, sedimentation may occur all the way upstream to Chongqing, blocking navigation to this major industrial center, exacerbating flood problems there, and perhaps reducing water quality. Operating the reservoir at a lower level reduces power generation and might still reduce navigation at Chongqing because of shallower water. The government projects that a deeper reservoir would increase business at Chongqing's port by four times and lower shipping costs by a third. Competing scenarios for operating the reservoir remain under study now that the dam is complete.

A more catastrophic aspect of sedimentation in the reservoir involves the potential for landslides that generate huge waves and flooding. Landslides along the river's length within the gorges made boat travel along the Chang Jiang hazardous for centuries. The 830 kilometers between Yichang and Chongqing included a thousand rapids formed where landslide debris steepened or temporarily dammed the river. Future landslides have the potential to generate such a large wave in the Three Gorges Dam reservoir that the dam fails or the giant wave flows over the dam and creates huge floods downstream. Imagine dropping the bathroom sink into a bathtub full of water. The bathtub would probably not break, but a big wave would slosh over its sides. In a confined canyon river, this wave sloshing over a dam continues downstream for long distances without attenuating.

Water content within and between rock layers influences landslides. The water level in the Three Gorges reservoir is expected to fluctuate forty meters vertically. Water seeps into the banks as water levels rise, saturating them and increasing their weight. As the water recedes, this water does not seep out as rapidly as the water level falls, making the saturated banks more likely to fail in a landslide. Chinese engineers have already identified 197 possible landslide zones in the region to be flooded by the dam, and the government has spent $482 million on landslide control measures. Landslide sites are notoriously difficult to identify and control, however, and the possibility remains strong that landslides might trigger waves that either destroy the dam or cause downstream flooding by overtopping it.

China has an ominous record of dam failures. As of 1981, 3,200 dams had collapsed, almost 4 percent of all dams in China. An average of 110 dams collapsed each year during the 1970s. Typhoon rains in

10.4 The navigation way station at lower left uses a system of river signals originally designed early in the twentieth century by Cornell Plant. The white line at center represents the vertical range between high- and low-water levels in the reservoir behind Three Gorges Dam. Photograph taken in 2002.

August 1975 caused floods that took out the massive Banqiao and Shimantan dams as well as dozens of smaller dams. These floods killed at least eighty-six thousand people. China's top hydrologist, Chen Xing, warned that the hastily and shoddily constructed Banqiao and Shimantan dams would fail, but the government declared them to be indestructible "iron dams." The Chinese government responds to criticisms of Three Gorges Dam with similar stern denial, but it remains to be seen how well these assertions correspond to reality.

Going, Going . . . Gone

Although Three Gorges Dam may not be immune to catastrophic failure, it can form an impassable "iron barrier" to anadromous fish that move between the ocean and the upstream reaches of the Chang Jiang at some point during their life cycle. Chinese sturgeon (*Acipenser sinensis*) and Yangtze paddlefish (*Psephurus gladius*) migrate long distances along the Chang Jiang, and they historically reached the size of a human or larger, but the Yangtze sturgeon (*Acipenser dabryanus*) may be the most impressive of the anadromous fish. Yangtze sturgeon can reach four meters and more than 450 kilograms. Fossils dating back 140 million years attest to the giant fish's ancient lineage. Sturgeon spend much of their adult lives in the Yellow Sea, but migrate up to three thousand kilometers as they return to the upper Chang Jiang and its tributaries to spawn.

Seasonal changes in flow signal fish when it is time to migrate, but these seasonal changes will be greatly reduced as flow in the river is regulated by dams and diversions. The Chinese government decided that building fish passages around Three Gorges Dam was too expensive, so the Yangtze sturgeon will be stocked from a captive breeding operation just below the dam. Similar attempts either to provide fish passage around a dam or to breed fish in a hatchery have been dismal failures for salmon and other species in the United States. Sturgeon disappeared from China's Qiantang River after Xinanjiang Reservoir was built in the 1950s. Sturgeon in the Chang Jiang are already at risk from overfishing, with about one hundred thousand tons of fish caught each year. This sounds impressive, but it is only 20 percent of the five hundred thousand tons a year caught during the early 1950s. Fish catches were halved during the few years between 1954 and 1970, largely as a result of overfishing. Flow regulation and barriers to migration will only exacerbate the pressure on fish populations.

Ecologists and environmentalists object to Three Gorges Dam because they fear the destruction of river ecosystems already stressed by water pollution, overfishing, and habitat loss. The Chang Jiang ranks eleventh in the world in drainage area but is fifth in fish biodiversity. Chinese scientists at the Wuhan Institute of Hydrobiology predict that changes associated with the dam will severely reduce the numbers of more than half of the 160 species present along the Chang Jiang. Many of these fish live in Dongting and Poyang lakes. The spring floods signal them to leave the lakes and spawn in the river, but flow regulation below the dam will curtail these floods. The cold water released from the base of the dam is also likely to be below the minimum temperature of 18°C that many fish require to lay their eggs.

Other unique and endangered species along the Chang Jiang include the baiji dolphin (*Lipotes vexillifer*), finless porpoise (*Neophocaena phocaenoides*), and Yangtze alligator (*Alligator sinensis*). Baijis were the last surviving species of a lineage dating back seventy million years and one of only six species of freshwater dolphins. These animals described as "pandas in water" were given the highest level of protection by the Chinese government, but baiji numbers went into freefall from the combined effects of accidental catches during fishing, river traffic, habitat loss, and pollution. Three Gorges Dam proved to be the final blow. When the baiji went extinct in 2006, the world lost an entire genus.

The gorges also provide habitat for forty-seven species of rare plants. Thirty-six of these species are found only in Three Gorges. Chen Weilie of the Institute of Botany at the Chinese Academy of Sciences noted in 1994 that land clearing had gone beyond the endurance of the local ecology and that many plants faced extinction. Inundation of existing farm and urban lands under the Three Gorges reservoir creates further pressures on adjacent lands as displaced people settle in new areas. Botanists are responding to this pressure by creating three national protection zones for plants in the region, but the effectiveness of these zones will depend on enforcement of protective regulations.

About half of the world's species of specialist river birds live in Asia, but scientists have done minimal research on these birds, which are vanishing before they are even studied. Ancient poems describe the howling of apes and chattering of birds along the Chang Jiang, but these sounds have not been heard in centuries. The relentless pressure of human population growth has steadily squeezed out Temminick's cat (*Catopuma temminickii*), South China sika deer (*Cervus nippon kopschi*), monkeys, Siberian cranes (*Grus leucogeranus*), and many other species.

Siberian cranes are the most endangered species of crane. The world's largest flock of these cranes, some three thousand birds, depend on the frogs, mollusks, insects, and fish they can catch along the shores of Poyang Lake downstream from Three Gorges Dam. The cranes winter along the lake after flying more than five thousand kilometers from their breeding grounds in western Siberia. As the regulation of river flow below the dam changes the natural seasonal fluctuations of water in Poyang Lake, the seasonal wetland habitat along the lake that the cranes depend on may shrink beyond a point where it can support many cranes. The cranes will likely not survive without such migratory feeding grounds.

Turtles are also vanishing from the river corridor. Scientists have identified about ninety species of river turtles in Asia, an unusually high level of diversity. Yet by 2003 an estimated 82 percent of these species were endangered by overharvest. The hemorrhaging of species from turtles to fish to river birds reflects the lack of research and concern about freshwater conservation or processes. The Chinese have done no environmental flow assessments in connection with proposed dams or water diversion projects. They are rushing headlong into massive river engineering—just as people did along the Ob, the Nile, the Ganges, the Mississippi, and the Murray-Darling—and giving little consideration to protecting river ecosystems.

Loss of History and Place

Another objection to the massive new dam has been the loss of human connection to the landscape. The Three Gorges preserve a long record of past cultures, and the rising waters of the reservoir inundate many irreplaceable archaeological sites. Human remains dating back to the upper Paleolithic (about 45,000 to 8,000 BC) have been found along the upper Chang Jiang. Approximately thirteen hundred archaeological sites are known in the area to be flooded. Some five hundred of these will be dismantled and moved elsewhere or protected in place. The rest will be abandoned to the rising waters.

Shibaozhai Temple in Qutang Gorge is among the sites to be protected in place. Built in the seventeenth century, Shibaozhai perches on a tall, narrow outcropping of bedrock bordered in front by the river and behind by a narrow valley that is a former course of the river. The temple's extravagant red and yellow enhance the drama of the setting. The twelve floors of Shibaozhai rise symmetrically upward to a tiny

10.5 Rebuilding above the projected high-water level of the Three Gorges Dam reservoir in 2002.

top story reached by a ladder. Wooden beams supporting the floors are levered into shallow pits gouged in the bedrock wall. The whole structure leans into the wall as though shrinking back from the forces of the river. Now the surrounding waters threaten Shibaozhai as never before, but engineers plan to build a wall around the temple to hold the reservoir water at bay.

Individual historical sites may be protected by walls, but it is not feasible to wall off the many cities being inundated by the rising waters behind the dam. The massive resettlement of people being displaced disrupts the deep-rooted associations with place that are present in many communities along Three Gorges, where 1.3 million people were displaced from their traditional homes. Perhaps only in China could such a widespread displacement be conducted in such a short time, with so little overt objection. Entire cities of more than a hundred thousand people were demolished and rebuilt above the high-water level.

The true human cost of resettlement is impossible to quantify. Environmental deterioration increases as these communities move upslope. As the vice president of the Chinese Academy of Sciences is quoted in an article in *Science,* "The land in the region is already overused, and soil erosion on the mountain slopes by the reservoir is serious" (Xiong 1998). Most of the agricultural villages being flooded by the reservoir cannot simply be reproduced elsewhere. The people have to leave farming because there is no available farmland in China where they can resettle. These individuals, with little or no formal schooling, instead take unskilled factory jobs or other positions in cities. They lose their ancestral connection to place, which in some cases extends back thousands of years.

Set against these losses, in many cases people forcibly displaced by rising reservoir waters along the Chang Jiang gained an improved material standard of living. The traditional agricultural villages along the river were picturesque, they but lacked electricity, running water, and sanitation. The new cities of unadorned concrete blocks look ugly compared with the traditional houses, but they have the basic amenities of industrialized society.

Nitrogen Stored and Nitrogen Flushed

One of the effects of replacing traditional agricultural villages with more industrialized cities may be the supplanting of nitrogen and sewage pollution with other forms of contamination. The farm fields terraced into the steeply sloping valley walls within the gorges leak nitrogen down into the Chang Jiang. Excess nitrogen in rivers has become a major scientific story within the past few years as scientists have grown concerned about eutrophication, toxic algal blooms, and hypoxia in coastal regions, as well as soil impoverishment and fluxes of greenhouse gases to the atmosphere as nitrogen is released from soils.

Human activities cause many of these changes in nitrogen distribution. Burning fossil fuels releases nitrogen oxide as waste, making nitrogen that had been sequestered in the fossil fuels available to living organisms. Together with food production, which releases excess nitrogen from fertilizers, energy production has greatly increased the availability of nitrogen around the world. Human processes now dominate the global supply of biologically available nitrogen. This human-released nitrogen goes into the atmosphere and into rivers, which dominate nitrogen delivery to coastal regions. Asia is a big player in this scenario because the continent releases the greatest amount of nitrogen during food production. Asia, Europe, and North America together account for nearly 90 percent of the contemporary human increase in biologically available nitrogen.

Rivers distribute changed amounts of nitrogen across the landscape and into coastal regions. The meeting of land and water along a river corridor forms a biogeochemical hot spot that has higher rates of chemical reactions than surrounding areas. River corridors are chemically very complex, and river basins differ in their ability to retain nitrogen. Retention depends in part on the percentage of the basin covered by wetlands, but also on factors such as climate. The more slowly water moves through, the more nitrogen is stored within the basin. Greater wetland coverage thus equates to more nitrogen storage.

The largest nitrogen fluxes today occur in rivers draining industrialized countries, where human-induced increases in nitrogen and losses of wetlands are testing the rivers' capacity to hold nitrogen. Scientists estimate that the input of biologically available nitrogen increased from 68 billion kilograms a year in 1860 to 118 billion kilograms a year in the early 1990s. This number is predicted to reach 150 billion kilograms a year by 2050. Contemporary nitrogen loading onto the continental landmass is highest in the drainages of the Mississippi, Danube, Chang Jiang, and Ganges, mainly from fertilizer use and livestock wastes.

Riverine export of nitrogen to coastal areas is likely to increase more rapidly in the future as terrestrial storage points for nitrogen become increasingly saturated and as continued destruction of riverine and wetland landscapes reduces biological use of nitrogen and increases its losses to rivers. This increased riverine export of nitrogen will create larger and more intense zones of coastal eutrophication. Along with global warming trends, it will expand the anoxic plumes that currently plague the Gulf of Mexico at the Mississippi's mouth and over 40 percent of the estuaries in the United States. Continuing increases in human population and land use in the Chang Jiang basin, along with altered river flows, may also create eutrophication in the Chang Jiang estuary.

Although traditional farming practiced within the gorges does not produce most of the nitrogen moving down the Chang Jiang, the river is far from pristine. Agricultural practices have increased the river's naturally high sediment load for centuries, as well as contributing nitrogen and phosphorus from fertilizer runoff. Less than 5 percent of the thirty billion tons of urban sewage produced annually in China is treated. Chinese scientists estimated in 1993 that each year more than three thousand industrial and mining enterprises released more than a billion tons of wastewater containing some 590 pollutants into the Chang Jiang. All of these contaminants currently move downstream attached to the suspended sediment and are distributed along the length of the river and its estuary. As Three Gorges reservoir traps sediment, it will also trap and concentrate contaminants. Tremendous amounts of garbage and industrial waste were left behind as cities below the projected high-water mark were demolished. The Chinese government set aside $4.8 billion to address these problems, but little was done as the reservoir filled. Cities being newly built above the high-water mark will likely continue to discharge sewage into the reservoir, but it will no longer be diluted as it is carried downstream.

As in the example of shoddy construction, Gezhouba Dam provides a foreshadowing of water pollution problems at Three Gorges Dam. As Gezhouba Dam closed in 1989 and water levels began to rise, people in the town of Yemingzhu admired the new lake that was said to shine like a mirror. By 1993 the mirror was covered with oil, sewage, and floating garbage. Oil accumulated on the surface so thickly that farmers could periodically skim off a few jars, pour it into their tractors, and drive off. Matches thrown on the water ignited fires. Despite being rated unsuitable for drinking, as of 1993 fifty thousand tons of drinking water were withdrawn daily from the reservoir. The national government and individual citizens of China recognize that rapid economic expansion is severely exacerbating environmental problems such as water pollution, but protection is hampered by limited environmental regulations and lack of enforcement of existing laws.

Thirsty China

The enormous experiment of Three Gorges Dam is now under way, with consequences to be determined during the decades to come. The dam will not be the final manipulation of the river. Increasingly the river's water will be diverted to slake the thirst of the north. The Chang Jiang is sometimes called the equator of China because it divides the country into a dry north and a humid south. Although 81 percent of the country's water resources are in the south, 64 percent of its arable land is in the north. The North China Plain is already dry, and its climate is growing progressively warmer and drier. Increasing water demand on northern rivers is causing them to cease flowing; before 1991, the Huang He failed to flow to the sea for up to 40 days each year. This reached 227 days in 1997, and the dry portion of the riverbed extended more than 650 kilometers. Northern rivers are massively polluted and of limited use for drinking water, and groundwater in northern China is being pumped faster than it is being recharged, limiting its future use for drinking water. Groundwater levels have dropped seventy meters in places. The cities of northern China are growing by millions despite the region's water scarcity. As a result of these dire constraints on water quantity and quality, the four hundred million people of northern China look to the Chang Jiang for water as well as hydropower.

Diverting water north from the Chang Jiang all the way to Beijing seems like a tremendous undertaking, yet navigation canals already cover this distance. By the sixth century the fertile agricultural plains of the lower Chang Jiang far outpaced rice farming in northern China, but there were no efficient ways to carry grain to the northern centers of military and political power. The Grand Canal, stretching 1,858 kilometers from Hangchow to Beijing, was progressively assembled from smaller canals over several hundred years before finally being completed by the Mongols in the late thirteenth century.

The Grand Canal also provides historical examples of the unintended consequences of such flow alteration. The canal interrupted and backed up drainage of the Huai River system, which flows to the coast midway between the Chang Jiang and the Huang He. Interruption of the Huai's flow deprived a strip of coastal land 125 to 170 kilometers wide of its normal input of groundwater. This allowed saltwater to penetrate inland in the subsurface, causing salinization and loss of soil fertility in once-productive agricultural lands.

The Huang He scenario of periods when the river ceases to flow may also occur on the Chang Jiang in the future. A more immediate threat is changes in the type and quality of habitat in the Chang Jiang's estuary at Shanghai on the Yellow Sea. Since the 1970s, monthly mean flow of the Chang Jiang to the sea has already declined dramatically during the dry portion of the year. The decline is a response to multiple changes upstream, including climatic warming in the headwater areas since the 1960s that has brought a continuous retreat of the Himalayan glaciers. Increased evaporation from lakes along the middle reaches of the Chang Jiang has lowered water levels and decreased water area in the lakes. Three Gorges Dam will exacerbate declining flows at the mouth; scientists from the Chinese Academy of Sciences predict that lower water discharges during October and November will allow the annual saltwater intrusion into the estuary to appear earlier in the year and last longer. Coastal subsidence caused by overextraction of groundwater and saltwater intrusion inland are already critical problems. The downstream effects of Three Gorges Dam will further disrupt the estuarine ecosystem.

The delta region of the Chang Jiang forms a large bowl more than 1,600 kilometers wide and long that slopes toward Lake Tai in the southwest. Sediment moving past Shanghai has built approximately 30,500 square kilometers of coastal plain over the past four thousand years. As long as this delta keeps growing or maintains its size, it buffers inland areas from wave erosion and saltwater intrusion. But smaller dams that trap and store sediment caused a measurable decrease in deposition on the delta by 2002, as well as associated erosion of the delta front where it faces the ocean tides and currents. Chinese scientists predict that completion of Three Gorges Dam will bring a 40 percent decrease in sediment discharge during the first fifty years and a 50 percent decrease during the next century.

Potential adverse effects of increasing subsidence or erosion of delta sediments are complicated by the presence of contaminants. Concentrations of heavy metals are relatively moderate in the Chang Jiang estuary compared with other estuaries in Asia and Europe. These smaller concentrations reflect the diluting effects of huge volumes of water and sediment as well as a shorter period of industrialization in the drainage basin. These diluting effects will be lost as water and sediment are stored behind dams or diverted from the river. Contaminant concentrations are already elevated at sites such as sewage outlets, and patterns of abundance of polycyclic aromatic hydrocarbons (PAHs) suggest significant petroleum contamination in the estuary.

Traveling the region in 1915, Jane Tracy wrote of how easily the yellow water of the river could be distinguished from the water of the surrounding sea, which took its name of Yellow Sea from the Chang Jiang's massive inflows. Three Gorges Dam and planned water diversions may cause regional or even global climate change as this outflow to the Yellow Sea is altered. Climate models predict that decreased outflows from the Huang He and Chang Jiang will alter circulation patterns in the Yellow Sea and the Sea of Japan, causing warmer sea surface temperatures and atmosphere over Japan. Along the Chang Jiang, as on the Ob, the Nile, the Danube, and the Ganges, river engineering undertaken by one country has rapid and readily detectable effects on other countries.

A dragon curls powerfully around one of the columns in the Huangling Temple near Xiling Gorge. Another column holds a faded stain from the high-water mark of the 1870 flood. Dragons represent the negative forces of evil and destruction. They also represent the positive forces of fertility, prosperity, creativity, and historical flow and continuity. As the Chinese grapple with the Chang Jiang at Three Gorges Dam and elsewhere along the river's length, they attempt to balance their ability to subdue the river's destructiveness during massive floods like that of 1870 against their own destruction of the river's nourishing supply of water, sediment, and habitat for a wide variety of life forms.

Three Gorges Dam is a metaphor for China's changing society. Determination to build the dam represents the engineering and technology culture of the Communist Party and its disregard for the rights of individuals as well as the importance of history, culture, and environment. Those opposing the dam represent the majority of intellectuals, independent thinkers seeking a means to express their own opinions in authoritarian China. The success or failure of the dam, now that it is built, may be crucial in determining which side of this struggle prevails in China during the decades to come.

Rivers have been perceived as eternal. As Chinese poet Tu Fu wrote twelve hundred years ago, "The state is shattered; / Mountains and rivers remain."

The Chang Jiang will remain, regardless of the development of the Three Gorges Dam. But the characteristics of what remains, and the effect the dam has on people and animals throughout the wide reach of this great river, will help to determine the world's future direction.

Interlude

More than two months after the water droplet entered the headwaters of the Chang Jiang, it leaves the river and flows into the East China Sea at the western margin of the Pacific Ocean. It is summer now, and the sun's heat evaporates the water droplet. Moving air masses carry it eastward across the Pacific, but the droplet cycles repeatedly between the ocean and the atmosphere, falling as rain and then evaporating once more. The Chang Jiang droplet gradually is carried east and slightly north with the North Pacific Current despite continually moving up and down. The North Pacific is a warm, shallow current that traverses the Pacific Ocean from Japan nearly to British Columbia at about 45° north. Near the west coast of North America, the North Pacific Current meets the southward-flowing California Current or the northward-flowing Aleutian Current. The Aleutian Current, which the Chang Jiang water droplet joins, helps to keep winter temperatures in southern Alaska milder than they otherwise would be.

The Chang Jiang water droplet reaches the coast of southeastern Alaska by October, where it is evaporated and carried to the northeast by a low-pressure cell. As temperatures in the air mass decline, the droplet freezes and falls as snow on the Canadian headwaters of the Mackenzie River.

ELEVEN

The Mackenzie: River on the Brink

On a physical relief map of North America, the Rocky Mountains form a wedge that clearly separates the eastern and western portions of the continent. From its widest extent in the southwestern United States, the mountain front slants back toward the northwest, so that the lines of longitude 110° and 120° west are in the midst of the mountains in the United States but cross a vast flat interior in the Northwest Territories of Canada. The Mackenzie is the great river of this vast interior, skirting the eastern edge of the mountain front as it recedes toward 135° west at the river's mouth. As it flows north to the Arctic Ocean, the Mackenzie collects sediment from western headwaters in the Rockies and water from countless ponds and lakes scattered across the plains in one of the world's largest wetland complexes.

The Mackenzie easily ranks as one of the world's great rivers. It is twelfth in both drainage area and length, yet it remains relatively unknown. The 1.7 million square kilometers of the Mackenzie's drainage basin have an average population density of less than one person per square kilometer, one of the lowest among the world's big rivers, and the Mackenzie does not have the fame of the Yukon River with its legendary nineteenth-century gold rushes. Obscurity has been good for the Mackenzie ecosystem, and it remains one of the least altered major rivers. Among boreal rivers, the Mackenzie contrasts with the highly polluted and heavily altered Ob-Irtysh River. Yet, like the relatively

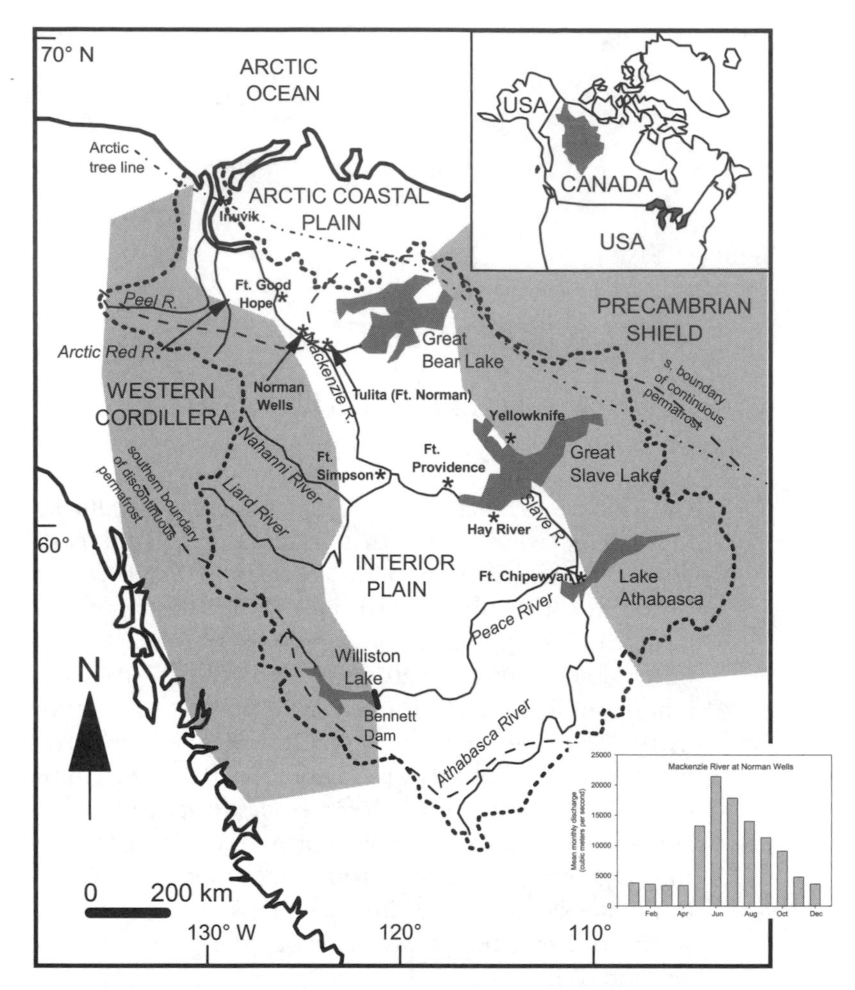

11.1 Schematic map of the Mackenzie River basin, showing drainage basin outline (short dashed line), physiographic regions, southern limits of continuous and discontinuous permafrost (long dashed lines), Arctic tree line (dash-dot line), and place-names mentioned in text. (After Droppo et al. 1998, figure 1.) Inset figure shows mean monthly discharge at Norman Wells.

pristine Amazon and Congo rivers, the Mackenzie is now at the center of controversial plans that would substantially alter the environment of its drainage basin. Thirty years ago the indigenous people living within the basin helped to dodge the bullet of a natural gas pipeline down the length of the remote river. Now the pipeline proposals have resurfaced, and the inhabitants are pushing for the economic benefits they expect from such a project. This time the bullet of hydrocarbon development will be much harder to dodge, and the nuclear explosion of climate change is unavoidable.

Origins

Maps show the Mackenzie River beginning at the southwestern outlet of Great Slave Lake, but the lake lies near the midpoint of the river's drainage basin. The Peace and Athabasca rivers feed Lake Athabasca, which empties into the Slave River, which in turn feeds Great Slave Lake. Only after it flows from this lake does the river assume the name Mackenzie after Alexander Mackenzie, the first European who explored its length, in 1789.

Much of the Mackenzie drainage basin is covered by dark green conifer forests. Small pale green or brown ponds form a watery mosaic between rivers that wind irregularly across the flat terrain. The forest is continuous yet sparse, a matchstick woodland of short, thin trees with widely spaced branches that lean precariously in all directions as the frozen ground beneath them thaws.

The mainstem Mackenzie River begins beyond 60° north, well beyond the equatorial bulge and the enduring warmth of the sun. The sun is a distinctly seasonal presence here where the planet slopes in toward its narrowest circumference, but even the endless hours of summer daylight cannot thaw the ground that winter has frozen. When the subsurface temperature remains below freezing throughout the year, permanently frozen ground develops. This permafrost is overlain by an active layer anywhere from two centimeters to a few meters deep that thaws each summer. Where temperature conditions are marginal the permafrost is discontinuous, its lateral continuity interrupted by patches that do not freeze beneath rivers and lakes. Only the fringes of the Mackenzie headwaters—the uppermost portions of the Athabasca and Peace River drainages—lie beyond the southern limits of discontinuous permafrost. Where the annual temperature averages −6°C or colder, continuous permafrost forms an ubiquitous subsurface layer that prevents infiltration and helps maintain the hundreds of thousands of lakes that characterize the far northern regions. The northern quarter of the Mackenzie drainage, north of Great Bear Lake, lies within the zone of continuous permafrost.

Permafrost gives rise to its own distinctive suite of surface features, from the leaning trees to the watery mosaics of thermokarst ponds. When the ground freezes each winter it also contracts, causing tension cracks to open at the surface. These cracks do not immediately close the following spring. Some meltwater from the active layer can flow into each crack, where it freezes at depth and forms an ice vein. Because the initial crack now forms a weaker zone, subsequent cold winters crack the ground in the same place. The ice vein gradually grows into an ice wedge that can be ten meters deep and three meters wide at the top. Ice wedges commonly form in groups connected in a polygonal pattern that resembles giant mud cracks when seen from above. The center of each ice wedge is depressed, and the margins are raised. When changing thermal regime causes the ice wedges to melt, the overlying sediments slump, creating a stand of tilted trees. Meltwater collects in the surface depression over each ice wedge. As the surface depression deepens, a thermokarst pond can form.

Only specially adapted plants can survive in the thin rooting depths available above the permafrost. Black spruce (*Picea mariana*), one of the most characteristic trees of permafrost regions, can grow where the active layer is only twenty-five centimeters thick. Plants also influence the permafrost. Spruce forests effectively intercept snowfall and reduce snow depth on the ground, inducing colder ground temperatures and deeper permafrost. Dense thickets of shrubby willows, on the other hand, trap snow, and the insulating effect of snow cover limits permafrost depths.

Permafrost also requires special adaptations from people who live above it. The town of Hay River stands just upstream from Great Slave Lake. Houses in the town sit on raised platforms that allow air to circulate beneath. Although this raises the cost of heating the house, it maintains the thermal regime of the ground and prevents the subsidence caused by thawing permafrost. The houses also lack underground utility lines. Water is delivered to each house by truck, and above-ground septic tanks are pumped out at regular intervals. The only high-rise in town, a sixteen-story tower, has a distinct lean from subsidence despite being built on a site not underlain by permafrost.

Hay River is currently prosperous with income from recently discovered diamond mines. The diamonds, which supply 13 percent of global production, come from kimberlite pipes. These pipes are literally conduits from Earth's interior that contain minerals coming from depths of approximately a hundred kilometers. The pipes are more easily weathered and eroded than the surrounding bedrock, so that a depression forms where each pipe reaches the surface. In the Far North, these depressions fill with water and blend in with the thousands of other lakes, making it difficult to find the diamonds.

The Mackenzie drainage has a long history of natural resources extracted for profit with great difficulty. Before diamonds, there were furs, gold, radium, and oil. Each period of extraction required an elaborate system of transport to carry the resource to distant southern markets. At present, tugs busily move barges up and down the Hay River past the town during ice-free seasons. The town is the northern terminus of the railroad, and PetroCanada has large oil storage tanks here from which supplies are sent north on barges. One tug can push six barges containing about eleven thousand tons of freight, but the barges come back upstream from the Arctic Ocean badly battered by ice. The banks of Hay River near the town are lined with barges being painted and repaired after their latest journey.

Among the barges rusting along the riverbanks are "radium barges" left from the uranium boom of the 1940s. The radioactive ore followed a long and convoluted journey from a remote mine on the eastern side of Great Bear Lake to more densely settled regions. Sacks of ore moved across lakes, down rivers, and around rapids by barge, jet boat, and truck. Some four thousand kilometers from its source, the ore was refined at Port Hope and then used to treat cancer and, later, to supply uranium for the Manhattan Project.

The city of Yellowknife, on the North Arm of Great Slave Lake, forms the center of another type of mining within the Mackenzie drainage. Alexander Mackenzie named Yellowknife for the copper knives used by the indigenous Dene people living here. The ancient (3,800 to 2,500 million years) rocks surrounding Yellowknife have been repeatedly uplifted, eroded, and intruded by superheated, mineral-bearing fluids from Earth's interior. A two hundred kilometer radius around the city includes deposits of a dozen metals. Gold mining began in 1934 and continued in fits and starts for the next few decades. From the 1930s to the 1970s the gold mines emitted arsenic, sulfur dioxide, and other toxins into the air, and sewage into the waters around Yellowknife, until pollution control measures and a decrease in mining reduced the flow of contaminants. Limited mining continues today, fluctuating with the prices of metals.

On a map of fish consumption advisories in the Mackenzie River basin attributable to mercury during 2002 and 2003, most sites occur in the highly mineralized zone along the southeastern edge of the basin. Some of the consumption advisories result from naturally occurring mercury concentrated in the mineralized zone, but many of them result from abandoned mines, tailing piles left from processing oil sands, and pulp mill loadings (which also contain dioxins and furans) in the

Peace River drainage. Mercury is one of the invisible, yet highly destructive, contaminants that can remain as a legacy of even brief resource extraction around Great Slave Lake.

Bennett Dam and Inland Deltas

Entering Great Slave Lake, the waters of the Hay River change from turbid brown to a clearer green. The horizon opens out so dramatically that the lake seems to be an ocean. Covering twenty-eight thousand square kilometers, Great Slave, known to the Dene people as Tucho, is the tenth largest lake in the world. It is also one of the deepest, reaching depths of 640 meters in the eastern arm. The lake takes its name from the Slave(y) tribe in whose territory it lies. The Slaves are one of several groups of Dene who were living in the Mackenzie region at the time of first European contact. Archaeologists continue to debate whether people first entered the region as early as twenty-five thousand years ago or as recently as twelve thousand years ago. Whatever the date of first human entry into North America, several lines of evidence indicate three major migrations. The first migration gave rise to the Native American tribes inhabiting most of the Americas south of Canada. The second migration gave rise to the Athabaskans, who in Canada refer to themselves as Dene. The third migration gave rise to the Inuit who inhabit the northern margins of Alaska and Canada, including the Mackenzie delta.

One of the attractions for the Dene and Inuit migrating into the Mackenzie drainage must have been the wealth of birds that could be hunted. All four major North American flyways—the Pacific, Central, Mississippi, and Atlantic—funnel into the Mackenzie valley. Many of the birds migrating along the river corridor travel to the great deltas of the river network.

Recent changes in the huge delta formed where the Peace and Athabaska rivers enter Lake Athabaska illustrate some of the unforeseen consequences of damming rivers in the Mackenzie drainage. The W. A. C. Bennett Dam on the Peace River was completed in 1968. Williston Reservoir upstream from the dam filled over the next four years, storing 3.8 billion cubic meters of water. This storage, along with operation of the dam for hydroelectric power generation, changed the hydrograph of the Peace River from one of low flow during winter with large spring snowmelt and summer rainfall peaks to a much more uniform flow throughout the year because of water retention during summer. Annual variability in flow after the dam went into operation was only 14 percent of what it had been before the dam. The loss of flood peaks meant that cobbles brought into the Peace River by tributary channels could no longer be mobilized during high flows. As a result, alluvial fans gradually constricted the Peace River at many of the tributary junctions. Balsam poplar (*Populus balsamifera*), river alder (*Alnus incana*), and willows (*Salix* spp.) constricted the channel by growing lower on the banks or covering gravel bars formerly scoured of vegetation by high flows. Secondary channels abandoned after the loss of high flows were completely filled by vegetation.

The thirty-nine hundred square kilometers of the Peace-Athabaska delta support a variety of plant communities and are used for at least part of the year by 215 species of birds, 45 species of mammals, and 20 species of fish. Eighty percent of the delta lies within Wood Buffalo National Park, which supports the world's largest herd of free-roaming bison (*Bison bison*). This biological diversity was supported by seasonally changing flows. Flooding maintained a mosaic of vegetation types, removed accumulated dissolved salts from lakes in the delta, and replenished many small ponds that together created extensive shoreline habitat.

Although Bennett Dam is nearly eleven hundred kilometers upstream, it now controls about half the flow of the Peace River at Lake Athabasca. Historical peak river flows of 4,000 to 9,000 cubic meters per second dropped to less than 280 cubic meters per second after the dam closed. Peak water levels dropped 3 to 3.6 meters below previous levels, and seasonal flooding of the delta largely ceased. The shorelines of perched basins dropped by about 40 percent as first these smaller basins, and then larger lakes fed by the annual flood, began to dry out. The diversity of plant communities decreased as species that thrived in recently flooded areas were supplanted by meadow and willow communities. Marshes became too shallow for overwintering, and muskrats (*Ondatra zibethicus*) disappeared from many perched basins. Waterfowl production and walleye (*Stizostedion vitreum*) spawning declined.

The Canadian government realized that it needed to take remedial measures to preserve the Peace-Athabaska delta. Engineers built weirs on major distributary channels with the intent of re-creating the raised water levels produced by the Peace River floods. Scientists and engineers still argue about the effectiveness of the weirs in the first years of the twenty-first century, and it has not yet proved feasible to reengineer the delta to its previous hydrological regime.

Scientists and local people also observed a general drying of the Slave River delta after completion of Bennett Dam. This may also be tied to changed flow regime on the Peace River, a principal tributary of the Slave. As distributary channels of the Slave River continually shift back and forth across the delta, they create bars that support populations of horsetail (*Equisetum* spp.). These stands of horsetail are very important to maintaining high biological productivity in the delta. The depressions between distributary channels grew drier over the thirty years after completion of Bennett Dam, so that horsetail gave way to willows and alder. The widespread and largely unforeseen effects of Bennett Dam provide a cautionary tale as new proposals to alter flows in the Mackenzie drainage basin come forward.

The Icy Mackenzie and Its Fish

Crossing the expanse of Great Slave Lake, the shore continually materializes ahead from a thin, dark, liquid line. The shoreline remains distant as the water flows into the Mackenzie River, which here is wide and shallow.

The first settlement downstream from Great Slave Lake is Fort Providence, a former fur-trading post now mostly occupied by Dene. During the hours of darkness, lights shine from the navigation markers along the riverbanks below Fort Providence like inland lighthouses. The Canadian Coast Guard has been responsible for navigation on the river since 1955. Their vessels, the *Eckaloo, Dumit, Miskanaw, Tembah,* and *Nahidik* all mean "pathfinder" in various Dene dialects. The annual task of these ships is literally to rediscover the path down the river and mark it with buoys as the ice breaks up during May and June. Unlike the dredged Danube and Mississippi, navigation on the Mackenzie remains an adventure.

The presence of thick winter ice along the Mackenzie and its tributaries gives rise to several phenomena unique to cold-region rivers. Large amounts of water are essentially put into storage as the river freezes each autumn. As rivers freeze, slushy masses of floating frazil ice typically grow and glom on to each other until a bridge or arch forms across the channel. New ice floating down the channel accumulates against this edge, reducing the cross-sectional area of the channel and increasing the hydraulic roughness of the flow. Because water is being conveyed downstream less efficiently, upstream water levels increase and those downstream from the forming ice decrease. Nearly a third of the Mackenzie's autumn flow is stored as ice. When this ice melts the following spring, it accounts for almost 20 percent of the river's high flow. Ice jams during spring breakup also create backwater upstream that accounts for most of the large floods recorded at individual communities along the Mackenzie and its tributaries the Liard and Peel rivers.

River ice holds clues to its own origin in isotopic ratios that reflect different contributions from surface water and groundwater. Isotopes can also provide clues about the life history of fish swimming beneath the ice in the turbid brown waters of the Mackenzie. Like the Ob, and unlike the big rivers of middle and lower latitudes, the Mackenzie has no really large fish or freshwater mammals. The diversity of fish is also low relative to other large rivers. Fish surveys undertaken starting in the 1970s identified fifty-three species that use the Mackenzie drainage for at least some portion of their life cycle.

Alexander Mackenzie described Dene people fishing in the river they called the Deh Cho (Mackenzie River) during his 1789 trip down and then back upriver. The Dene fished for species such as broad whitefish (*Coregonus nasus*), inconnu (*Stenodus leucichthys*), and lake whitefish (*Coregonus clupeaformis*) in the eddies of the big river during midsummer. Subsistence fishing continues today. Commercial fishing in the Mackenzie drainage has mostly been limited to Great Slave Lake, Lake Athabasca, and the Mackenzie delta. Even these limited fisheries have had mixed results, with only twenty to forty years of success before catches of the target species declined and other species such as lake trout (*Salvelinus namaycush*) were largely eliminated.

Fish population dynamics create one of the challenges to sustainable commercial fishing in the Mackenzie drainage. Fish in the far northern regions have slow growth, late maturity, and long life spans. Lake trout in Great Bear Lake may take twenty years to mature and can live forty years. Fisheries scientists interpret late maturity and long life spans as adaptations to a capricious environment. For long-lived fish, loss of some year classes through reproductive failure during bad years may have less effect on the population as a whole. Fish such as lake trout, Arctic char (*Salvelinus alpinus*), and some whitefish species appear to conserve energy by spawning only every second or third year. Although unexploited lakes and streams can have dense populations of fish, this represents the cumulative production of many years. Intense fishing quickly eliminates older fish and destroys the breeding stock.

Some of the energy that the fish conserve by skipping spawning seasons presumably goes into the extensive annual migrations characteristic of many fish in the Mackenzie drainage. Lake whitefish leave Lake Athabasca in summer to spawn 250 kilometers up the Athabasca River.

Walleye (*Stizostedion vitreum*) can travel more than 300 kilometers between successive spawnings as they move between spring spawning sites in the Peace-Athabasca delta and summer feeding sites in Lake Athabasca. Arctic cisco (*Coregonus autumnalis*) move from the Liard River out to the coast off Point Barrow, Alaska, in a journey estimated at 4,800 kilometers. These migrations use a lot of energy but bring advantages in being able to select the best spawning, feeding, and overwintering areas within a fish's range. The mass seasonal migrations also make the fish more vulnerable to humans and other predators. Bennett Dam is the only human-created barrier to long-distance migration in the Mackenzie drainage.

Charting fish migration patterns in the turbid waters of the remote Mackenzie drainage presents many challenges for fishery scientists. One approach uses strontium in the otoliths of fish to infer their life histories. An otolith is the ear bone of a fish and helps the animal with balance, orientation, and sound detection. Otoliths form through deposition of alternating layers of calcium carbonate and the protein otolin from the lymphatic fluid surrounding the otolith. Environmental trace elements such as strontium become incorporated into otoliths as they grow. Because the strontium concentration of water increases proportionally with salinity, fluctuations in strontium concentration in the layers of an otolith indicate whether a fish spent its entire life in freshwater or saltwater or alternated between the two. Another study compared isotopes in the muscle of broad and lake whitefish with isotopes from their potential food of plants and aquatic microorganisms at different locations within the Mackenzie drainage to indicate what the fish ate and where they ate it. In general, the big rivers serve as migration routes. Spawning, rearing, and summer feeding habitats are mostly small tributaries or the eddies of large streams.

Migration and feeding patterns of fish in the Mackenzie drainage all evolved fairly recently, because until nearly ten thousand years ago much of the drainage was covered with a sheet of ice up to two kilometers thick. An estimated twenty-eight fish species colonized the Mackenzie drainage from their glacial refuge in the Mississippi River drainage. The drainages were apparently temporarily linked during deglaciation, when the weight of the massive ice sheet warped Earth's crust and changed regional drainage patterns across low-elevation landscapes, and when huge meltwater lakes along the margins of the ice sheet inundated low drainage divides. A northwestern outlet of Glacial Lake Agassiz briefly flowed into the Mackenzie basin about 11,000 years ago, providing a dispersal corridor for fish. It must have been a wild ride at times. Geologists interpret valley morphology and sediment deposits as recording a catastrophic flood 11,250 years ago that lowered the level of Lake Agassiz by forty-six meters and discharged more than 2.4 million cubic meters of water per second for at least seventy-eight days.

At the start of the twenty-first century, scientists are still mapping the extent over which individual fish species use the Mackenzie drainage. A study in 2000 found four species of salmon previously unknown there. Another study in 2002 found bull trout (*Salvelinus confluentus*) eight hundred river kilometers north of their previously mapped locations. Accurate knowledge of species distributions is particularly important as global warming begins to alter the ranges of species such as salmon. The present distribution of Pacific salmon (*Oncorhynchus gorbuscha*) in the Arctic is most likely limited by low water temperature. Increasing numbers of Pacific salmon in the region might affect indigenous populations of salmonid such as Arctic char.

Land of Lakes

Downstream from Fort Providence, the forests spread thick and unbroken along both banks of the Mackenzie. Above the water surface lie cobbles and boulders that in places are aligned and tightly packed into a pavement like a cobblestone street. Farther up the banks, sand and silt several meters thick support willows and aspen backed by darker spruce. Some of these finer sediments come from deposition during high flows on the Mackenzie. Some are left over from the great glacial outburst floods. The floodwaters from Lake Agassiz carried with them a great deal of sand, silt, and clay, and these sediments are ideal for storing water that can freeze and become permafrost.

The bright green deciduous trees and thickly growing black spruce of the riparian fringe cover a slight rise, the natural levees that run along the river. Beyond the levee the land drops down to a flat plain pockmarked with irregular ponds and lakes that resemble the scraps left when cutting out cookies. Among the lakes sinuous rivers wind tortuously, leaving abandoned meanders scattered behind. This is a sodden landscape patterned in shades of gray and green. From the air, the dwarfed trees look like patches of different mosses growing on a log. In places the land is softly corrugated, like a rumpled blanket, with vegetation growing to different heights. Elsewhere in the world these patterns would reflect the history of rivers cutting down and moving

11.2 *Above*, vegetation succession along the banks of the Mackenzie River. Shrubby willows beside the channel are backed by aspen and black spruce. *Below*, aerial view of irregular patches of vegetation associated with permafrost processes, west of Mackenzie River near Fort Simpson. Thermokarst ponds are at rear of view.

sideways, but that is probably not so here, where ground ice can throw up strange ridges and polygons as it grows and then melt away to leave depressions. Permafrost began accumulating in this region about thirty-seven hundred years ago. As the permafrost collapses because of climate warming, fires, or a cyclical evolution of buildup and collapse, it creates a hummocky topography rich in microenvironments able to support diverse communities of plants.

Scattered across the landscape are the straight gashes of human-made scars. There are no roads in these scars, but the different color and type of the vegetation make them immediately apparent. During the 1970s, energy companies ran seismic lines across substantial portions of the Mackenzie drainage as they searched for natural gas and oil reserves.

Seismic surveys allow geologists to map the trend of rock units below the surface and to search for accumulations of oil and gas. Images of the subsurface are created by measuring the time it takes for reflected or refracted impulses from a sound wave initiated at the surface to return to the surface at another location. Seismic surveyors set out lines and grids of sensors, initiate a sound wave with dynamite or strong vibrations, and then pick up the returning impulses along the line or grid. This can be done with minimal disruption of the surface environment if dynamite is used to generate a sound wave just below the surface and sensors are set out and then retrieved using helicopters. Or the surveyors can cut a dense grid into the existing forest and use a vibroseis vehicle to drive up and down the grid, creating vibrations as the vehicle moves. The vibroseis generates higher-resolution images of the subsurface, but of course it is much more disruptive.

Low-elevation aerial photographs taken after the 1970s surveys indicate local melting of the permafrost where the forest cover was cleared along the grids, and accelerated erosion as each melted line became in effect a small channel that changed drainage patterns across the flat landscape and eventually created an alluvial fan of sediment where the line intersected a larger opening in the forest. Clearing the grids was not necessary to obtain the seismic data, but there was probably little environmental oversight to force the exploration companies to use less destructive methods.

The Middle Mackenzie

As the river approaches Camsell Bend, the faint outlines of mountains appear. These mountains—the Camsell Range and other subsets of the

Mackenzie Mountains—form part of the great Western Cordillera, the northern extension of the Rocky Mountains, where peaks over twenty-four hundred meters high reflect folding and faulting of the underlying bedrock between approximately one hundred million and fifty million years ago. Mountain rivers flow steeply down valleys cut deep into the bedrock, and relatively little precipitation is stored in groundwater. Paralleling the relationship between the Missouri and the Mississippi, Cordilleran rivers such as the Liard bring most of the sediment into the Mackenzie drainage system as they steadily carry the disintegrating rock of the mountains toward the sea. Eastward, the Cordillera gives way to the flat or gently rolling terrain of the Interior Plains. Younger, undeformed sedimentary rocks that in places contain coal, oil, and gas underlie the plains. River channels meander broadly across the flatlands, and deep soils store abundant water, helping to support up to 75 percent wetland cover. The Interior Plains merge almost imperceptibly into the Canadian Shield to the east. The shield is the ancient root of Canada, a complex mixture of sedimentary, metamorphic, and igneous rocks folded and faulted into mountain ranges and then eroded

11.3 The Mackenzie riverscape near Camsell Bend. The dwarfed vegetation and clear air promote long views, and the shifting clouds overhead are reflected in the calm water of the river.

down to the nearly flat terrain of today. Rivers draining the shield are really chains of lakes linked by channels that follow faults or joints in the bedrock. Wetlands occupy up to a quarter of the land surface, and the thin soils have little ability to store groundwater. Forming a northern fringe along all of these terrains, the Arctic Coastal Plain consists of thick deposits of fairly young glacial, marine, and river sediments overlying sedimentary rocks.

Vegetated islands and sandbars increasingly interrupt the river downstream from Camsell Bend. Driftwood litters a broad, shallow sandbar at the junction of the North Nahanni River. Some of the wood temporarily stored here is on a long journey that will take it far to the north and east. The wood originates from trees of the boreal forest undercut by bank erosion. The Liard River contributes a disproportionately large share of wood that will be carried progressively downstream, past treeline in the Mackenzie delta and out into the Beaufort Sea, where the East Greenland Current will carry some of it as far as Greenland.

The continuous forest and abundant wetlands covering much of the Mackenzie drainage can produce a misleading impression of a wet climate. During winter the region is dominated by cold, dry air moving south from the highest latitudes of the Arctic. Low-pressure cells occasionally bring moisture from the Pacific Ocean or the Beaufort Sea, but there is limited opportunity for air masses to take on water by evaporation as they pass over these cold water bodies studded with pack ice. Annual precipitation is only 30 to 40 centimeters through most of the Mackenzie Valley, although it rises to 160 centimeters in mountainous parts of the western basin. Precipitation falling from winter air masses also may not make it into the surface water or groundwater supply. Of the snow falling on the town of Inuvik in the Mackenzie delta, 40 to 60 percent is lost to sublimation, in which snow crystals vaporize directly back into the air without first melting. Sublimation rates are likely even higher in the drier near-surface environment of inland areas.

The Arctic high pressure retreats northward during the brief summer, allowing low-pressure cells from the northern Pacific to bring in moisture as they move through Alaskan valleys into the Yukon and then down the Mackenzie. Air drawn from farther south in the Pacific Ocean can also break through the mountainous Cordillera in central British Columbia and points south, then move north into the Mackenzie by way of the Liard valley or the plains east of the Cordillera. These low-pressure systems produce sequences of moisture pulses that climatologists refer to as atmospheric rivers, which contribute a quarter to a half of the total water vapor budget of the Mackenzie drainage.

Although daily summer rainfalls are typically less than half a centimeter, convective storms can bring up to five centimeters of rain. By November any precipitation that falls comes in the form of snow. The snowpack reaches a maximum depth that is locally up to half a meter in March.

Winds from the Pacific bring more than just moisture to the Canadian Arctic. Scientists were surprised to discover during the 1980s that many of the apparently pristine northern environments were highly contaminated with heavy metals and persistent synthetic chemicals. Although some of these contaminants come from local sources such as mining, the primary source seems to be long-range transport from Eurasia. Contaminants are transported through the atmosphere in the gas phase and as aerosols. Deposited on the snowpack, vegetation, and soils, the contaminants are washed off, buried, or revolatilized into the atmosphere. Revolatilization is reduced by cold temperatures, so that the contaminants become less mobile at higher latitudes. Lake sediments along a south-north transect in Arctic Canada reveal that PCB deposition began during the 1930s and 1940s in the low Arctic but was delayed until the 1950s to 1960s in the high Arctic.

A pervasive white haze and scent of wood smoke signal more obvious aerial contaminants along the Mackenzie during summer. The Canadian government allows wildfires to burn themselves out in these vast, uninhabited forests unless they threaten infrastructure. These are mostly small, slow-burning fires started by summer lightning storms, but they generate so much smoke that they can shut down navigation on the river for hours. Bush pilots and gold prospectors around Yellowknife caused extensive forest fires during the 1930s and 1940s, but lightning has been starting fires for centuries. Scientists estimate that each square kilometer of the Mackenzie drainage has been burned repeatedly since deglaciation, so that the forest ecosystems have adapted to the existing fire regime.

Black spruce and jack pine (*Pinus banksiana*), for example, depend on fire to release their seeds. Black spruce can grow up to thirty centimeters a year after fire because the fire creates a nutrient flush in this cold environment of slow organic decay and can also lower the permafrost table. Frequent fire and constantly changing moisture conditions also produce highly diverse forests in which trees of different sizes and shapes grow within a small area. This diversity attracts a wide range of songbird species, each specializing in particular kinds of trees and particular portions of the forest canopy. The boreal forests that are spread across Canada and throughout the Mackenzie basin support a higher diversity of breeding birds than anywhere else in North America. Scientists estimate that three hundred bird species breed in the boreal forest and, depending on the species, anywhere from 50 percent to more than 90 percent of the global population converges there during the nesting season. One of the concerns with global warming is that warmer air temperatures and increased lightning will lead to more extensive fires, altering permafrost levels and forest ecology.

Farther downstream, smoke hazing the air comes from the ground itself. At the Smoking Hills, smoke from slow-burning coal seams drifts out of the riverbank. Sediments near the coal are baked to a brick red. These are very slow burning indeed; Alexander Mackenzie described the same sights here in 1789.

Downstream from the Smoking Hills, the town of Tulita perches on a terrace above the Mackenzie where the Great Bear River enters. Except for Norman Wells and Inuvik, most of the towns in the Mackenzie drainage basin originated as fur-trading posts that reflected the search for new sources of furs and the hope of finding a Northwest Passage. The French began this search in Canada when Jacques Cartier sailed up the St. Lawrence River as far as present-day Montreal in 1535, looking for the Pacific Ocean.

On June 3, 1789, Alexander Mackenzie left Fort Chipewyan on Lake Athabasca intending to find the Northwest Passage and reach the Pacific Ocean. His party of a dozen followed the Mackenzie River all the way downstream to the Arctic Ocean, which they reached on July 12. The journey does not compare with the great contemporary epics of exploration in terms of physical danger or lives lost, but Mackenzie's group struggled with dangerous rapids in the upper drainage, cold and wet weather, insects, and the uncertainty of unknown territory ahead. Mackenzie also fought disillusion as his northwest-trending river gradually turned north and then remained northward-flowing. Mackenzie christened the waterway the River of Disappointment when he realized that he would not reach the Pacific. Ironically, the nonexistent Northwest Passage that Mackenzie and hundreds of other explorers struggled so mightily to find is now coming into existence as a result of global warming and the melting of Arctic sea ice.

Oil

A few kilometers downstream from the Mackenzie's junction with the Great Bear River lies the town of Norman Wells. People have known

11.4 Natural gas flaring off an oil well at Norman Wells.

of the energy potential of the Mackenzie basin for more than a century. Alexander Mackenzie described oil seeps and burning coal seams. But it was at Norman Wells that industrial development began in the Mackenzie drainage, when an oil field was developed during the 1920s. In 2005 Norman Wells was a community of eight hundred residents, fifty-five of whom worked for Imperial Oil as it pumped an oil field ten kilometers wide and fifteen kilometers long. A Devonian-age coral reef underlies the Mackenzie here, and the carbonate rocks host high-quality oil that is mostly found beneath the river itself. Imperial has 383 mainland wells and wells on two natural islands and six artificial islands built in the river. The oil comes from vertical depths of 180 to 200 meters, although Imperial also does a lot of lateral drilling. The oil comes out of the ground at 20°C, but to minimize disruption to the permafrost, it is then chilled to −4°C before being sent south via the Embridge pipeline. The pipeline follows the Mackenzie to Fort Simpson and then continues south. A proposed new gas pipeline will follow the same route if built. Imperial estimates another fifteen to twenty years worth of oil at Norman Wells, but the company is vigorously exploring other potential sites in the Mackenzie drainage.

Although fur trappers were the first to commercially exploit the resources of the Mackenzie drainage, hydrocarbon removal has had the greatest effects on the region, from seismic lines to new airstrips. The Canol Project, an uncompleted effort to build a pipeline designed to carry oil from Norman Wells to naval bases in Alaska during World War II, made it much easier to enter the Mackenzie drainage, for example. Landing strips with weather and radio stations built during the project began to be used after the war, giving rise to further development in once remote areas, just as the roads being built into the Amazon and the Congo rain forests today accelerate forest removal.

The Mackenzie Delta

Despite increasing development after World War II, the lower portions of the Mackenzie drainage remain largely unpopulated. In much of the world, following a river downstream entails a progression from remote, sparsely populated headwaters to the more densely settled lower regions of the river. On the Mackenzie this progression is reversed as the river flows toward the Arctic, and only small, isolated towns interrupt the forests along the riverbanks.

The Mackenzie crosses the Arctic Circle at 66°30′ north. At this point, the broad river flows placidly through a predominantly horizontal landscape formed by an immensity of sky and water with a thin line of vegetated sediment at the horizon. The small settlement of Tsiigehtchic appears on the left bank where the Arctic Red River enters the Mackenzie. Here the Dempster Highway ferry crosses the Mackenzie. Finished in 1979, the Dempster partly follows a former dogsled trail and is the only road of any kind in the Mackenzie drainage downstream from Fort Good Hope.

Beyond Separation Point the river officially enters the Mackenzie delta, an area of thirteen thousand square kilometers that is the second largest Arctic delta in the world. Separation Point was named by Sir John Franklin when his 1826 expedition in search of a Northwest Passage divided into two groups to follow different distributary channels down the delta. The Mackenzie River grows perceptibly wider beyond Separation Point, although the effect is diminished by the numerous islands that effectively form banks when viewed from river level. Rapid erosion of some of the banks exposes layers of black sediment contorted by the growth of ice lenses. As the river cuts into the bank the ice melts more rapidly, and whole chunks of bank spill into the shallow water

11.5 Bank collapsing along a channel in the Mackenzie Delta as ice underlying the surface melts.

just as Alexander Mackenzie described, creating half-submerged forests of trees that are still green but tilted like a jumble of pickup sticks. Shallow lakes perch on the islands.

More than twenty-five thousand lakes spread across the delta. There are so many lakes, of such varied shape and size, that the landscape appears unfinished, as though the water had only ponded and had not yet begun to drain downhill. One lake might be clear and dark while the next, just a short distance away, is an opaque bright green. These colors likely reflect the degree of connection to river water.

Most of the delta lakes are less than ten hectares in area and four meters deep. As the minor channels of the delta shift back and forth, they leave depressions that can become lakes. Thawing of the underlying permafrost also creates depressions. Sills of sediment separate the lakes from the nearest channels, and high river discharges overtop these sills. Some lakes are continuously connected to the river, whereas others are connected only when spring floods are higher than normal. Lakes with differing amounts of river inflow also have differing water chemistries and aquatic communities. Frequently flooded lakes have turbid water that limits the growth of rooted aquatic plants and algae, whereas those with a higher sill can have clear water and abundant aquatic plants. The clearwater lakes contain a total mass of plants and animals comparable to that of floodplain lakes in temperate and tropical regions despite the low temperatures and short growing season, which implies very high growth rates.

Each lake and river channel on the delta is fringed by paler green deciduous vegetation, with darker green spruce forming a ragged-topped stand beyond. The combination of wetlands, river channels, and forests supports high densities of migratory birds. Thousands of Canada geese (*Branta canadensis*), white-fronted geese (*Anser albifrons*), brant geese (*Branta* spp.), pintail ducks (*Anas acuta*), and other birds breed here during summer. Muskrats, pike (*Esox lucius*), and whitefish also thrive in the lakes and rivers, and the river's abundance supports seals, whales, herring, and arctic trout in the Beaufort Sea.

The oil and gas industry became interested in the Mackenzie delta when large quantities of hydrocarbon reserves were discovered in the Beaufort Sea near the delta during the late 1960s. Oil and gas consultants and the federal government undertook numerous studies designed to identify potential pipeline crossings, as well as associated hazards and environmental impacts. As a result, much more is known about the Mackenzie delta than about other Arctic systems.

The delta developed in an estuary of the Beaufort Sea created between eleven and thirteen thousand years ago during the last glaciation. Except for the active channels, permafrost extends up to 395 meters deep beneath portions of the delta. This is indeed the frozen north, but frozen here does not mean static. Flows through the delta can reach thirty-four thousand cubic meters of water per second during May and June. The force of all this water, combined with huge blocks of ice jostling each other and forming jams along the river margins, creates constant erosion and sedimentation. Geologists estimate that 128 million tons of sediment reach the delta annually from the Mackenzie and the tributary Peel River. Approximately 103 million tons are at least temporarily deposited on the delta, of which 43 million tons remain; the rest flow from the delta into Mackenzie Bay and the Beaufort Sea.

The main distributary channels of the delta have remained fairly stable since they were originally mapped in 1826 by the Franklin expedition, despite all this water and sediment moving through and all the localized erosion and deposition. Most scientists attribute this large-scale stability to the presence of permafrost that substantially increases the resistance of the channel banks to erosion.

Flow through these channels is extremely complicated. The West Channel of the delta opens earlier than the Middle and East channels because the spring breakup on the rivers tributary to the southwestern end of the delta generally occurs one to two weeks earlier than in other parts of the region. Jams of floating logs and cakes of ice further complicate the hydraulics, and up to 95 percent of the delta is flooded during spring. Extensive ice jams or logjams can also back water upstream, producing flood levels four meters higher than low-water levels at Inuvik on the East Channel. Water levels across the delta rise continuously during the winter, despite constant inflows, as smaller channels freeze and impede flow. Flow that does make it all the way through the delta is likely to be impounded as liquid under the landfast ice in the Beaufort Sea or to be incorporated into the landfast ice until the following summer.

This land of frozen ground and rivers choked with masses of ice is the home of the Inuvialuit, a group of Inuit living along the portion of the Arctic coast centered on the delta of the Mackenzie, which they call Kuukpak, the "great river." This region supported the richest and densest concentration of humans in Arctic Canada before European contact, thanks to abundant driftwood from the Mackenzie, seals and whales, and two great caribou herds east and west of the Mackenzie River. Both herds winter in the boreal forests and migrate onto the Arctic tundra in spring, and they provided the Inuvialuit with meat and with hides for warm winter clothing.

Although Inuvialuit began trading for furs with the Europeans during the 1820s, the real change in their life came with the arrival of American whaling ships in 1889. The whalers brought not only modern tools but epidemics of disease that reduced the original Inuvialuit population of 2,500 to about 250 people.

The remaining Inuvialuit replaced subsistence hunting with cash-paying jobs operating heavy equipment or loading freight on airplanes during construction of the DEW Line in the 1950s. The Distant Early Warning (DEW) Line was a string of radar sites from Greenland across the Canadian Arctic to western Alaska designed to give the United States early warning of a Soviet attack. The DEW Line was eventually abandoned, creating a bust like those that had followed earlier resource booms. The pattern was repeated when Imperial Oil started seismic work in the Mackenzie delta in 1958; oil drilling and jobs boomed in 1974; and everything went bust by 1979. The Inuvialuit were left with no traditional lifestyle, but no real way to join the economy of the rest of Canada.

Today Inuvik forms a regional center for the Inuvialuit. Inuvik is the northernmost town along the Mackenzie, a town of thirty-four hundred people originally established in 1955 as an administrative center for the western Arctic. To go down the Mackenzie is to descend into more and more intense cold in winter. Annual air temperature averages −4°C at Fort Simpson, −6°C at Norman Wells, and −9°C at Inuvik. Many of the buildings at Inuvik are brightly colored, a legacy from the days when this was a naval base and the "southerners" needed color to psychologically survive the long, dark polar winters.

Beyond Inuvik, the waters of the Mackenzie join the Arctic Ocean at the Beaufort Sea. The coastal waters are directed eastward toward the maze of large islands that lie between the mouth of the Mackenzie and Greenland. Like other pristine rivers such as the Amazon, the plume of Mackenzie River water is deficient in nitrogen and phosphorus and consequently does not produce the blooms of phytoplankton and hypoxia supported by excess nutrients coming down rivers such as the Mississippi or the Danube.

Other specifics of the Mackenzie's water chemistry reflect the chemical composition and weathering of rocks in the drainage basin. Concentrations of polycyclic aromatic hydrocarbons (PAHs) are much higher than those found in other pristine rivers of the Arctic and are comparable to levels in regions polluted by oil and gas development. In the Mackenzie, however, the PAHs come from weathering of natural oil seeps, bitumen deposits, and peat. Scientists have documented fish pathologies typical of high levels of PAHs. This means that the Mackenzie system is at greater risk from increased levels of PAHs associated with oil and gas development because there is no buffer; the system is already at levels that are dangerously high to aquatic organisms.

Changing Climate

At present the Mackenzie is among the world's great rivers least affected by human activities. Although the Bennett Dam earns the basin a rating of moderately impacted on a 2005 compilation of flow regulation around the world, the mainstem and other tributaries are little affected by flow regulation. This chapter is subtitled "River on the Brink," however, for at least two reasons: the probable effects of climate change, and the proposed gas pipeline that would parallel the great river for a substantial portion of its length.

Worldwide, the Mackenzie River basin is one of the regions with the greatest documented warming over the last thirty years. Temperatures in the center of the basin have increased about half a degree each decade. This may not sound substantial, but it has been enough to create perceptible changes. Indigenous people who live closely tied to the land are particularly cognizant of the changes accompanying this warming. Dene living above the Arctic Circle speak of drier summers in which the usual light rain is replaced by thunderstorms. The lightning ignites forest fires, but the rain cannot sustain normal water levels. Higher summer temperatures cause berries to dry up, so grizzly bears stay out longer in autumn to forage for food. Melting the upper permafrost increases landslides and soil flowage even where the terrain is not steep, because saturated soil resting over still-frozen deeper layers can flow on nearly level ground.

These perceptions are substantiated by measurements. In the prairie provinces of Canada the southern edge of permafrost has receded about a hundred kilometers northward during the past century as a result of warmer air temperatures and larger areas burned by wildfires. Scientists established a twelve hundred kilometer transect in the Mackenzie drainage during the late 1980s as part of a program to monitor permafrost. It was clear by 2004 that the depth of thaw penetration is increasing at most sites. Interannual variations in air temperature increased starting in the mid-1970s, supporting another prediction of climate change: not only will the means change, but the extremes will grow larger. Long-term records from rivers draining into the Arctic Ocean indicate a slight trend toward increased flow during the last decade of the twentieth century, as predicted by models of climate change.

Although the Arctic Ocean contains only 1 percent of the world's total ocean water, it receives 11 percent of the world's annual river runoff. Arctic rivers have very low variability from year to year; extreme annual values differ from the mean by less than 20 percent. Long-term trends in these normally consistent rivers play an important role in controlling the production of sea ice in the Arctic Ocean. Because formation of sea ice is a critical part of the controls on temperature and salinity that determine the density of seawater and thus the density-driven circulation of the world's oceans, ultimately what affects the Arctic Ocean can substantially influence global climate in ways not yet well understood. A team of scientists predicted in 2005 that within a century present rates of warming will produce an Arctic Ocean free of ice during summer. Such a situation has not existed for at least a million years.

Simulations of climate change suggest that carbon dioxide levels could double by 2040, causing temperatures in the Mackenzie drainage to warm by 2–5°C in summer and 4–8°C in winter. These changes in air temperature affect everything, as illustrated by permafrost. Warmer air temperatures create a greater energy flux to the soil, deepening the active layer. Porous sediments such as the silt and sand underlying the surface in much of the Mackenzie valley store a lot of water when frozen in permafrost. Thawing greatly reduces the strength and volume of the sediments, and this induces landslides, increases sediment contribution to rivers, and reduces or eliminates the ability of the sediments to support buildings and transport corridors. The predicted changes in air temperature in the Mackenzie valley will eradicate permafrost in the Fort Simpson area, cause substantial thinning from the top down in Norman Wells, and increase the thickness of the active layer that melts each summer at the Arctic coast.

The majority of the world's wetlands occur between 50° and 70° north, including many wetlands in the Mackenzie drainage. Thawing of permafrost changes the hydrology, temperature, and vegetation dynamics of these wetlands. Because wetlands worldwide emit about 20 percent of the methane released to the atmosphere, warming-induced changes in wetlands will also significantly alter emissions of this greenhouse gas. The position of the water table and temperature of the peat control wetland methane emissions by controlling oxygen levels and biochemical reactions among microbes living in the wetlands. Evidence thus far indicates that warming will result in higher methane emissions. Collapse scars over melted permafrost, for example, have particularly high rates of methane emission.

Lake dynamics on the edge of the Arctic Ocean provide a second example of the extremely complicated ramifications of warmer air temperatures. Photosynthesis by bottom-dwelling plants contributes about half of the organic carbon available to organisms in the lake. (Floating algae and plant detritus coming into the lake from the surrounding catchment provide the other half.) Increased levels of atmospheric carbon dioxide would retard degassing of carbon dioxide from the lake, raise the partial pressure of carbon dioxide in the lake water, and ultimately increase the amount and proportion of carbon contributed to the lake by photosynthesis of bottom-dwellers. Such changes are likely to ripple through the finely adjusted ecosystem of the lakes in ways that scientists cannot yet predict, though not for lack of trying.

The Mackenzie Basin Impact Study is a six-year, $950,000 effort to document and understand the effects of ongoing and predicted climate change in the Mackenzie drainage. Among the predictions: Warmer winter temperatures will result in more snow, making it more difficult for caribou to get at the ground lichens that form the mainstay of their winter diet. Warmer temperatures will also favor the parasitic nasal botflies that can drive the caribou to near starvation when the harassed animals take refuge in breezier areas that are barren of vegetation. Shallow wetlands favored by the twenty-six species of shorebirds that breed almost exclusively in the Arctic or subarctic will decrease. Numbers of the nearly extinct Eskimo curlew and of the long-billed dowitcher, which breeds only in the Mackenzie delta, will likely decline. Other birds that prefer drier conditions will expand their ranges northward. Peatlands will shift northward by eight hundred kilometers. Those at the southern edge of the current range will dry up, changing them from sinks to sources for atmospheric carbon. Minimum water levels in the rivers will decline as loss of wetlands reduces the drainage basin's capacity to store water.

How can we really predict the effects of climate change? Like the old song telling how one bone is connected to another, the connections just keep spreading. How will climate warming affect Arctic fish? Fish depend on access to spawning and rearing habitat. Habitat depends on flow regime, sediment concentration in the channels, bank stability, and water temperature. Flow regime depends on precipitation patterns and on how the precipitation runs off the surface or infiltrates and moves through the subsurface. The routes that precipitation travels to the stream channels are controlled by ground temperature (permafrost) and moisture content. Sediment concentration in the channels reflects rates of bank erosion, which depend on permafrost, riverside vegetation type and density, and amount and timing of water flowing in the channel and exerting erosive force against the channel banks, and so on. Although the Ob and other high-latitude rivers in Eurasia will be as affected by warming climate as the Mackenzie, it is on the Mackenzie that much of the basic research needed to understand and predict ecosystem responses to climate change is being carried out.

A Thread or a Razor Slash?

It is in this climate of uncertainty about the extent and intensity of side effects from global warming in the Mackenzie drainage that a lengthy natural gas pipeline is proposed. The pipeline is a ghost returning. When the costs of fossil fuels rose steeply during the 1970s, North

American energy companies responded with enthusiastic proposals for domestic energy development. At that time, however, the land claims of indigenous groups in the Mackenzie valley were still unsettled, and the indigenous peoples vigorously opposed energy development because they saw no economic benefits to themselves. Thirty years later the land claims have been settled, and many former pipeline opponents are highly supportive of the newest iteration of the proposal because they will profit from it.

The earlier version of the pipeline was dropped after the 1977 report of a government commission explained how any gas pipeline would be followed by an oil pipeline and an enormous infrastructure of roads, airports, maintenance bases, and new towns. Pipeline supporters in the 1970s compared the pipeline to a thread stretched across a football field. Opponents described it as a razor slash across the *Mona Lisa*.

The Canadian government lifted its moratorium on petroleum development in the Northwest Territories in 1994. This, along with rising energy prices, revived interest in the natural gas pipeline sufficiently by 2000 to give rise to two competing pipeline proposals. One proposed route from northern Alaska would go through the Yukon and then southward along the border between British Columbia and Alberta. This route would involve opening up the Arctic National Wildlife Reserve in Alaska and would require the pipeline to cross some substantial mountain ranges. The second route, which would go from Alaska to the Mackenzie delta and then south along the river, is estimated to be several billion dollars cheaper because it follows low-lying topography. Major energy producers have brought their deep pockets to the process, spending an estimated $350 million on exploration and preliminary studies. A formal application for the pipeline route was submitted in 2003 by Imperial Oil, ConocoPhillips Canada, Shell Canada Ltd., ExxonMobil Canada, and the Aboriginal Pipeline Group. The pipeline was approved in early January 2010 and the companies are now eagerly preparing to develop the area, although they also seek further financial backing from the Canadian government.

The proposed gas pipeline would run 1,220 kilometers along the Mackenzie River valley, with compressor stations at regular intervals to maintain the flow of gas. Total development associated with the project would bring into the basin an estimated 4,000 kilometers of feeder pipelines, 62,000 kilometers of seismic lines, and thousands of kilometers of new roads. Although technologies for oil and gas extraction and transport have improved, the environmental impact studies conducted back in the 1970s remain highly relevant to the latest iteration of the proposed pipeline. These studies highlighted several potential primary impacts, each with secondary impacts.

The most immediate direct effect of pipeline construction is likely to be disruption of the permafrost in the Mackenzie valley as removal of vegetation cover increases depth of soil thawing. Melting of permafrost decreases bank stability along rivers, causing more lateral movement of the river and more sediment entering the channel from bank collapse. Aquatic ecologists find that naturally turbid rivers and lakes in the Mackenzie drainage support less abundant and diverse invertebrate fauna than do clear waters. The clearwater systems are thus particularly vulnerable to increased sedimentation tied to the pipeline directly through trench construction and transport corridors, or indirectly through disruption of permafrost. Sediment on the banks of some streams in the Mackenzie drainage contains relatively high amounts of nutrients and heavy metals as a result of local geology, so increased bank erosion will pollute streams such as the Great Bear River. Most vegetation in the soggy environments of the Mackenzie valley is also very sensitive to changes in drainage or water level. Such changes will accompany pipeline development either through thawing of the permafrost or through construction of road and pipeline embankments. Dust and herbicides kill roadside vegetation along existing unpaved roads, and this would substantially increase with pipeline development.

Contamination associated with existing energy development provides examples of the biological disruption that can result from pipeline construction. Noting the high probability of an accidental oil spill from a pipeline near a river in the Mackenzie drainage because of bank erosion, lateral channel movement, and ice jams, scientists examined sites of accidental oil spills in Inuvik and Yellowknife Bay. Documented effects of oil spills were rapid—within hours—and lethal to many aquatic organisms. Crude oil can directly damage tissues by interfering with cell structure or function. Viscous whole oil can also impair movement, feeding, and respiration of aquatic organisms. The rate of recovery for the entire lake or river remains unknown, but it may be slower than in temperate environments because winter in the Mackenzie drainage is such a long period of limited chemical and biological activity.

Coastal oil spills are another possible consequence of oil and gas pipeline development. Ice abrasion limits the establishment of littoral communities in all but the most protected areas of the Arctic coastline, so sublittoral communities would likely be more damaged by an oil spill than might happen in temperate latitudes. Dense algal blooms form on and within the undersurface of ice in the Arctic Ocean during spring. Increasing levels of nutrients in the water column and increasing levels of diffuse light filtering through the ice stimulate these blooms. Invertebrates congregate under the ice to eat the algae, feeding intensively during the brief period of phytoplankton productivity and accumulating nutrient reserves that see them through the rest of the year. Spilled oil also concentrates under the ice. Because oil is toxic to algal species such as phytoplankton, it can prevent or inhibit the formation of the normal spring algal bloom. A spill under the ice retains its toxicity longer than spills in warmer regions, and higher concentrations of the soluble components of the oil ultimately dissolve in the seawater.

Scientists predict that terrestrial animals including Dall sheep, grizzly bears, waterfowl, some raptors, and caribou will be sensitive to pipeline disruptions such as aircraft noise, fires that destroy the climax forest caribou use, harassment during winter when food and energy supplies are low, interference with migration routes, and opening of large areas to hunters. Waterfowl migration patterns will be disrupted by construction and blasting and by aircraft flying low, and the birds will lose breeding and migration staging habitat as wetlands and rivers are altered. Fish in the Mackenzie drainage are most sensitive during May 15 to June 30 and September 15 to November 15, when most species are migrating or spawning or both. Fisheries biologists identify siltation or removal of spawning gravel, blockage of fish migration routes, destruction of rearing areas, and chemical contamination from spills as the pipeline's greatest threats to fish.

Probably the most damaging indirect effect of pipeline development will be fragmentation of a large portion of Canada's intact boreal forest, which supports an estimated three to five billion birds. These birds are the boreal warblers, vireos, and flycatchers that stop over in the United States on their way to the tropics of Central and South America. They are the white-throated sparrows that overwinter in the United States. They are the waterfowl, the shorebirds such as sandpipers and yellowlegs, and the endangered whooping cranes, which breed on boreal lakes. Less than 10 percent of Canada's boreal forest is permanently protected, and two hectares of it are lost each minute to timber harvest and clearing for oil and gas exploration and development.

The future of the Mackenzie drainage basin provides a catch-22. No longer willing or able to survive using a traditional subsistence lifestyle, the Dene and Inuit peoples living in the drainage now aggressively promote oil and gas development. It is not clear how the indigenous peoples can escape the severe social problems that now beset them without participating in the employment and income provided by oil and gas development. Development, however, even if held to high environmental standards, will certainly change the remote and undeveloped character that is a unique aspect of the region. Energy development will also degrade the existing terrestrial and aquatic ecosystems in ways ranging from direct impacts to the variety of indirect effects that will follow greater air and ground access to the region and higher population levels. The ecological losses and massive contamination of the Ob drainage, which is intensively developed for energy production, provide a worst-case scenario for the future of the Mackenzie.

River Health

The potential effects of pipeline development on the Mackenzie River might seem inconsequential to someone who views as wasted land a river corridor unpopulated or not obviously used by humans. A river can simply provide a downstream conduit for water: even a concrete-lined canal or a pipe can do this. In a healthy river corridor the river is free to adjust to changes in water and sediment entering the channel. These physical adjustments create diverse habitat for aquatic and riverside organisms, which in turn helps to support communities of diverse species and age structure. In a healthy river corridor, organisms can migrate—upstream and downstream, between channel and floodplain, between hill slopes and valley bottoms, and between surface and subsurface—because the river remains fully connected to the adjacent landscape. A healthy river corridor provides many ecosystem services vital to humans and other animals. Among the most obvious are clean water, clean air, fertile soil, flood control, productive marine ecosystems, places to fish, paddle, hunt, and watch birds, and the habitat and food for species that humans eat, from fish to the ducks that use river-margin wetlands and the deer that graze in riverside woodlands.

Bands of riverside vegetation promote infiltration and flow of subsurface water from adjacent lands into river channels and thereby filter and cleanse water. Flowing rivers can also absorb a lot of organic waste and naturally dilute and process it. Microbes in small headwater creeks fix large amounts of nitrogen into biologically usable forms, providing a critical nutrient for living organisms and limiting the buildup of greenhouse gases in the atmosphere. Plants within and along the channel convert carbon dioxide into oxygen through photosynthesis. Floods that spill over the channel banks and across the floodplain deposit fine sediment and organic matter from which fertile soils develop. The slower passage of floodwaters across the broad, hydraulically rough valley bottom flushes and revitalizes wetlands, recharges groundwater, and damps out flood waves. Fish and other organisms that live some part of their life cycle in a river and some part in the ocean effectively transfer nutrients between the two environments. Many marine fisheries would not exist if fish did not have access to inland spawning and rearing habitat. The dissolved and suspended nutrients carried into the ocean at the mouths of rivers fuel algal populations that feed invertebrates, fish, marine mammals, and seabirds.

Whether or not we as humans remain aware of it, we cannot exist without these river services. And a river cannot provide these services unless it is fully connected to the subsurface, valley bottom, hill slopes, ocean, headwaters, and atmosphere of both its own drainage basin and, as discussed in the interludes of this book, the entire world. The river must have naturally fluctuating supplies of water, sediment, nutrients, and organisms. It must have room to move in adjusting to changes in the supply of water and sediment. It must have a riverside corridor of native plants adapted to its dynamic changes through time and space. And it must have complete, intact communities of aquatic and riverside species of plants and animals that promote so many of the processes that provide clean water, clean air, and fertile soil—the basis for human life.

Humans remain extraordinarily fortunate that so many rivers around the world continue to provide some level of these ecosystem services despite the numerous changes we have imposed with little thought for long-term loss of river health. Like the three spirits who visit Ebenezer Scrooge in Charles Dickens's *Christmas Carol*, this book's portraits of ten of the world's large rivers demonstrate what has been, what is, and what may be. The question for humanity, as for Scrooge after the ghostly visits, is how individuals, societies, and the global community will react to this knowledge—along the Mackenzie and along all the other great rivers threatened by rapid and widespread contemporary or proposed development.

TWELVE

An Ebbing Flow

Rivers should be such a small, insignificant component of Earth. Taken together, all the freshwater in rivers, lakes, and glaciers forms less than 3 percent of the total volume of the planet's water. Yet its distribution and quality are vital to humans and many other organisms. Withhold fresh, potable water, and we suffer. More than a billion people today do not have access to clean drinking water, and more than five thousand, mostly children, die each day from water-related diarrheal diseases. Lack of suitable drinking water leads to human rights debates over water quality and quantity. Is water a need that can be sold, as well as used and polluted by some? It takes more than 378,000 liters of water to manufacture one car. Many growing industries are water-intensive. Or is water a right that is guaranteed to all and must be protected by national governments and the international community?

People are also vulnerable to water shortages through their crops. Approximately 85 percent of the water consumed by humans goes to agriculture. Damage from droughts averages $6 billion a year in the United States. On the flip side of scarcity, superabundance of water from floods and storms also causes immense suffering. Approximately two hundred major floods during the year 2004 resulted in 173,474 deaths, 51.5 million people displaced, and $7.6 billion in damages. People spend enormous amounts of energy and money manipulating water supplies precisely because the quantity and quality of water are so vital to human existence. These manipulations are one of the most fundamental components of the rapidly occurring changes in Earth's environments.

One of the most obvious alterations of the hydrologic cycle is the storage of water in dams. Worldwide, an estimated forty thousand dams higher than fifteen meters, and more than eight hundred thousand smaller dams, store approximately nine thousand cubic kilometers of water. This is equal to five times the volume of water in all the rivers of the world. What does it mean to store five times the flow of the world's rivers? Storing water in a reservoir changes the amount and timing of evaporation into the atmosphere, which in turns changes the dynamics of local air masses, including their temperature, humidity, and ability to release precipitation. Water lost to evaporation from reservoirs reduces the net amount of water that flows downstream in rivers.

Storing water changes the timing and magnitude of high and low flows in rivers and the seasonal cues that river plants, insects, and fish rely on for their reproductive cycles. Fishery biologists have documented river after river on which the abundance and diversity of fish have plummeted when dams block migration routes, destroy seasonal spawning cues, and reduce fish habitat. Plant ecologists document a consistent scenario downstream from dams of disrupted patterns of seed dispersal, plant migration, and the flood disturbance that creates new germination sites. As riparian zones grow narrower, plant populations become disconnected from one another, exotic species invade, and the abundance and diversity of native plants decline.

Storing water behind dams also changes the chemistry of the water. An adequate supply of dissolved oxygen is important to a river's ability to purify itself of excess nutrients and organic matter, which commonly result from agriculture and the wastes produced by human communities. The availability of dissolved oxygen depends on several factors, including the amount of organic material in the river, photosynthesis by river plants, respiration by other organisms living in the river, and diffusion between the river water and the atmosphere. Smaller amounts of water moving more slowly down rivers below dams can be depleted of dissolved oxygen. More important, water released from a dam commonly comes from the bottom of the reservoir. As microbes and algae living in the upper, sunlit portion of the water column in a reservoir die, their decaying bodies sink through the water column. Decomposition depletes the dissolved oxygen deeper in the reservoir. Reservoir water is consequently low in dissolved oxygen when it is released into a river from a dam.

Chronic depletion of dissolved oxygen leads to problems such as increased incidence of human disease and disappearance of fisheries because many species of fish cannot tolerate low oxygen levels. Depletion of dissolved oxygen also enhances release of nutrients, heavy metals, and toxic compounds. Microbes living in the upper fraction of the sediments at the base of a reservoir consume what little oxygen is present, creating conditions under which ferric iron is transformed to ferrous iron that dissolves into the water and releases metals that had been adsorbed onto the ferric iron compound. As these metals are released into solution they become available for uptake by living organisms. Mercury is one of the deadliest of these metals.

Storing water, of course, also means storing sediment. At present, dams store 25 to 30 percent of the sediment that formerly was carried to the oceans. The Aswan High Dam alone reduced the Nile's annual sediment load from 100 million tons a year to almost zero. Human enterprises that cause soil erosion have increased the transport of sediment by the world's rivers by approximately 2.3 billion tons a year, but the storage in reservoirs has more than compensated for this increase, reducing sediment flux to the oceans by 1.4 billion tons a year. Over 100 billion tons of sediment and 1 billion to 3 billion tons of carbon are now stored in reservoirs built mostly within the past fifty years.

What does all this sediment storage mean? First, it means coastlines retreat as sediment that formerly replenished deltas and estuaries no longer reaches the coast, allowing older sediment to subside as it compacts under its own weight or to be eroded by waves and tides. As of 2001, more than half of the world's population lived within two hundred kilometers of a coastline. As coastlines erode and retreat, agricultural lands, fisheries, and settlements are lost. Coastal areas become increasingly vulnerable to damage during storms. Reductions in sediment supply affect the bottom-dwelling animals in coastal estuaries as well as coral reefs and sea grass communities. Fisheries decline. These changes result partly from altered coastal configuration and partly from loss of the nutrients carried with the sediment. More than 90 percent of the total flux of vital elements such as phosphorus, nickel, manganese, iron, and aluminum from rivers to the world's oceans is transported in sediment. More than 40 percent of the total organic carbon transported to the oceans comes down rivers in particulate form.

Take away enough sediment, and the vitality of the entire coastal zone is diminished or destroyed. Yet in Asia 31 percent of the sediment from the continent is now retained in reservoirs, just as 30 percent of the sediment that once entered the Mediterranean and Black seas now rests in the bottom of reservoirs built along the rivers. The chapters in this book chronicle declines in coastal environments in the Gulf of Mexico, the Black Sea, the Mediterranean Sea, the Arctic Ocean, and the Sundarbans. Each of these examples can be traced upstream to changes in water and sediment discharge from a great river.

As the ten examples in this book illustrate, many other changes occur along river corridors when massive amounts of water and sediment are stored in dams. As many as 90 percent of the floodplains historically present along rivers in Europe and North America are already cultivated and therefore functionally extinct as floodplain ecosystems. The remaining natural floodplains are disappearing faster and faster in the developing world as river flows are changed by dams and land use. The most imminently threatened floodplains are those in southeastern Asia, Sahelian Africa, and relatively undeveloped portions of North America.

As floodplain wetlands and flooded forests disappear, and as complex river channels are simplified and straightened, the species that depend on river habitats also disappear. Although small in area, freshwater habitats are disproportionately rich in species, supporting 10 percent more species than terrestrial habitats per unit area, and 150 percent more than marine habitats. Of all animal species, 12 percent, including 41 percent of all recognized fish species, live on the less than 1 percent of Earth's surface that is freshwater. As freshwater habitat has been lost over the past few decades, at least 35 percent of all freshwater fish species have become extinct, threatened, or endangered. Forty freshwater fish species disappeared from North America between 1880 and 1990, with a sharp increase in the rate of extinction during the past thirty years. The future does not look any brighter. Projected future declines in aquatic biodiversity are about five times greater than the rate for terrestrial fauna and three times the rate for coastal marine areas. Scientists modeling losses in river flow associated with changing climate and water withdrawal estimate up to 75 percent loss of local fish diversity by 2070 as a result of extinction, with disproportionately greater loss in poor countries.

We do not even know what we are losing. Two hundred new species of fish have been described annually during the past ten years. Although fish are among the best studied freshwater groups, many species will go extinct before they are discovered, as a result of habitat degradation, introduced exotic species, pollution, hybridizing, and harvesting faster than the fish can reproduce. Scientists know even less about microbial, invertebrate, and insect communities in most rivers.

CHAPTER TWELVE

Toward Healthy Watersheds

Five alterations appear consistently among most of the rivers profiled in this book. The first is loss of longitudinal connectivity, which alters the ability of fish and other aquatic organisms to move upstream and downstream as well as attenuating the downstream transport of sediment and nutrients. Loss of flow variability forms a second consistent alteration, as seasonal floods diminish in volume and duration and thereby limit the amount and diversity of habitat available to aquatic and riverside organisms. Loss of flow variability is intimately connected to a third form of alteration: loss of lateral connectivity as floodplains and secondary channels are either physically altered or hydrologically disconnected from the main channel. Loss of complexity also characterizes most of the great rivers. Removal of wood, disconnection from secondary channels and floodplain habitats, and channelization that replaces meanders and alternating pools and riffles all simplify channels and thus reduce channel stability and habitat diversity. Scientists studying eight large rivers in Europe and the United States, including the Mississippi and the Danube, found that more than a quarter of the native fish species in these rivers complete their life cycles exclusively in river channels and that many of them are imperiled. This indicates that, important as the river-floodplain connection is to many fish, the characteristics of the main channel are also critical. The fifth consistent alteration is local contamination of water and stream sediments so severe that it compromises the ability of aquatic and riparian organisms to survive.

The good news is that river ecosystems are resilient enough to recover at least some function when the extent or intensity of these alterations is reduced. Restoration efforts to date have had at least partial success in rejuvenating river health. The difficult news is that the people living within a river basin must recognize that alterations have occurred and must develop a consensus that the alterations are damaging enough to warrant correction. Recognizing alteration can be particularly difficult if it occurs over decades or is too subtle to be readily apparent without specialized scientific studies.

At this point in the book, I expect readers to ask, *Is there reason for optimism?* and *What can I do?* Because of the many examples I came across while writing this book, I believe each individual can make a difference. What does it take to make a difference? The first requirement is to care enough not to turn away in detachment or despair.

No matter what your age, chances are that you will not live long enough to see abrupt and complete catastrophe. The botos and the sturgeons will not go extinct all at once. The tremendous carbon storage and plant diversity of the boreal wetlands will not vanish overnight. The dead zone in the Gulf of Mexico may grow steadily larger, but it will probably not extend along the entire east coast of North America every year during the next decade. The changes will be swift and far-reaching, yet they will not be readily perceived if we do not care enough to watch.

Instead of waiting for even more obvious signs of river decline, or despairing over the magnitude of the problems, think about how you would like rivers to look and function. Envision the river system nearest your home as a vibrant corridor with a diverse main channel and connected floodplains that are periodically inundated by large floods. Envision the diversity of plants and animals that flourish in the uncontaminated water and the joy that people living nearby draw from fishing, bird-watching, canoeing, and simply spending time in this river corridor. Then think about how you can work toward this vision, from pulling old tires out of the river to replanting native grasses and trees along the bottomlands or starting school field trips and Adopt-A-Mile or Riverkeeper programs.

The second requirement for making a difference is to be aware. Seek to know as much as you can about the world around you. Each of us lives in a watershed. What is the state of that watershed? What were the rivers like twenty, fifty, one hundred, and five hundred years ago? How have they changed through time? What are the most biologically rich components of the local watershed, and what processes maintain this biological wealth? What are the most imminent threats to the health of the watershed? What actions must be taken to protect and restore ecological health? So many questions might seem like a lot to ask of any individual, yet those who live in democracies like to cite the adage that the price of freedom is eternal vigilance. The price of health, wholeness, and sustainability is also eternal vigilance, but vigilance widened beyond the political sphere to include the ecosystems we depend on for life.

Eternal vigilance implies not just awareness but also willingness to act on knowledge. The third requirement is to make choices in your private life that help to preserve watershed health. Rivers flow through everything we eat, wear, use, and discard, thanks to use of river water for drinking, agriculture, manufacturing, transport, and power generation. While writing this book, I read Jared Diamond's *Collapse*, which discusses in detail how individual choices can make a difference. Diamond cites examples of consumers using market forces to pressure industry to enforce sustainable practices of the type certified by the Forest Stewardship Council for wood products or the Marine Stewardship Council for marine fisheries. Consumers can exert buying power either directly when choosing what brand of gasoline to put in a car or through major middlemen when they refuse to buy diamonds or gold not certified as "conflict-free." Corporations must operate according to their bottom line, but consumers can demand that the bottom line include sustainable use of resources.

Similarly, many governments will not spend money on watershed conservation unless public pressure from individuals and groups of voters creates a strong political incentive. The fourth requirement is to publicly voice support for restoring and protecting watershed health. Many avenues exist for supporting rivers, from writing to elected representatives and participating in local advisory councils and public hearings to joining local watershed alliances, national environmental groups, or international environmental organizations and voting for candidates who actively support watershed protection and restoration. No matter what your particular talents, you can become an advocate for rivers and make a critical difference in the way governments treat watersheds.

Toward the end of her life, Rachel Carson asserted that human rights include the right to know about environmental contamination and the right to protection against it. An environmental bill of rights includes the right of every human and every living organism to clean water and healthy watersheds.

How can watershed health be judged? The best indicators are physical integrity and ecological integrity. Physical integrity implies that any segment of river and adjacent floodplain can adjust to continuing changes in water and sediment entering that segment from upstream and from adjacent hill slopes. The heart of this definition is the ability to undergo continual adjustments, because water and sediment supply do change, even in heavily regulated rivers, and because the adjustments maintain the diversity and complexity of habitat that aquatic and riverside communities of plants and animals depend on. Ability to adjust implies that the river is not completely channelized, confined within artificially stabilized banks or levees, or regulated by dams.

Physical integrity and river restoration and protection are not all-or-nothing scenarios. Substantial health can be restored to rivers such as the Danube and the Mississippi by restoring some of the natural flood peaks and seasonal low flows, by allowing the rivers to migrate back and forth across a limited portion of their historical floodplains, or by re-creating some of the historical complexity of the streambanks. The string of beads along a river can start with small and widely spaced beads that grow in size and number as opportunities arise to acquire land or manage riverfront areas differently.

The key to protecting and restoring the physical integrity of rivers is to abandon the attempt to create static, fixed rivers so confined within engineered banks that channel configuration never changes, and so regulated that they have a nearly constant level of flow throughout the year and between years. Repeated experience from rivers around the world demonstrates that attempts to create static rivers never succeed for more than a short time or along a short distance. Floods overtop the artificial levees, or depletion of sediment supply undermines stabilized banks, even as the attempts to engineer static rivers have encouraged denser land use and settlement adjacent to the river channel and thus raised the risk of loss of human life and damage to property during floods. Common sense and scientific understanding of rivers indicate that humans cannot engineer static rivers and that continuing to try destroys watershed health and costs more in the long run than watershed protection.

Ecologists define the ecological integrity of a river as the ability to sustain ecological populations and communities as well as processes such as nutrient dynamics and sediment transport. As with physical integrity, the central component of ecological integrity involves adjustments to continuing variability. That these adjustments must be self-sustaining, rather than dependent on continual human manipulation, entails that organisms must be free to move upstream and downstream, between the channel and adjacent floodplains, and between the main channel and tributaries or secondary channels in response to changes in factors such as river flow, water chemistry or temperature, food supply, competition, or predation. Ability of organisms to move implies that the river ecosystem includes a diversity of habitats that are physically connected to one another rather than fragmented by dams or levees. Again, this is not all-or-nothing. Rivers can have some level of beneficial ecological integrity without a complete return to historical conditions before any dams or channelization.

CHAPTER TWELVE

A World of Rivers

Earth and all its ecosystems form a round river. Materials and energy are continually cycled between the reservoirs of groundwater, the oceans, glaciers, and lakes, the atmosphere, and the bodies of organisms. Rivers form the links between these reservoirs and the pathways for movement and exchange. They occupy the low points in the landscape and integrate everything that occurs in the watershed, even while altering the landscape by continually transporting rock weathered from the continents toward the oceans.

Rivers remain mesmerizing to many people, even when they have been engineered and altered. Rivers express the primal force of water. River currents flow through human history and through time, extending back far beyond humans. The tremendous energy and flow of rivers is literally the source of life for most societies. The flows of water narrow and expand, pass downstream, and yet remain always present . . . thus far. Flows are ebbing on many of the world's greatest rivers, diverted for human use for drinking water, irrigated crops, or hydrologically generated electric power. As the examples of specific rivers in this book demonstrate, these ebbing flows are unsustainable. Whole river ecosystems, coastal and marine ecosytems, and ultimately human communities decline with the loss of healthy rivers. This book is not solely about the importance of maintaining watershed health; it is also about how to maintain the health of humanity. We live in a world of rivers.

Selected Bibliographic Sources

All River Chapters

Nilsson, C., C. A. Reidy, M. Dynesius, and C. Revenga. 2005. Fragmentation and flow regulation of the world's large river systems. *Science* 308:405–8.

All Interlude Chapters

Elphick, J., ed. 1995. *The atlas of bird migration: Tracing the great journeys of the world's birds*. New York: Random House.

Neumann, G. 1968. *Ocean currents*. Amsterdam: Elsevier.

Chapter One. A Round River

Epigraph of John Muir is from E. W. Teale, ed., *The wilderness world of John Muir* (Boston: Houghton Mifflin, 1954), 322; the second Muir quotation, in the text, is from Muir, *My first summer in the Sierra* (Boston: Houghton Mifflin, 1911), 157.

Burbank, D. W., J. Leland, E. Fielding, R. S. Anderson, N. Brozovic, M. R. Reid, and C. Duncan. 1996. Bedrock incision, rock uplift, and threshold hillslopes in the northwestern Himalayas. *Nature* 379:505–10.

Leopold, A. 1966. The round river. In *A Sand County almanac*, by A. Leopold. New York: Oxford University Press.

Schulze, R. E. 2004. River basin responses to global change and anthropogenic impacts. In *Vegetation, water, humans and the climate*, ed. P. Kabat, M. Claussen, P. A. Dirmeyer,

J. H. C. Gash, L. B. de Guenni, M. Meybeck, R. A. Pielke, C. J. Vörösmarty, R. W. A. Hutjes, and S. Lütkemeier, 339–74. Berlin: Springer.

Chapter Two. The Amazon: Rivers of Blushing Dolphins

Best, R. C., and V. M. F. da Silva. 1989. Amazon river dolphin, boto *Inia geoffrensis* (de Blainville, 1817). In *Handbook of marine mammals*, ed. Sam H. Ridgway and Richard Harrison, 4:1–23. San Diego: Academic Press.

De Almeida, A. L. O., and J. S. Campari. 1995. *Sustainable settlement in the Brazilian Amazon*. Oxford: Oxford University Press.

Dunne, T., L. A. K. Mertes, R. H. Meade, J. E. Richey, and B. R. Forsberg. 1998. Exchanges of sediment between the flood plain and channel of the Amazon River in Brazil. *Geological Society of America Bulletin* 110:450–67.

Feddema, J. J., K. W. Oleson, G. B. Bonan, L. O. Mearns, L. E. Buja, Gerald A. Meehl, and W. M. Washington. 2005. The importance of land-cover change in simulating future climates. *Science* 310:1674–78.

Goulding, M. 1989. *Amazon: The flooded forest*. New York: Sterling.

Goulding, M., R. Barthem, and E. Ferreira. 2003. *The Smithsonian atlas of the Amazon*. Washington, DC: Smithsonian Books.

Heckenberger, M. J., A. Kuikuro, U. T. Kuikuro, J. Russell, J. Christian, M. Schmidt, C. Fausto, and B. Franchetto. 2003. Amazonia 1492: Pristine forest or cultural parkland? *Science* 301:1710–14.

Jordan, C. F., ed. 1989. *An Amazonian rain forest*. Paris: UNESCO.

Junk, W. J., ed. 1997. *The central Amazon floodplain: Ecology of a pulsing system*. Berlin: Springer.

Kolk, A. 1996. *Forests in international environmental politics: International organizations, NGOs and the Brazilian Amazon*. Utrecht: International Books.

Lucas, R. M., P. J. Curran, M. Honzak, G. M. Foody, I. Amaral, and S. Amaral. 1996. Disturbance and recovery of tropical forest: Balancing the carbon account, in *Amazonian deforestation and climate*, ed. J. H. C. Gash, C. A. Nobre, J. M. Roberts, and R. L. Victoria, 383–98. Chichester: John Wiley.

Mahar, D. J. 1989. *Government policies and deforestation in Brazil's Amazon region*. Washington, DC: World Bank.

Meade, R. H. 2007. Transcontinental moving and storage: The Orinoco and Amazon rivers transfer the Andes to the Atlantic. In *Large rivers*, ed. A. Gupta, 45–63. Chichester: John Wiley.

Mertes, L. A. K., T. Dunne, and L. A. Martinelli. 1996. Channel-floodplain geomorphology along the Solimões-Amazon River, Brazil. *Geological Society of America Bulletin* 108:1089–1107.

Parker, E. 1985. The Amazon caboclo: An introduction and overview. In *The Amazon caboclo: Historical and contemporary perspectives*, ed. E. P. Parker, xvii–li. Studies in Third World Societies Publication 32. Williamsburg, VA: College of William and Mary.

Richey, J. E., C. Nobre, and C. Dreser. 1989. Amazon river discharge and climate variability: 1903 to 1985. *Science* 246:101–3.

Wallace, A. R. 1969 (1853). *A narrative of travels on the Amazon and Rio Negro.* New York: Haskell House.

Welcomme, R. L. 1995. Relationships between fisheries and the integrity of river systems. *Regulated Rivers: Research and Management* 11:121–36.

Interlude

Crane, K., and J. L. Galasso. 1999. *Arctic environmental atlas.* New York: Hunter College Office of Naval Research, Naval Research Laboratory.

Lumpkin, R., and S. L. Garzoli. 2005. Near-surface circulation in the tropical Atlantic Ocean. *Deep-Sea Research I* 52:495–518.

Moore, W. S., J. L. Sarmiento, and R. M. Key. 1986. Tracing the Amazon component of surface Atlantic water using 228Ra, salinity and silica. *Journal of Geophysical Research* 91 (C2): 2574–80.

Chapter Three. The Ob: Killing Grandmother

Bityukov, V. P. 1990. Downstream of the Novosibirsk hydroelectric station on the Ob River. *Hydrotechnical Construction* 23:587–91.

Crane, K., and J. L. Galasso. 1999. *Arctic environmental atlas.* Office of Naval Research, Naval Research Laboratory. New York: Hunter College.

Gebhardt, A. C., B. Gaye-Haake, D. Unger, N. Lahajnar, and V. Ittekkot. 2004. Recent particulate organic carbon and total suspended matter fluxes from the Ob and Yenisey rivers into the Kara Sea (Siberia). *Marine Geology* 207:225–45.

Holmes, R. M., B. J. Peterson, V. V. Gordeev, A. V. Zhulidov, M. Meybeck, R. B. Lammers, and C. J. Vorosmarty. 2000. Flux of nutrients from Russian rivers to the Arctic Oean: Can we establish a baseline against which to judge future changes? *Water Resources Research* 36:2309–20.

Kempe, F. 1992, *Siberian odyssey: A voyage into the Russian soul.* New York: Putnam's Sons.

Kenna, T. C., and F. L. Sayles. 2002. The distribution and history of nuclear weapons related contamination in sediments from the Ob River, Siberia, as determined by isotopic ratios of plutonium and neptunium. *Journal of Environmental Radioactivity* 60:105–37.

Klubnikin, K., C. Annett, M. Cherkasova, M. Shishin, and I. Fotieva. 2000. The sacred and the scientific: Traditional ecological knowledge in Siberian river conservation. *Ecological Applications* 10:1296–1306.

Loucks, D. P., ed. 1998. *Restoration of degraded rivers: Challenges, issues and experiences.* Dordrecht: Kluwer.

Medvedev, Z. 1976. Two decades of dissidence. *New Scientist* 70:264–67.

———. 1979. *Nuclear disaster in the Urals*. Trans. G. Saunders. New York: W. W. Norton.

Nuttall, Mark, and Terry V. Callaghan, eds. 2000. *The Arctic: Environment, people, and policy*. Amsterdam: Harwood.

Osintsev, S. P. 1995. Heavy metals in the bottom sediments of the Katun' River and the Ob' upper reaches. *Water Resources* 22:36–42.

Peterson, B. J., R. M. Holmes, J. W. McClelland, C. J. Vorosmarty, R. B. Lammers, A. I. Shiklomanov, I. A. Shiklomanov, and S. Rahmstorf. 2002. Increasing river discharge to the Arctic Ocean. *Science* 298:2171–73.

Pryde, P. R. 2002. Radioactive contamination. In *The physical geography of northern Eurasia*, ed. M. Shaghedanova, 448–62. Oxford: Oxford University Press.

Ruban, G. I. 1997. Species structure, contemporary distribution and status of the Siberian sturgeon, *Acipenser baerii. Environmental Biology of Fishes* 48:221–30.

Serreze, M. C., D. H. Bromwich, M. P. Clark, A. J. Etringer, T. Zhang, and R. Lammers. 2003. Large-scale hydro-climatology of the terrestrial Arctic drainage system. *Journal of Geophysical Research* 108 (D2): ALT 1-1–1-28.

Shahgedanova, M., ed. 2002. *The physical geography of northern Eurasia*. New York: Oxford University Press.

Shvartsev, S. L., and O. G. Savichev. 1997. The ecological and geochemical state of the major tributaries of the Ob in its middle course. *Water Resources* 24:707–13.

Shvartsev, S. L., O. G. Savichev, G. G. Vertman, R. F. Zarubina, N. G. Nalivaiko, N. G. Trifonova, Y. P. Turov, L. F. Frizen, and V. V. Yankovskii. 1996. The ecological and geochemical state of the water in the middle course of the Ob River. *Water Resources* 23:673–82.

Vasiliev, O. F. 1998. Water quality and environmental degradation in the Tom River basin (western Siberia): The need for an integrated management approach. In *Restoration of degraded rivers: Challenges, issues and experiences*, ed. D. P. Loucks, 283–92. Dordrecht: Kluwer.

Walker, H. Jesse. 1998. Arctic deltas. *Journal of Coastal Research* 14:718–38.

Wenyon, Charles. 1896 (1971). *Across Siberia on the great post-road*. New York: Arno Press.

Zhulidov, A. V., V. V. Khlobystov, R. D. Roberts, and D. F. Pavlov. 2000. Critical analysis of water quality monitoring in the Russian Federation and former Soviet Union. *Canadian Journal of Fisheries and Aquatic Sciences* 57:1932–39.

Interlude

Crane, K., and J. L. Galasso. 1999. *Arctic environmental atlas*. New York: Hunter College Office of Naval Research, Naval Research Laboratory.

Chapter Four. The Nile: Lifeline in the Desert

Abu-Zeid, M. A., and A. K. Biswas, eds. 1996. *River basin planning and management.* Calcutta: Oxford University Press.

Baker, S. 1869. *The Albert N'yanza: Great basins of the Nile and explorations of the Nile sources.* London: Macmillan.

Collins, R. O. 1990. *The waters of the Nile: Hydropolitics and the Jonglei Canal, 1900–1988.* Princeton, NJ: Markus Wiener.

Collins, R. O. 2002. *The Nile.* New Haven, CT: Yale University Press.

Crisman, T. L., L. J. Chapman, C. A. Chapman, and L. S. Kaufman, eds. 2003. *Conservation, ecology and management of African fresh waters.* Gainesville: University Press of Florida.

Goldschmidt, T. 1996. *Darwin's dreampond: Drama in Lake Victoria.* Trans. S. Marx-Macdonald. Cambridge, MA: MIT Press.

Goudie, A. S. 2005. The drainage of Africa since the Cretaceous. *Geomorphology* 67:437–56.

Hillel, D. 1994. The mighty Nile. In *Rivers of Eden: The struggle for water and the quest for peace in the Middle East.* New York: Oxford University Press.

Howell, P. P., and J. A. Allan, eds. 1994. *The Nile: Sharing a scarce resource.* Cambridge: Cambridge University Press.

Hurst, H. E. 1951. Long-term storage capacity of reservoirs. *Transactions of the American Society of Civil Engineers* 116:770–808.

Kashef, A.-A. I. 1981. Technical and ecological impacts of the High Aswan Dam. *Journal of Hydrology* 53:73–84.

McCauley, J. F., G. G. Schaber, C. S. Breed, and M. J. Grolier. 1982. SIR-A images reveal major subsurface drainages in the eastern Sahara: Applications to Mars. *NASA Technical Memorandum* 85127:311–13.

Moghraby, A. I. 1982. The Jonglei Canal—needed development or potential ecodisaster? *Environmental Conservation* 9:141–48.

Mohamed, Y. A., B. J. J. M. van den Hurk, H. H. G. Savenije, and W. G. M. Bastiaanssen. 2005. Impact of the Sudd wetland on the Nile hydroclimatology. *Water Resources Research* 41:W08420.

Revenga, C., S. Murray, J. Abramovitz, and A. Hammond. 1998. *Watersheds of the world: Ecological value and vulnerability.* Washington, DC: World Resources Institute.

Pitcher, T. J., and P. J. B. Hart, eds. 1995. *The impact of species changes in African lakes.* London: Chapman and Hall.

Sene, K. J., E. L. Tate, and F. A. K. Farquharson. 2001. Sensitivity studies of the impacts of climate change on White Nile flows. *Climatic Change* 50:177–208.

Speke, J. H. 1864 (1967). *What led to the discovery of the source of the Nile.* London: Frank Cass.

Stanley, J.-D., F. Goddio, T. F. Jorstad, and G. Schnepp. 2004. Submergence of ancient Greek cities off Egypt's Nile delta—a cautionary tale. *GSA Today* 14:4–10.

Tafesse, T. 2001. *The Nile question: Hydropolitics, legal wrangling, modus vivendi and perspectives*. Münster: Lit Verlag.

Walling, D. E. 1996. Hydrology and rivers. In *The physical geography of Africa*, ed. W. M. Adams, A. S. Goudie and A. R. Orme, 103–21. Oxford: Oxford University Press.

Waterbury, J. 1979. *Hydropolitics of the Nile Valley*. Syracuse, NY: Syracuse University Press.

Interlude

Skliris, N., and A. Lascaratos. 2004. Impacts of the Nile River damming on the thermohaline circulation and water mass characteristics of the Mediterranean Sea. *Journal of Marine Systems* 52:121–43.

Chapter Five. The Danube: Remnants of Beauty

Aarts, B. G. W., F. W. B. van den Brink, and P. H. Nienhuis. 2004. Habitat loss as the main cause of the slow recovery of fish faunas of regulated large rivers in Europe: The transversal floodplain gradient. *River Research and Applications* 20:3–23.

Avis, C., J. Van Wetten, J. Seffer, G. Tinchev, and P. Weller. 2000. Danube River basin: Wetlands and floodplains. In *The root causes of biodiversity loss*, ed. Alexander Wood, Pamela Stedman-Edwards, and Johanna Mang, 183–212. London: Earthscan Publications.

Bell, R. G., J. B. Stewart, and M. T. Nagy. 2002. Fostering a culture of environmental compliance through greater public involvement. *Environment* 44:34–44.

Bloesch, J. 2002. The unique ecological potential of the Danube and its tributaries. *Archives of Hydrobiology Supplement* 141:175–88.

Bloesch, J. 2003. Flood plain conservation in the Danube River Basin, the link between hydrology and limnology. *Archives of Hydrobiology Supplement* 147:347–62.

Bloesch, J., and U. Sieber. 2003. The morphological destruction and subsequent restoration programmes of large rivers in Europe. *Archives of Hydrobiology Supplement* 147:363–85.

Botterweg, T., and D. W. Rodda. 1999. Danube River basin: Progress with the environmental programme. *Water Science and Technology* 40:1–8.

Chovanec, A., F. Schiemer, H. Waidbacher, and R. Spolwind. 2002. Rehabilitation of a heavily modified river section of the Danube in Vienna (Austria): Biological assessment of landscape linkages on different scales. *International Review of Hydrobiology* 87:183–95.

Chovanec, A., J. Waringer, R. Raab, and G. Laister. 2004. Lateral connectivity of a fragmented large river system: Assessment on a macroscale by dragonfly surveys (Insecta: Odonata). *Aquatic Conservation: Marine and Freshwater Ecosystems* 14:163–78.

Damacija, B., I. Ivancev-Tumbas, J. Zejak, and M. Djurendic. 2003. Case study of petroleum contaminated area of Novi Sad after NATO bombing in Yugoslavia. *Soil and Sediment Contamination* 12:591–611.

Fitzmaurice, J., 1996. *Damming the Danube: Gabčikovo and post-Communist politics in Europe*. Boulder, CO: Westview.

Gouder de Beauregard, A.-C., G. Torres, and F. Malaisse. 2002. Ecohydrology: A new paradigm for bioengineers? *Biotechnolology, Agronomy, Society, and Environment* 6:17–27.

Guieu, C., J.-M. Martin, S. P. C. Tankéré, F. Mousty, P. Trincherini, M. Bazot, and M. H. Dai. 1998. On trace metal geochemistry in the Danube River and western Black Sea. *Estuarine, Coastal and Shelf Science* 47:471–85.

Guti, G. 2002. Significance of side-tributaries and floodplains for Danubian fish populations. *Archives of Hydrobiology Supplement* 141:151–63.

Habersack, H. M., and H.-P. Nachtnebel. 1995. Short-term effects of local river restoration on morphology, flow field, substrate and biota. *Regulated Rivers: Research and Management* 10:291–301.

Hirzinger, V., H. Keckeis, H. L. Nemschkal, and F. Schiemer. 2004. The importance of onshore areas for adult fish distribution along a free-flowing section of the Danube, Austria. *River Research and Applications* 20:137–49.

Hohensinner, S., H. Habersack, M. Jungwirth, and G. Zauner. 2004. Reconstruction of the characteristics of a natural alluvial river-floodplain system and hydromorphological changes following human modifications: The Danube River (1812–1991). *River Research and Applications* 20:25–41.

International Union for the Conservation of Nature and Natural Resources. 1992. *Conservation status of the Danube delta*, vol. 4, *Environmental status reports*. Berkshire, UK: IUCNNR.

Lancelot, C., J.-M. Martin, N. Panin, and Y. Zaitsev. 2002. The north-western Black Sea: A pilot site to understand the complex interaction between human activities and the coastal environment. *Estuarine, Coastal and Shelf Science* 54:279–83.

Lenhardt, M., P. Cakić, and J. Kolarevic. 2004. Influences of the HEPS Djerdap I and Djerdap II dam construction on catch of economically important fish species in the Danube River. *Ecohydrology and Hydrobiology* 4:499–502.

Lucas, C. 2001. The Baia Mare and Baia Borsa accidents: Cases of severe transboundary water pollution. *Environmental Policy and Law* 31:106–11.

Nachtnebel, H.-P. 2000. The Danube River basin environmental programme: Plans and actions for a basin wide approach. *Water Policy* 2:113–29.

Reichert-Facilides, D. 1998. Down the Danube: The Vienna Convention on the law of treaties and the case concerning the Gabčíkovo-Nagymaros Project. *International and Comparative Law Quarterly* 47:837–54.

Schneider, E. 2002. The ecological functions of the Danubian floodplains and their restoration with special regard to the Lower Danube. *Archives of Hydrobiology Supplement* 141:129–49.

Tamas, S., N. Nicolescu, T. Toader, and F. Corduneanu. 2001. Floodplain forests of the Danube delta. In *The floodplain forests in Europe: Current situation and perspectives*, ed. E. Klimo and H. Hager, 221–32. Leiden: Brill.

Tockner, K., C. Baumgartner, F. Schiemer, and J. V. Ward. 2000. Biodiversity of a Danubian floodplain: Structural, functional and compositional aspects. In *Biodiversity in wetlands: Assessment, function and conservation*, ed. B. Gopal, W. J. Junk, and J. A. Davis, 1:141–59. Leiden: Backhuys.

Interlude

Daskalov, G. M. 2003. Long-term changes in fish abundance and environmental indices in the Black Sea. *Marine Ecology Progress Series* 255:259–70.

Kideys, A. E. 2002. Fall and rise of the Black Sea ecosystem. *Science* 297:1482–84.

Meinesz, A. 2001. *Killer algae: The true tale of a biological invasion*. Chicago: University of Chicago Press.

Chapter Six. The Ganges: Eternally Pure?

Adel, M. M. 2002. Man-made climatic changes in the Ganges basin. *International Journal of Climatology* 22:993–1016.

Agnihotri, N. P., V. T. Gajbhiye, M. Kumar, and S. P. Mohapatra. 1994. Organochlorine insecticide residues in Ganges River water near Farrukhabad, India. *Environmental Monitoring and Assessment* 30:105–12.

Ahmad, S., M. Ajmal, and A. A. Nomani. 1996. Organochlorines and polycyclic aromatic hydrocarbons in the sediments of Ganges River (India). *Bulletin of Environmental Contamination and Toxicology* 57:794–802.

Alley, K. D. 2002. *On the banks of the Ganga: When wastewater meets a sacred river.* Ann Arbor: University of Michigan Press.

Bandyopadhyay, J. 2002. Water management in the Ganges-Brahmaputra basin: Emerging challenges for the 21st century. In *Conflict management of water resources*, ed. M. Chatterji, S. Arlosoroff, and G. Guha, 179–218. Hampshire, UK: Ashgate.

Biswas, Asit K., and Juha I. Uitto, eds. 2001. *Sustainable development of the Ganges-Brahmaputra-Meghna basins*. Tokyo: United Nations University Press.

Chander, D. V. R., C. Venkobachar, and B. C. Raymahashay. 1994. Retention of fly ash-derived copper in sediments of the Pandu River near Kanpur, India. *Environmental Geology* 24:133–39.

Ives, J., and D. C. Pitt, eds. 1988. *Deforestation: Social dynamics in watersheds and mountain ecosystems*. London: Routledge.

Kannan, K., R. K. Sinha, S. Tanabe, H. Ichihashi, and R. Tatsukawa. 1993. Heavy metals and organochlorine residues in Ganges River dolphins from India. *Marine Pollution Bulletin* 26:159–62.

Kumar, K. S., K. Kannan, O. N. Paramasivan, V. P. S. Sundaram, J. Nakanishi, and S. Masunaga. 2001. Polychlorinated dibenzo-p-dioxins, dibenzofurans, and polychlorinated biphenyls in human tissues, meat, fish, and wildlife samples from India. *Environmental Science and Technology* 35:3448–55.

Markandya, A., and M. N. Murty. 2004. Cost-benefit analysis of cleaning the Ganges: Some emerging environmental and development issues. *Environment and Development Economics* 9:61–81.

Mirza, M. M. Q., ed. 2004. *The Ganges water diversion: Environmental effects and implications*. Dordrecht: Kluwer.

Morrison, P., and P. Morrison. 2002. No one checked: Natural arsenic in wells. *American Scientist* 90:123–25.

Rahman, M. M., M. Q. Hassan, M. S. Islam, and S. Z. K. M. Shamsad. 2000. Environmental impact assessment on water quality deterioration caused by the decreased Ganges outflow and saline water intrusion in south-western Bangladesh. *Environmental Geology* 40:31–40.

Sainju, M. M. 2002. Some issues related to conflict management of water resources in Nepal. In *Conflict management of water resources*, ed. M. Chatterji, S. Arlosoroff, and G. Guha, 251–58. Hampshire, UK: Ashgate.

Sarkar, S. K., A. Bhattacharya, and B. Bhattacharya. 2003. The river Ganga of northern India: An appraisal of its geomorphic and ecological changes. *Water Science and Technology* 48:121–28.

Sharma, C. B., and N. C. Ghose. 1987. Pollution of the river Ganga by municipal waste: A case study from Patna. *Journal of the Geological Society of India* 30:369–85.

Siddiqi, N. A. 1997. *Management of resources in the Sundarbans mangroves of Bangladesh*. Intercoast Network, Mangrove edition, Coastal Resources Center. Narragansett Bay: University of Rhode Island.

Singh, M. 2001. Heavy metal pollution in freshly deposited sediments of the Yamuna River (the Ganges River tributary): A case study from Delhi and Agra urban centres, India. *Environmental Geology* 40:664–71.

Smith, B. D., B. Bhandari, and K. Sapkota. 1996. *Aquatic biodiversity in the Karnali and Narayani river basins–Nepal*. Kathmandu: IUCN–World Conservation Union.

Subramanian, V., N. Madhavan, R. Saxena, and L. C. Lundin. 2003. Nature of distribution of mercury in the sediments of the river Yamuna (tributary of the Ganges), India. *Journal of Environmental Monitoring* 5:427–34.

Chapter Seven. The Mississippi: Once and Future River

Balogh, S. J., D. R. Engstrom, J. E. Almendinger, M. L. Meyer, and D. K. Johnson. 1999. History of mercury loading in the Upper Mississippi River reconstructed from the sediments of Lake Pepin. *Environmental Science and Technology* 33:3297–3302.

Barry, J. M. 1997. *Rising tide: The great Mississippi flood of 1927 and how it changed America*. New York: Simon and Schuster.

Brown, A. V., K. B. Brown, D. C. Jackson, and W. K. Pierson. 2005. Lower Mississippi River and its tributaries. In *Rivers of North America*, ed. Arthur C. Benke and Colbert E. Cushing, 231–81. Amsterdam: Elsevier.

Changnon, S. A., ed. 1996. *The great flood of 1993: Causes, impacts, and responses*. Boulder, CO: Westview.

Changnon, S. A. 1998. The historical struggle with floods on the Mississippi River basin. *Water International* 23:263–71.

Clay, F. M. 1983. *History of navigation on the lower Mississippi*. U.S. Army Engineer Water Resources Support Center, Navigation History NWS–83-8. Washington, DC: U.S. Government Printing Office.

Costner, P., and J. Thornton. 1989. *We all live downstream: The Mississippi River and the national toxics crisis*. Seattle: Greenpeace, Vision Press.

Delong, M. M. 2005. Upper Mississippi River basin. In *Rivers of North America*, ed. A. C. Benke and C. E. Cushing, 327–73. Amsterdam: Elsevier.

Dickens, C. 1842 (1972). *American notes for general circulation*. Harmondsworth, UK: Penguin Books.

Fremling, C. R. 2005. *Immortal river: The Upper Mississippi in ancient and modern times*. Madison: University of Wisconsin Press.

Fremling, C. R., J. L. Rasmussen, R. E. Sparks, S. P. Cobb, C. F. Bryan, and T. O. Claflin. 1989. Mississippi River fisheries: A case history. In *Proceedings of the International Large River Symposium*, ed. D. P. Dodge. Canadian Special Publication of Fisheries and Aquatic Sciences 106, Ottawa: Department of Fisheries and Oceans.

Goldstein, R. M., K. Lee, P. Talmage, J. C. Stauffer, and J. P. Anderson. 1999. *Relation of fish community composition to environmental factors and land use in part of the Upper Mississippi River basin, 1995–97*. U.S. Geological Survey Water-Resources Investigations Report 99–4034. Washington, DC: U.S. Geological Survey.

Goolsby, D. A., W. A. Battaglin, G. B. Lawrence, R. S. Artz, B. T. Aulenbach, R. P. Hooper, D. R. Keeney, and G. J. Stensland. 1999. *Flux and sources of nutrients in the Mississippi-Atchafalaya River basin*. Topic 3 report on the integrated assessment on hypoxia in the Gulf of Mexico. NOAA Coastal Ocean Program Decision Analysis Series 17. Silver Spring, MD: NOAA Coastal Ocean Program.

Hoops, R. 1993. *A river of grain: The evolution of commercial navigation on the Upper Mississippi River*. College of Agricultural and Life Sciences Report R3584. Madison: University of Wisconsin.

Keeney, D. R. 2002. Reducing nonpoint nitrogen to acceptable levels with emphasis on the Upper Mississippi River basin. *Estuaries* 25:862–68.

Kesel, R. H. 2003. Human modifications to the sediment regime of the Lower Mississippi River floodplain. *Geomorphology* 56:325–44.

Manning, R. 2004. *Against the grain: How agriculture has hijacked civilization.* New York: North Point Press. McCall, E. 1984. *Conquering the rivers: Henry Miller Shreve and the navigation of America's inland waterways.* Baton Rouge: Louisiana State University Press.

Meade, R. H., ed. 1996. *Contaminants in the Mississippi River, 1987–92.* Circular 1133. Washington, DC: U.S. Geological Survey.

Meade, R. H., T. R. Yuzyk, and T. J. Day. 1990. Movement and storage of sediment in rivers of the United States and Canada. In *Surface water hydrology,* ed. M. G. Wolman and H. C. Riggs, 255–80. Boulder, CO: Geological Society of America.

Merritt, R. H. 1984. *The Corps, the environment, and the Upper Mississippi river basin.* Washington, DC: Office of the Chief of Engineers, U.S. Government Printing Office.

Mitsch, W. J., J. W. Day, J. W. Gilliam, P. M. Groffman, D. L. Hey, G. W. Randall, and N. Wang. 2001. Reducing nitrogen loading to the Gulf of Mexico from the Mississippi River basin: Strategies to counter a persistent ecological problem. *BioScience* 51:373–88.

Pike, Z. M. 1810. *An account of expeditions to the sources of the Mississippi, and through the western parts of Louisiana, to the sources of the Arkansaw, Kans, La Platte, and Pierre Jaun, rivers.* Philadelphia: C. and A. Conrad.

Reuss, M. 1998. *Designing the bayous: The control of water in the Atchafalaya basin, 1800–1995.* Alexandria, VA: Office of History, U.S. Army Corps of Engineers.

Richter, B. D., R. Mathews, D. L. Harrison, and R. Wiggington. 2003. Ecologically sustainable water management: Managing river flows for ecological integrity. *Ecological Applications* 13:206–24.

Scarpino, P. V. 1985. *Great River: An environmental history of the Upper Mississippi, 1850–1950.* Columbia: University of Missouri Press.

Schramm, H. L. 2004. Status and management of Mississippi River fisheries. In *Proceedings of the 2nd International Symposium on the Management of Large Rivers for Fisheries,* vol. 1, ed. R. L. Welcomme and T. Petr, 301–34. RAP Publication 2004/16. Bangkok, Thailand: FAO Regional Office for Asia and the Pacific.

Schoolcraft, H. R. 1953. *Schoolcraft's narrative journal of travels: Through the northwestern regions of the United States, extending from Detroit through the great chain of American lakes to the sources of the Mississippi River, in the year 1820.* Edited by M. L. Williams. East Lansing: Michigan State University Press.

Turner, R. E. 1997. Wetland loss in the northern Gulf of Mexico: Multiple working hypotheses. *Estuaries* 20:1–13.

Twain, M. 1883 (1992). *Life on the Mississippi.* New York: Book of the Month Club.

Tweet, R. D. 1983. *History of transportation on the upper Mississippi and Illinois rivers*. U.S. Army Engineer Water Resources Support Center, Navigation History NWS–83-6. Washington, DC: U.S. Government Printing Office.

U.S. Geological Survey. 1999. *Ecological status and trends of the Upper Mississippi River system, 1998*. Long Term Resource Monitoring Program 99-T001. Washington, DC: U.S. Government Printing Office.

Upper Mississippi River Conservation Committee. 2000. *A working river and a river that works*. Rock Island, IL: Upper Mississippi River Conservation Committee.

Wang, M., and D. D. Adrian. 1998. Wetland loss in coastal Louisiana. *International Journal of Sediment Research* 13:1.

Winger, P. V., and P. J. Lasier. 1998. Toxicity of sediment collected upriver and downriver of major cities along the Lower Mississippi River. *Archives of Environmental Contamination and Toxicology* 35:213–17.

Chapter Eight. The Murray-Darling: Stumbling in the Waltz

Allison, G. B., P. G. Cook, S. R. Barnett, G. R. Walker, I. D. Jolly, and M. W. Hughes. 1990. Land clearance and river salinisation in the western Murray basin, Australia. *Journal of Hydrology* 119:1–20.

Arthington, A. H. 1996. The effects of agricultural land use and cotton production on tributaries of the Darling River, Australia. *GeoJournal* 40:115–25.

Arthington, A. H., and B. J. Pusey. 2003. Flow restoration and protection in Australian rivers. *River Research and Applications* 19:377–95.

Boulton, A. J., and L. N. Lloyd. 1992. Flooding frequency and invertebrate emergence from dry floodplain sediments of the River Murray, Australia. *Regulated Rivers: Research and Management* 7:137–51.

Bowling, L. C., and P. D. Baker. 1996. Major cyanobacterial bloom in the Barwon-Darling River, Australia, in 1991, and underlying limnological conditions. *Marine and Freshwater Research* 47:643–57.

Braaten, R., and G. Gates. 2003. Groundwater-surface water interaction in inland New South Wales: A scoping study. *Water Science and Technology* 48:215–24.

Chen, X. Y. 1995. Geomorphology, stratigraphy and thermoluminescence dating of the lunette dune at Lake Victoria, western New South Wales. *Palaeogeography, Palaeoclimatology, Palaeoecology* 113:69–86.

Crabb, P. 1988. Managing the Murray-Darling basin. *Australian Geographer* 19:64–84.

Dyer, F. J. 2002. Assessing the hydrological changes to flood plain wetland inundation caused by river regulation. In *The structure, function and management implications of fluvial sedimentary systems*. IAHS Publication 276, 245–53. Wallingford, UK: International Association of Hydrological Sciences.

Fisher, T. 1996. Fish out of water: The plight of native fish in the Murray-Darling. *Fish and Fisheries Worldwide* 24:17–24.

Goss, K. F. 2003. Environmental flows, river salinity and biodiversity conservation: Managing trade-offs in the Murray-Darling basin. *Australian Journal of Botany* 51:619–25.

Harris, G. 1995. Eutrophication—are Australian waters different from those overseas? *Water* 2:9–12.

Herczeg, A. L., H. James Simpson, and E. Mazor. 1993. Transport of soluble salts in a large semiarid basin: River Murray, Australia. *Journal of Hydrology* 144:59–84.

Hillman, T. J., and Quinn, G. P. 2002. Temporal changes in macroinvertebrate assemblages following experimental flooding in permanent and temporary wetlands in an Australian floodplain foreset. *River Research and Applications* 18:137–54.

Humphries, P. L., A. J. King, and J. D. Koehn. 1999. Fish, flows and flood plains: Links between freshwater fishes and their environment in the Murray-Darling River system, Australia. *Environmental Biology of Fishes* 56:129–51.

Jenkins, K. M., and A. J. Boulton. 2003. Connectivity in a dryland river: Short-term aquatic microinvertebrate recruitment following floodplain inundation. *Ecology* 84:2708–23.

Jensen, A. 1998. Rehabilitation of the river Murray, Australia: Identifying causes of degradation and options for bringing the environment into the management equation. In *Rehabilitation of rivers: Principles and implementation*, ed. L. C. de Waal, A. R. G. Large, and P. M. Wade, 215–36. Chichester, UK: John Wiley.

Langford-Smith, T., and J. Rutherford. 1966. *Water and land: Two case studies in irrigation*. Canberra: Australian National University Press.

Leslie, D. J. 2001. Effect of river management on colonially-nesting waterbirds in the Barmah-Willewa Forest, south-eastern Australia. *Regulated Rivers: Research and Management* 17:21–36.

Low, T. 2002. *Feral future: The untold story of Australia's exotic invaders*. Chicago: University of Chicago Press.

MacNally, R., and G. Horrocks. 2002. Habitat change and restoration: Responses of a forest-floor mammal species to manipulations of fallen timber in floodplain forests. *Animal Biodiversity and Conservation* 25:41–52.

McGinness, H. M., M. C. Thoms, and M. R. Southwell. 2002. Connectivity and fragmentation of flood plain—river exchanges in a semiarid, anabranching river system. In *The structure, function and management implications of fluvial sedimentary systems*. IAHS Publication 276, 19–26. Ottawa: International Association of Hydrological Sciences.

Pigram, J. J., and W. F. Musgrave. 2002. Sharing the waters of the Murray-Darling basin: Cooperative federalism under test in Australia. In *Conflict prevention and resolution in water systems*, ed. A. T. Wolf, 261–81. Cheltenham, UK: Elgar.

Robertson, A. I., P. Bacon, and G. Heagney. 2001. The responses of floodplain primary production to flood frequency and timing. *Journal of Applied Ecology* 38:126–36.

Rolls, E. 1981. *A million wild acres*. Ringwood, Victoria, Australia: Penguin Books.

Schulze, D. J., and K. F. Walker. 1997. Riparian eucalypts and willows and their significance for aquatic invertebrates in the river Murray, South Australia. *Regulated Rivers: Research and Management* 13:557–77.

Sheldon, F., M. C. Thoms, O. Berry, and J. Puckridge. 2000. Using disaster to prevent catastrophe: Referencing the impacts of flow changes in large dryland rivers. *Regulated Rivers: Research and Management* 16:403–20.

Sturt, C. 1849. *Narrative of an expedition into Central Australia, performed under the authority of Her Majesty's Government, during the years 1844, 5, and 6, together with a notice of the Province of South Australia, in 1847.* London: T. and W. Boone.

Thoms, M. C., and F. Sheldon. 2000. Water resource development and hydrological change in a large dryland river: The Barwon-Darling River, Australia. *Journal of Hydrology* 228:10–21.

Walker, K. F. 1992. The river Murray, Australia: A semiarid lowland river. In *The rivers handbook: Hydrological and ecological principles*, ed. P. Calow and G. E. Petts, 1:472–92. Oxford: Blackwell Science.

Walker, K. F. and M. C. Thoms. 1993. Environmental effects of flow regulation on the lower Murray River, Australia. *Regulated Rivers Research and Management* 8:103–19.

Young, R. W., A. R. M. Young, D. M. Price, and R .A. L. Wray. 2002. Geomorphology of the Namoi alluvial plain, northwestern New South Wales. *Australian Journal of Earth Sciences* 49:509–23.

Chapter Nine. The Congo: River That Swallows All Rivers

Amarasekera, K. N., R. F. Lee, E. R. Williams, and E. A. B. Eltahir. 1997. ENSO and the natural variability in the flow of tropical rivers. *Journal of Hydrology* 200:24–39.

Crisman, T. L., L. J. Chapman, C. A. Chapman, and L. S. Kaufman, eds. 2003. *Conservation, ecology and management of African fresh waters*. Gainesville: University Press of Florida.

Gauthier, G., and M. Deliens. 1999. Cobalt minerals of the Katanga Crescent, Congo. *Mineralogical Record* 30:255–67.

Goodall, J., G. McAvoy, and G. Hudson. 2005. *Harvest for hope: A guide to mindful eating*. New York: Warner Books.

Goudie, A. S. 2005. The drainage of Africa since the Cretaceous. *Geomorphology* 67:437–56.

Hopkins, B. 1992. Ecological processes at the forest-savanna boundary. In *Nature and dynamics of forest-savanna boundaries*, ed. P. A. Furley, J. Proctor, and J. A. Ratter, 37–62. New York: Chapman and Hall.

Lévêque, C. 1997. *Biodiversity dynamics and conservation: The freshwater fish of tropical Africa*. Cambridge: Cambridge University Press.

Lowe-McConnell, R. H. 1975. *Fish communities in tropical freshwaters: Their distribution, ecology and evolution*. London: Longman.

Pacini, N., and D. M. Harper. 2000. River conservation in central and tropical Africa. In *Global perspectives on river conservation: Science, policy and practice*, ed. P. J. Boon, B. R. Davies, and G. E. Petts, 155–78. Chichester, UK: John Wiley.

Quammen, D. 2006. *The long follow: J. Michael Fay's epic trek across the last great forests of central Africa*. New York: Random House.

Salopek, P. 2005. Who rules the forest? *National Geographic* 208 (3): 74–95.

Stanley, Henry M. 1890. *In darkest Africa*. New York: Harper.

Chapter Ten. The Chang Jiang: Bridling a Dragon

Barber, M., and G. Ryder, eds. 1993. *Damming the Three Gorges: What dam builders don't want you to know*. London: Earthscan.

Chen, X., Y. Zong, E. Zhang, J. Xu, and S. Li. 2001. Human impacts on the Changjiang (Yangtze) River basin, China, with special reference to the impacts on the dry season water discharges into the sea. *Geomorphology* 41:111–23.

Green, P. A., C. J. Vörösmarty, M. Meybeck, J. N. Galloway, B. J. Peterson, and E. W. Boyer. 2004. Pre-industrial and contemporary fluxes of nitrogen through rivers: A global assessment based on typology. *Biogeochemistry* 68:71–105.

Lin, B. 1999. Sedimentation in lock approaches of Three Gorges Project (TGP). In *River sedimentation: Theory and applications*, ed. A. W. Jayawardena, J. H. W. Lee, and Z. Y. Wang, 377–83. Rotterdam: Balkema.

Liu, J. G., P. J. Mason, N. Clerici, S. Chen, A. Davis, F. Miao, H. Deng, and L. Liang. 2004. Landslide hazard assessment in the Three Gorges area of the Yangtze River using ASTER imagery: Zigui-Badong. *Geomorphology* 61:171–87.

Lu, J. Y., and X. Y. Hu. 1999. Study on the influence of the Three Gorges Project on the middle and lower Yangtze River. In *River sedimentation: Theory and applications*, ed. A. W. Jayawardena, J. H. W. Lee, and Z. Y. Wang, 421–25. Rotterdam: Balkema.

Thibodeau, J. G., and P. B. Williams, eds. 1998. *The river dragon has come! The Three Gorges Dam and the fate of China's Yangtze River and its people*. London: Sharpe.

Tracy, J. A. F. 1930. *See China with me*. Boston, MA: Stratford.

Xiong, L. 1998. Going against the flow in China. *Science* 280:24–26.

Yang, Z., H. Wang, Y. Saito Saito, J. D. Milliman, K. Xu, S. Qiao, and G. Shi. 2006. Dam impacts on the Changjiang (Yangtze) River sediment discharge to the sea: The past 55 years and after the Three Gorges Dam. *Water Resources Research* 42:W04407.

Yin, H., and C. Li. 2001. Human impacts on floods and flood disasters on the Yangtze River. *Geomorphology* 41:105–9.

Chapter Eleven. The Mackenzie: River on the Brink

Brunskill, G. J. 1986. Environmental features of the Mackenzie system. In *The ecology of river systems*, ed. B. R. Davies and K. F. Walker, 435–62. Dordrecht: Junk.

Brunskill, G. J., D. M. Rosenberg, N. B. Snow, G. L. Vascotto, and R. Wagemann. 1973. *Ecological studies of aquatic systems in the Mackenzie-Porcupine drainages in relation to proposed pipeline and highway developments*, vol. 1. Report 73–40. Calgary: Information Canada.

Carson, M. A., J. N. Jasper, and F. M. Conly. 1998. Magnitude and sources of sediment input to the Mackenzie delta, Northwest Territories, 1974–94. *Arctic* 51:116–24.

Davies, B. R., and K. F. Walker, eds. 1986. *The ecology of river systems*. Dordrecht: Junk.

Dyke, L. D., and G. R. Brooks, eds. 2000. *The physical environment of the Mackenzie valley, Northwest Territories: A base line for the assessment of environmental change*. Bulletin 547. Ottawa: Geological Survey of Canada.

Grescoe, T. 1997. Temperature rising. *Canadian Geographic* 117:36–41.

Harlan, R. L. 1974. *Hydrogeological considerations in northern pipeline development*. Report R57–37/1975. Calgary: Information Canada.

Headley, J. V., P. Marsh, C. J. Akre, K. M. Peru, and L. Lesack. 2002. Origin of polycyclic aromatic hydrocarbons in lake sediments of the Mackenzie delta. *Journal of Environmental Science and Health* A37:1159–80.

Howland, K. L., W. M. Tonn, J. A. Babaluk, and R. F. Tallman. 2001. Identification of freshwater and anadromous inconnu in the Mackenzie River system by analysis of otolith strontium. *Transactions of the American Fisheries Society* 130:725–41.

Liblik, L. K., T. R. Moore, J. L. Bubier, and S. D. Robinson. 1997. Methane emissions from wetlands in the zone of discontinuous permafrost: Fort Simpson, Northwest Territories, Canada. *Global Biogeochemical Cycles* 11:485–94.

Mackenzie River Basin Board. 2003. *Highlights of the Mackenzie River Basin Board's State of the Aquatic Ecosystem Report*. Fort Smith, Northwest Territories, Canada.

McCart, P. J. 1986. Fish and fisheries of the Mackenzie system. In *The ecology of river systems*, ed. B. R. Davies and K. F. Walker, 493–515. Dordrecht: Junk.

McDonald, T. H., ed. 1966. *Exploring the Northwest Territory: Sir Alexander Mackenzie's journal of a voyage by bark canoe from Lake Athabasca to the Pacific Ocean in the summer of 1789*. Norman: University of Oklahoma Press.

Nijssen, B., G. M. O'Donnell, A. F. Hamlet, and D. P. Lettenmaier. 2001. Hydrologic sensitivity of global rivers to climate change. *Climatic Change* 50:143–75.

Parkinson, Dennis. 1973. *Effects of oil spillage on microorganisms in northern Canadian soils*. Report R72–8573. Calgary: Information Canada.

Prowse, T. D., F. M. Conly, M. Church, and M. C. English. 2002. A review of hydroecological results of the Northern River Basins Study, Canada, part 1, Peace and Slave rivers. *River Research and Applications* 18:429–46.

Rempel, L. L., and D. G. Smith. 1998. Postglacial fish dispersal from the Mississippi refuge to the Mackenzie River basin. *Canadian Journal of Fisheries and Aquatic Sciences* 55:893–99.

Rouse, W. R., M. S. V. Douglas, R. E. Hecky, A. E. Hershey, G. W. Kling, L. Lesack, P. Marsh, M. McDonald, B. J. Nicholson, N. T. Roulet, and J. P. Smol. 1997. Effects of climate change on the freshwaters of Arctic and subarctic North America. *Hydrological Processes* 11:873–902.

Serreze, M. C., D. H. Bromwick, M. P. Clark, A. J. Etringer, T. Zhang, and R. Lammers. 2003. Large-scale hydro-climatology of the terrestrial Arctic drainage system. *Journal of Geophysical Research* 108 (D2): ALT 1-1–1-28.

Stein, J. N., C. S. Jessop, T. R. Porter, and K. T. J. Chang-Kue. 1973. *An evaluation of the fish resources of the Mackenzie River valley as related to pipeline development*, vol. 1. Report 73–1, Calgary: Information Canada.

Stewart, R. E., H. G. Leighton, P. Marsh, G. W. K. Moore, H. Ritchie, W. R. Rouse, E. D. Soulis, G. S. Strong, R. W. Crawford, and B. Kochtubajda. 1998. The Mackenzie GEWEX study: The water and energy cycles of a major North American river basin. *Bulletin of the American Meteorological Society* 79:2665–83.

Tallman, R. F., K. L. Howland, and S. Stephenson. 2005. Stability, change, and species composition of fish assemblages in the lower Mackenzie River: A pristine large river. *American Fisheries Society Symposium* 45:13–21.

Vincent, W. F., and J. E. Hobbie. 2000. Ecology of Arctic lakes and rivers. In *The Arctic: Environment, people, policy*, ed. M. Nuttall and T. V. Callaghan, 197–232. Amsterdam: Harwood.

Watson, G. H., W. H. Prescott, E. A. de Bock, J. W. Nolan, M. C. Dennington, H. J. Poston, and I. G. Stirling. 1973. Report 73–27. Calgary: Information Canada.

Wonders, W. C., ed. 2003. *Canada's changing north*, rev. ed. Montreal: McGill-Queen's University Press.

Yunker, M. B., S. M. Backus, E. G. Pannatier, D. S. Jeffries, and R. W. Macdonald. 2002. Sources and significance of alkane and PAH hydrocarbons in Canadian Arctic rivers. *Estuarine, Coastal and Shelf Science* 55:1–31.

Chapter Twelve. An Ebbing Flow

Arthington, A. H., and R. L. Welcomme. 1995. The condition of large river systems of the world. In *Condition of the world's aquatic habitats*, ed. N. B. Armantrout, 44–75. New Delhi: Oxford and IBH.

Barlow, M., and T. Clarke. 2002. *Blue gold: The fight to stop the corporate theft of the world's water.* New York: New Press.

Diamond, J. 2005. *Collapse: How societies choose to fail or succeed.* New York: Viking.

Galat, D. L., and I. Zweimüller. 2001. Conserving large-river fishes: Is the highway analogy an appropriate paradigm? *Journal of the North American Benthological Society* 20:266–79.

Graf, W. L. 2001. Damage control: Restoring the physical integrity of America's rivers. *Annals of the Association of American Geographers* 91:1–27.

Green, P. A., C. J. Vörösmarty, M. Meybeck, J. N. Galloway, B. J. Peterson, and E. W. Boyer. 2004. Pre-industrial and contemporary fluxes of nitrogen through rivers: A global assessment based on typology. *Biogeochemistry* 68:71–105.

Hughes, R. M., J. N. Rinne, and B. Calamusso. 2005. Historical changes in large river fish assemblages of the Americas: A synthesis. *American Fisheries Society Symposium* 45:603–12.

Postel, S. 2005. *Liquid assets: The critical need to safeguard freshwater ecosystems.* Worldwatch Paper 170. Washington, DC: Worldwatch Institute.

Syvitski, J. P. M., C. Vörösmarty, A. J. Kettner, and P. Green. 2005. Impact of humans on the flux of terrestrial sediment to the global coastal ocean. *Science* 308:376–80.

Tockner, K., and J. A. Stanford. 2002. Riverine flood plains: Present state and future trends. *Environmental Conservation* 29:308–30.

Vörösmarty, C., D. Lettenmaier, C. Leveque, M. Meybeck, C. Pahl-Wostl, J. Alcamo, W. Cosgrove, H. Grassl, H. Hoff, P. Kabat, F. Lansigan, R. Lawford, and R. Naiman. 2004. Humans transforming the global water system. *EOS, Transactions of the American Geophysical Union* 85 (48): 509–20.

Index